T0192530

The Island of South Georgia

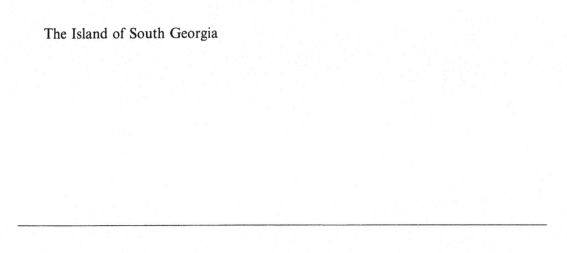

The Island of
South Georgia

ROBERT HEADLAND

CAMBRIDGE
UNIVERSITY PRESS

CAMBRIDGE UNIVERSITY PRESS
Cambridge, New York, Melbourne, Madrid, Cape Town, Singapore, São Paulo, Delhi

Cambridge University Press
The Edinburgh Building, Cambridge CB2 8RU, UK

Published in the United States of America by Cambridge University Press, New York

www.cambridge.org
Information on this title: www.cambridge.org/9780521424745

First published 1984
Reprinted 1986
First paperback edition 1992
Re-issued in this digitally printed version 2009

A catalogue record for this publication is available from the British Library

Library of Congress Catalogue Card Number: 83–24008

ISBN 978-0-521-25274-4 hardback
ISBN 978-0-521-42474-5 paperback

Contents

List of Illustrations

Foreword

SIR REX HUNT CMG

I am delighted to introduce and commend to everyone this book on the Island of South Georgia. Until the dramatic events of 1982, few people had heard of South Georgia and fewer knew where it was. Thanks to President Galtieri's folly, it is now firmly on the map and in the public eye. Apart from scientific papers and reports of various expeditions, there is, however, little literature available; Bob Headland's book is the most comprehensive work on this fascinating island.

Bob has passed through Government House in Stanley several times since my arrival in early 1980, on his way to and from South Georgia. He has tramped the length and breadth of the Island, visited all the old whaling stations, and made a scientific and systematic investigation of the flora and fauna. His enthusiasm for South Georgia is infectious and he whetted my appetite to go down and see for myself. By courtesy of HMS *Endurance*, I was able to visit South Georgia in December, 1981. I was not disappointed and readily understood why Bob was hooked on the place. It is spectacularly beautiful, with awesome glaciers that sweep down from snowcapped mountains to the sea. The abundance of wildlife reminded me of the River Nile in Uganda before Idi Amin turned his gunmen loose on the hippos and the elephants. It could be a veritable tourists' paradise but, thank goodness, it is probably too remote and the elements too harsh for tourists to spoil it. For the adventurous visitor, who is not deterred by distance or difficulties, it is the experience of a lifetime.

I can think of no person better qualified than Bob Headland to have written this book. I congratulate him on his diligence and industry and wish the book deserved success.

Preface

This book is a result of an interest in South Georgia which was greatly stimulated during my first tour of duty on the island with the British Antarctic Survey from 1977. During this period I started to compile a bibliography of the region and to accumulate other relevant information. Another tour of duty there, which started in January 1982, was interrupted on 3 April when I was taken prisoner with some of my colleagues and a platoon of Royal Marines by invading Argentine forces. After being released and returning to the United Kingdom I found that the conflict in the region had greatly increased general interest in what had been, to many people, a virtually unknown part of the Earth. The only book with a reasonably comprehensive account of South Georgia prior to this was *South Georgia, The British Empire's Sub-Antarctic Outpost* by Dr L. Harrison Matthews, published in 1931 and long out of print. I am indebted to this for part of the stimulus to write the present work as well as its value as a source of much information. As did Dr Harrison Matthews, I have quoted several extracts from older sources in full as these are not easily accessible; they are of great relevance to the island's history, and they are best left in the original form.

The book attempts to be a detailed account of South Georgia, which is one of the few places on Earth where this may still reasonably be encompassed in one volume. Nevertheless many decisions concerning the amount of material to be included were needed. In it I have discussed many subjects about which my knowledge is far from expert. Fortunately most of the writing was done while I was at the Headquarters of the British Antarctic Survey in Cambridge where my colleagues provided much advice and assistance as well as saving me from error in many cases. The Scott Polar Research Institute, also in Cambridge, has been another very valuable source of information. As well as these two institutes the Falkland Islands Government Archives in Stanley, Government records at King Edward Point, Norse Hvalfangstmuseum in Sandefjord, the Royal Navy in London, Institute of Oceanographic Sciences in Wormley, Sea Mammal Research Institute in Cambridge, Trans-Antarctic Association in Cambridge, Whaling Museum in New Bedford, Public Record Office in Kew, Hakluyt Society in London, and the Foreign and Commonwealth Office in London have all provided valuable information and many have made illustrations available.

I particularly wish to express my thanks to: the Civil Commissioner (previously Governor) of the Falkland Islands and Dependencies, Sir Rex Hunt CMG, who kindly provided the foreword and to the Director of the British Antarctic Survey, Dr R.M. Laws CBE FRS, who reviewed the typescript, and to both for permitting me many of the facilities and some of the time to prepare this account; Mr W.N. Bonner, Head of Life Sciences Division of the Survey and a South Georgia specialist for many years, who read all the typescript during preparation, gave valuable advice and suggested the addition of some important details; and Dr R.I. Lewis Smith, Mr P.R. Stark and Dr D.W.H. Walton who read large parts of the typescript and offered suggestions for improvements. As well as these persons I wish to acknowledge the assistance of the following who provided advice on specific matters and many of the illustrations: Dr R.J. Adie, Mr M.J. Baker, Captain N.J. Barker, Dr W. Block, Mr S.G. Brown, Commander M.K. Burley, Mr D. Carse, Mr E.J. Chinn, Mr I.B. Collinge, Dr J.P. Croxall, Mr R.D.J. Edwins, Mr G.H. Elliot, Dr J.C. Ellis-Evans, Dr I. Everson, Dr F. Goldberg, Mr I.B. Hart, Dr J.A. Heap, Mr T.D. Heilbronn, Dr R.B. Heywood, Mr C.A. Holland, Mr A.G.E. Jones, Mr H.G.R. King, Mr H-K. Larsen, Mr D.W.S. Limbert, Dr T.S. McCann, Mr S.J. Martin, Dr D.W. Matthews, Mr S. Miller, Dr J. Mitton, Mr A.W. North, Miss C.M. Phillips, Mr A. Pinheiro-Torres, Dr J. Priddle, Mr P.A. Prince, Mr A.S. Rodger, Mr D.J. Sanders, Dr R. Schenker, Mr J. Smith, Dr M.J. Smith, Dr L. Sømme, Dr B.C. Storey, Mr A.F. Saunders, Mr J.J. Thompson, Mrs J.W. Thomson, Dr M.R.A. Thomson, Mr R.J. Timmis, Miss G.E. Todd, Mr R.W. Vaughan†, Dr C.C. West, Mr E. Wexelsen, Mr M.G. White, Dr T.M. Whitaker, and Dr D.D. Wynn-Williams.

The illustrations are derived from a large number of sources which have provided far more material than my own photographs do and greatly improve the book. The origin of each is given in its caption and I wish to record my thanks to the many institutions and persons who made this material available. I am greatly indebted to Mr C.J. Gilbert who prepared nearly all the photographs used for the illustrations from a large variety of prints, plates, glass and celluloid negatives, colour slides, and original material.

There are four points for the reader to bear in mind.

Seasons: most annual periods are expressed as seasons such as 1982/83. The austral summer is the period during which most activity occurs on the island and in the Antarctic generally. As this covers two calendar years this is the most efficient method of referring to it.

References: I have not attempted to provide detailed references throughout the book for, as much of the information is unpublished and space is limited, this would have been quite impracticable. Those given in the text are of substantial works covering all or part of the subject or are sources of direct quotation. All of these, as well as many others, are given in full in the bibliography.

Accuracy: some dates, numerical details, and other matters I have recorded are, in a few cases, not unequivocal. I have referred to original sources wherever possible but sometimes have had no option other than to assess which of several is most likely to be correct. Fortunately the differences are rarely substantial.

Errors: finally, I would appreciate advice of errors in this work as they are detected.

R.K. Headland   *Cambridge 1984*

Note added at first printing of the paperback edition
The publishing of the paperback edition of *The Island of South Georgia* has allowed some textual revisions to be incorporated. I am indebted to several colleagues and correspondents for indicating items for attention. Fortunately no major amendments and corrections have been required: I still request readers to inform me of any errors detected.

It was not practicable to make substantial additions in the text to describe recent circumstances. I have the opportunity, however, to mention briefly some items in this introduction. At King Edward Point a garrison has been stationed since 1982. This has had many consequences, one of which is to increase access to the island (mainly from the Falkland Islands by Royal Naval vessels), almost to that prevailing during the whaling days. Thus a greater number of expeditions have visited, undertaking scientific research, surveying, mountaineering, and other activities. Royal Naval and Air Force patrols keep the island and the South Sandwich Islands under more frequent surveillance. The greater naval presence has also resulted in continued improvement of the charts.

The British Antarctic Survey has maintained the Bird Island station continuously from the 1982/83 season and has had several summer parties at Husvik. Similarly other expeditions from Britain, Germany, The Netherlands, New Zealand, Norway, Sweden, Switzerland, and elsewhere have worked at various sites. An automatic meteorological observatory has been deployed on the south side of King Edward Cove; this communicates automatically with the Falkland Islands.

A major clean-up and salvage of the abandoned whaling stations took place in 1990–91. This was undertaken by a British Company associated with the Falkland Islands, with results far more beneficial than those of the 1982 attempt. It was funded by the Government of South Georgia and the South Sandwich Islands, and the Christian Salvesens Company. The possibility of serious contamination has been effectively eliminated – over 3000 tonnes of petroleum products were removed as well as other potentially dangerous substances.

The Commissioner, Mr W. H. Fullerton, has made funds available for a South Georgia Whaling Museum and has actively encouraged its establishment. Work towards this began in 1988–89 and has continued over subsequent summers. An exhibition of the operations of this major aspect of Antarctic history is to be established at Grytviken. An intriguing variety of industrial archaeological material remains at the seven whaling sites, much of this will be concentrated at the Museum. Interest in visiting South Georgia and other Antarctic regions has increased greatly during recent years and tourist vessels are becoming more frequent. A number of non-military personnel also benefit from the opportunities provided by naval vessels. Many of these visitors will benefit from the Museum.

The political status of South Georgia was amended in 1985; the term Falkland Islands Dependencies was dropped and the territory designated as part of 'South Georgia and the South Sandwich Islands' – *all territories between 20° W and 50° W from 50° S to 60° S* (which thus includes Shag Rocks). A Government Gazette for the territory was first issued in 1986. The administration is by the Commissioner, resident in the Falkland Islands (who is also the Governor of the Islands). Locally the Officer-in-Charge of the garrison is appointed Magistrate. A resident civilian Harbour Master has been appointed since 1989.

R. K. Headland,
Scott Polar Research Institute,
Lensfield Road,
Cambridge,
United Kingdom, CB2 1ER
24 December 1991

1

Geography, administration, and population

This chapter introduces South Georgia and gives a general description of its geography with an account of the government and habitation. Some of the subjects mentioned are discussed in greater detail in other chapters.

Location

South Georgia is an isolated island which lies in the Antarctic or Southern Ocean. It is the highest, most mountainous and second largest of the small number of islands which encircle the continent of Antarctica. South Georgia lies between latitudes 53°56′ and 54°55′ S and longitudes 34°45′ and 38°15′ W. It is roughly crescent shaped along a north-west to south-east axis, about 170 km long and from 2 to 30 km wide. The surface area is approximately 3755 km², well over half of which is permanently covered with ice and snow. Around South Georgia are several small islands, islets, and rocks. The major ones are Bird Island and the Willis Islands off the western extremity, Annenkov Island and the Pickersgill Islands off the south-western coast, Cooper Island at the south-eastern end, and several islands in the Bay of Isles. Two groups of rocks are generally regarded as associated with South Georgia: Clerke Rocks, approximately 65 km east south-east from Cooper Island; and Shag Rocks approximately 250 km west north-west from the Willis Islands with neighbouring Black Rock to their south-east.

South Georgia is situated approximately 550 km from the nearest of the South Sandwich Islands; 1030 km from the South Orkney Islands; 1450 km from Port Stanley in the Falkland Islands; 1550 km from Cape Dubouzet, the nearest point on continental Antarctica; 2050 km from Cape Horn; 2150 km from Cabo Virgenes, the nearest point of continental South America; and 4800 km from the Cape of Good Hope. It lies on a small continental block which forms part of the North Scotia Ridge between Tierra del Fuego and the South Sandwich Islands. Owing to the very distinct features of South Georgia and the region in which it lies, it is not possible to make any reasonably close comparisons between it and other islands. One may, however, regard it as having some similarities with Heard Island in the southern hemisphere and Jan Mayen Island or Spitsbergen in the northern hemisphere. Although lying at approximately the same distance from the equator as northern England, Denmark, Labrador, northern British Columbia, and Kamchatka, South Georgia is far more frigid than these places owing, principally, to the effects of the surrounding ocean. In the North Atlantic region, climates similar to that of South Georgia are found about 20° closer to the Pole.

Topography

South Georgia has been described as like 'the Alps in mid-ocean' or 'the Himalayas seen from Simla' as the mountains are a magnificent sight – especially in moonlight. It is the summit of a partly drowned mountain range which rises on a

South Georgia and the other parts of the Falkland Islands Dependencies showing their relation to the British Antarctic Territory (BAT), Falkland Islands and part of South America. (BAT is the land and ice-shelves within 20° W to 80° W and 60° S.) (Author.)

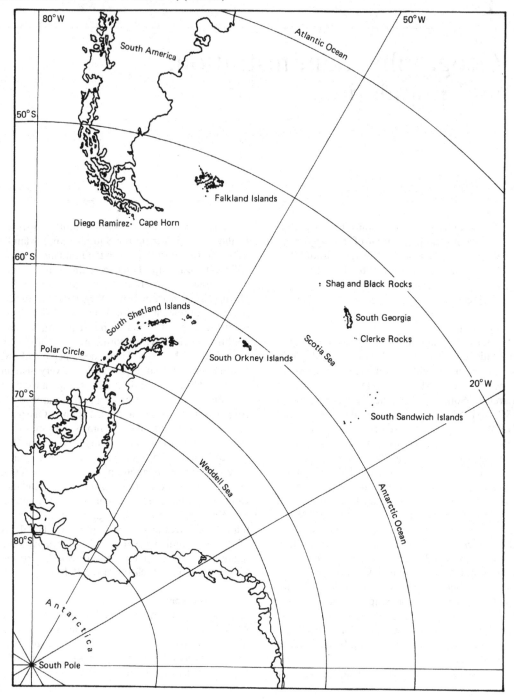

small continental shelf over 6000 m from the ocean floor. These mountains are dissected by large numbers of deep fjords, most of which contain glaciers. Two principal mountain chains, the Allardyce and Salvesen Ranges, effectively form the island's spine. The highest peak is Mount Paget, 2934 m above sea level whilst twelve other peaks exceed 2000 m, including Nordenskjöld Peak (2355 m), Mount Carse (2331 m), Mount Sugartop (2323 m) and Mount Paterson (2200 m). The highest peaks are concentrated mainly around the middle section of the crescent shaped island, where they provide a substantial barrier against the severe weather which reaches the south-west side of the island with the prevailing winds. The area in their lee has a comparatively less severe climate.

The coasts consist mainly of high sea cliffs which are interrupted by many fjords and glaciers, although some have restricted areas of flat ground behind them. The fjords, especially on the north-eastern coast, provide a variety of harbours and anchorages, some deep and clear, others with sunken rocks and reefs; many of them have glaciers at their heads. The south-western shores have far fewer fjords, many more glaciers and comparatively few safe anchorages. Permanent snow begins at about 200 m on the exposed south-west side and 400 m on the protected parts of the north-east side of the island. The peaks are lower and non-glaciated areas are more common towards the western extremity of the island where nearly all of the few anchorages on the south-western side occur. In contrast, the eastern end has much more precipitous mountains and deeper, more rugged fjords.

Plains of several square kilometres occur in only two places: Hestesletten at Cumberland East Bay and Salisbury Plain at the Bay of Isles. At the heads of many of the bays and fjords there are limited more or less level areas (some of which have provided the sites for the whaling stations which were established on the island). Apart from small areas with a hilly, undulating aspect, most of the topography of the parts of the island free from permanent ice or snow cover is high cliffs, crags and rocky outcrops (too steep to permit snow and ice to remain) separated by deep steep-sided valleys, often with unstable scree or talus slopes. Almost everywhere else is covered with snow fields, ice fields or glaciers which may be highly crevassed. These

constitute some of the most notable features of South Georgia. Some floating glacier fronts may be up to 50 m high, 250 m deep and over 1 km wide. Surface travel on the island is thus rendered particularly difficult.

Only two water-courses have been designated rivers: Hope River at Undine Harbour and Penguin River at Cumberland Bay; but many others exist. Virtually every bay has several streams emptying into it from adjacent valleys. Many waterfalls occur; these, together with the streams, are spectacular when in flood following the spring thaw, a bright day causing snow melting or a period of heavy summer rain. Melt streams from glaciers are also common. These have a diurnal variation, being at their maxima in the afternoon and minima in the mornings, (owing to the effect of solar heat on the melting of glaciers).

There are over two dozen lakes on the island, most of which are situated between Antarctic Bay and St Andrews Bay. Several of them have been formed where glaciers have dammed valleys. Gulbrandsen Lake, near Husvik, is the largest and most spectacular of these with icebergs floating in it and a series of over 12 major terraces on its shores (representing previous lake levels). Ponds, pools, and tarns are common throughout the island and may (especially during the thaw and periods of rainfall in summer) merge into swamps, bogs and mires near the coast. Another distinctive ground feature is found near many flatter areas of the coast behind beaches – elephant seal wallows. These are formed by the seals lying closely packed, in mud during their moulting periods. The wallows are often 1 m or more deep and become exceedingly foetid with skin, fur, faeces, combined with thin mud and the occasional dead seal. The wallows are at their worst during February and March, the height of the moulting season, when they emit rising clouds of noxious vapour. The author can testify to the extreme undesirability of falling into them.

Most of the island's beaches are shingle, although a few are sandy. Otherwise, where the shores are not cliffs or glaciers, wave-cut platforms are not uncommon and some are easily passable on foot at low tide. Glacier snouts, which reach the ocean or waters of the bays and fjords, can be very spectacular – especially when enormous pieces break off and crash into the water (a process,

A general map of South Georgia. (British Antarctic
Survey and author.)

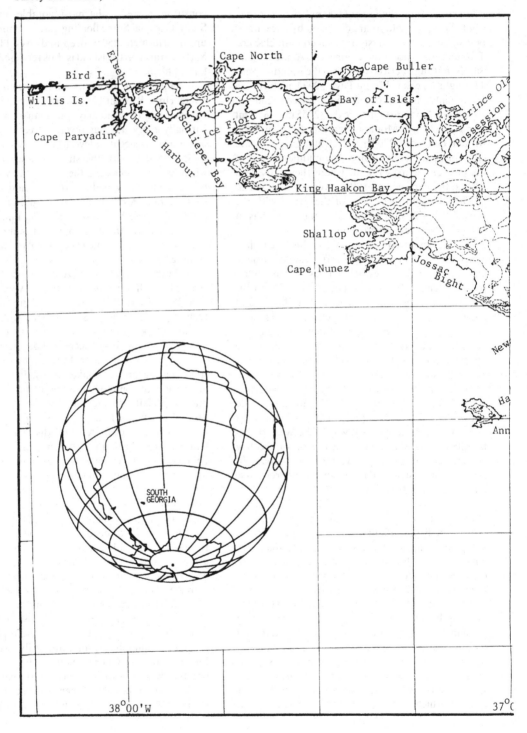

Cape North

Cape Buller

Bird I.

Willis Is.

Elsehul

Bay of Isles

Prince Ol

Possession

Cape Paryadin

Undine Harbour

Schlieper Bay

Ice Fjord

King Haakon Bay

A

Shallop Cove

Cape Nunez

Jossac
Bight

New

Ha

Ann

SOUTH
GEORGIA

38°00'W

37°0

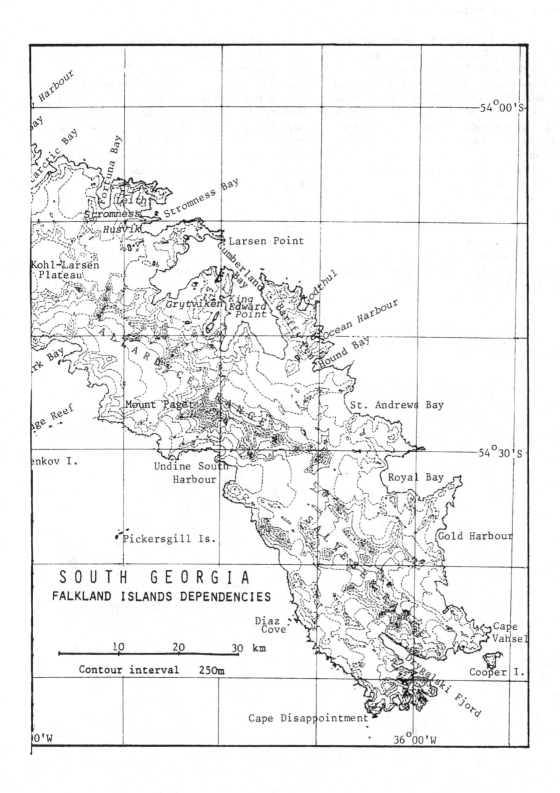

SOUTH GEORGIA
FALKLAND ISLANDS DEPENDENCIES

0   10   20   30 km

Contour interval   250m

termed calving). Beaches in the vicinity of large glaciers show the effects of this quite distinctly, with erratic boulders and large masses of ice deposited on them. Bergy bits, growlers, and brash ice (pieces of ice of different sizes from calving glaciers) infest many bays, especially after storms and during early summer when the glacial advances (formed during winter) break off. These may present hazards to navigation.

Vegetation, which is discussed in detail in Chapter 8, is mainly confined to coastal areas; some lichens may, however, be found on the highest exposed rocks. The largest plants on the island are grasses. The distribution of vegetation is profoundly affected by the shelter provided by the mountain ranges; the area from Royal Bay to Possession Bay is, by South Georgian standards, comparatively luxuriant.

Sovereignty over South Georgia

The sovereignty exercised over South Georgia and other parts of the Falkland Islands Dependencies (as well as over the Falkland Islands and parts of British Antarctic Territory) by the British Government is disputed by the Government of Argentina. In the Dependencies this appears to have begun in 1925 and has presented increasing difficulties. This culminated in 1982 when Argentine military forces invaded South Georgia and, after 3 weeks, were removed by the Royal Navy. Similar events took place in the Falkland Islands at that time. The diplomatic and political matters involved are discussed in Chapter 9, as they refer to parts of the history and other subjects in earlier chapters. Various aspects of political science and international law are also involved and are briefly described there, together with the war which followed the failure of diplomacy.

Part of the Allardyce Range showing Mount Sugartop with Hamberg Lakes and the overspill from the Hamberg Glacier. (Author.)

The events of 1982 mark a substantial change in South Georgia and the probable future for the island was both greatly affected and discussed in consequence. Thus the last chapter closes with a brief note on what might subsequently become of South Georgia.

Government

South Georgia is part of the Falkland Islands Dependencies, a British dependent territory. Constitutional responsibility to Parliament for the government of the Dependencies rests with the Secretary of State for Foreign Affairs. For reasons of administrative convenience the Government of the Colony of the Falkland Islands is empowered to legislate for the Dependencies which comprise the eleven South Sandwich Islands, Clerke Rocks, Shag Rocks and Black Rock as well as South Georgia. In 1962, however, the Colonial Office considered that they might later be designated as a separate colony but nothing resulted. The Governor and Commander-in-Chief of the Falkland Islands is also appointed to these offices for the Dependencies and his Executive Council is common to them. (Additionally he is the High Commissioner of British Antarctic Territory.)

A Magistrate, appointed by the Governor, has resided at South Georgia since 1909 and conducts the local administration. For various periods he has been assisted by a Customs Officer, a Police Constable, and other officers. From 13 November 1969 until early 1982, the Magistrate was also the Commander of the British Antarctic Survey scientific station at King Edward Point. Adminis-

The **Brøgger** Glacier in the south-eastern part of the island. (**B. C. Storey.**)

tration on South Georgia has required the Magistrate to function in a large number of capacities including: Administrative Officer, Deputy Post Master, Immigration Officer, Collector of Customs, Harbour Master, Receiver of Wrecks, Coroner, Gaoler, Registrar, and Health Officer. In recent years, most of the Magistrate's work has been connected with shipping during the summer and with the Post Office. A Customs House was established at Leith Harbour in 1958 for the convenience of the Magistrate and other Government officers visiting the station.

Legislation is enacted, and regulations, proclamations, notices and appointments are made for the Dependencies by the Government in the Falkland Islands. These are promulgated in the Falkland Islands Gazette published in Stanley. Falkland Islands law does not apply in the Dependencies unless specifically extended to them, when notification appears in the Gazette. In the case of

many earlier ordinances this was qualified by the phrase 'as far as circumstances permit and local conditions render necessary'. The basis of the law of the Dependencies was defined in 1908 as including the Common Law, Doctrines of Equity, and Statutes of general application which were in force in England on 22 May 1900 and provision for its administration was made in the same Act. In 1953, a revised edition of the laws of the Falkland Islands and Dependencies was published in two volumes which came into force, by proclamation, on 12 March of that year.

After the defeat of the Argentine invasion, described in Chapter 9, the *Falkland Islands and Dependencies (Interim Administration) Order 1982* was made by the Queen in Council at Windsor, which came into operation on 18 June 1982. This provided for the interim administration of the Colony and Dependencies by both a Civil Commissioner and a Military Commissioner. It defined their

The coast north from a pass between Royal Bay and Doris Bay. (Author.)

powers and suspended the office of Governor and Commander-in-Chief during its effect. The exercise of the functions of this office was vested in the Civil Commissioner while the order is in force and the previous Governor, Mr (later Sir) Rex Hunt, was appointed to the post. The Military Commissioner was given the right to attend and take part in the proceedings of the Executive Council but has no vote in it. At South Georgia most actions of Government have been conducted by the officer commanding the garrison at King Edward Point, who is appointed as Magistrate.

Armorial Ensigns were assigned to the Falkland Islands Dependencies by a Royal Warrant signed by the Queen on 11 March 1952. These are heraldically described as 'Per fesse barry wavy of six Argent and Azure and Argent on a Pile Gules a Torch enflamed proper: and for Supporters: On the dexter side a Lion Or and on the sinister side an Emperor Penguin proper the whole upon a Compartment divided per pale and representing dexter a Grassy Mount and sinister an Ice Floe; together with the Motto "Research and Discovery"'. As well as the Union Flag, the flag of the Dependencies is the Blue Ensign with the Shield of Arms of the Dependencies on the fly. The island's weather ensures only a brief life for any particular flag.

The judicial system operates with the Magis-

The Busen Peninsula across Stromness Bay from near Leith Harbour showing typical lowland scenery in the area protected by the Allardyce Range. (Courtesy Trans-Antarctic Expedition.)

trate holding the court of first instance in the Dependencies. Appeals lie with the Supreme Court of the Falkland Islands (this and the Falkland Islands Magistrates Court are common to the Dependencies). The most recent sitting on a criminal case in South Georgia took place at King Edward Point in 1962 and concerned the illicit production of alcohol. The defendant was found guilty and was sentenced to a term of imprisonment with labour – at about the time several Government buildings needed painting.

A gaol was built in the Customs Warehouse in 1914 and was approved as a 'Common Gaol' in 1915. The first prisoner broke out through parts of the structure then unfinished. He fled up the steep slopes of Mount Duse and stoned his pursuers. After one very cold night in the open he returned voluntarily to gaol and later served as an assistant to a biologist at Leith Harbour. Since its establishment, the gaol has probably provided more accommodation for members of expeditions visiting the island than for prisoners.

Revenue

Government moneys are presently derived from Posts and Telegraphs, taxation of income, port fees, the sale of the water, the whaling station leases, and some minor items. In normal circumstances the Dependencies budget is without deficit. During the whaling era, much larger sums were

The Armorial Ensign of the Falkland Islands Dependencies (with permission of the Civil Commissioner.)

RESEARCH AND DISCOVERY

raised from taxes on whale and seal products as well as shipping (some of which supported research on whaling and related matters). The majority of the Post and Telegraph revenue is raised from the sale of postage stamps for philatelic purposes and the stamp-issuing programme, managed by the Crown Agents, takes account of this. Officers of the British Antarctic Survey, serving for other than brief periods on South Georgia, pay income tax at Dependencies rates. Harbour fees and Customs duties are substantial sources of revenue; as many as 25 ships liable for these may visit the harbour in the course of a summer season. Water is also sold to ships, drawn from a dam built by the Grytviken whaling station. The aqueduct and other facilities have been maintained by the British Antarctic Survey and the water is considered to be of high quality.

The currency of South Georgia is the Falkland Islands Pound, which is on a par with the Pound Sterling. Falkland Islands monetary notes and coins together with Sterling are in circulation. The only actual use for these on the island is for the purchase of postal items, a small number of publications sold by the British Antarctic Survey and some charts. An unofficial currency once circulated on the island when the Compañía Argentina de Pesca issued tokens for 50 øre, 1 and 5 Norwegian crowns (Norse Kroner), with its insignia stamped on one side and the value on the reverse. They were issued by a 'Grytvikmynt' from 1908 to 1914. There are no commercial or banking facilities established on South Georgia. Arrangements for purchasing items not supplied by the British Antarctic Survey, Falkland Islands Dependencies Government, or HM Forces for their personnel, may be made occasionally with some visiting ships, or ordered from the merchants in Stanley and elsewhere. The only publicly available transport for merchandise is the Post Office. No supplies of food, fuel, equipment, etc. are available for sale on the island.

Government moneys are transmitted to the Post Office and Colonial Treasury in Stanley as appropriate. Fiscal estimates for, and the accounts of, the Dependencies are prepared by the Treasury. Presently, the only imports are food, fuel, scientific and personal equipment together with items required by HM forces especially since April 1982. There are now no exports.

Population and settlement

Permanent settlement of South Georgia began on 16 November 1904, with the foundation of the Grytviken whaling station. Prior to this, however, many sealers had overwintered and spent periods of a year or more on the island, especially in the early part of the nineteenth century. One scientific expedition, the German International Polar Year Expedition of 1882–83, remained for more than a year at Royal Bay where they established a station with some eight buildings, the ruins of which are still apparent. These early visitors are described in Chapters 3 and 4, with the island's history.

The vicinity of Grytviken and King Edward Point is by far the best site on the island for habitation. C.A. Larsen, the founder of Grytviken, selected it following two visits to South Georgia. It has a well sheltered harbour, a bay within a bay (which is of good size and depth), sufficient flat land suitable for the erection of buildings, and abundant supplies of fresh water. It occupies a central position, and it is in the most climatically protected region of the island. No other site on South Georgia has all these advantages.

King Edward Cove from Mount Hodges, the scientific station is on the left and the whaling station in the foreground, January 1981. (Courtesy British Antarctic Survey, C.J. Gilbert.)

Population increased very rapidly during the early years of this century after the foundation of Grytviken, when five other whaling stations were established and a maximum of eight floating factories worked at the island. The first building was established at King Edward Point in 1907; this was a meteorological station, erected by the Compañia Argentina de Pesca, which continued recording observations in accordance with the requirements of the company's lease of the whaling station site. Following these changes in occupation of the island, the Governor appointed a Stipendiary Magistrate to reside there in 1909. He conducted a census of the population on 31 December 1909. It showed a total population of 720, 472 of whom were residing ashore and 248 on floating factories anchored for the summer season. Of the population, three were female and there was one child. Distribution according to nationality showed an overwhelming preponderance of Norwegians (80.5%), followed by Swedes (8%), Britons (4.4%), Danes (2.25%), Finns (2.08%), Germans (1.25%), Russians (1%), Dutchmen (0.27%), Austrians (0.14%) and Frenchmen (0.14%). Scandinavians represented 92.83% of the total. The average age of the population was 28 years, with 65 below 18 years and 8 over 61. In 1919, six land-based whaling stations operated. With the administration outpost, there were seven centres of population, namely: Prince Olav Harbour, Leith Harbour, Stromness Harbour, Husvik Harbour, Grytviken, Ocean Harbour and King Edward Point. A floating factory, with a small shore depot, was also established at Godthul. Population size depended principally on the whaling industry. Thus it declined during the winter when no whaling was done and fell precipitously when whaling from land stations finished at South Georgia.

Since the end of the whaling era in early 1966, the population of South Georgia has varied from a winter minimum of about 12 to a summer maximum of about 50. In recent years, the wintering population has been generally 20 men, rising to 25 in summer (almost all of whom are British nationals). They have been almost entirely employees of the British Antarctic Survey, from late 1969 when the Survey took over the island's administration up to the Argentine invasion in early 1982. Exceptions have been the occasional summer visitors performing specialised functions. Subsequent to the

Argentine attack, a somewhat larger contingent of British forces has been in residence.

The first birth in the Falkland Islands Dependencies, indeed in the Antarctic, took place at Grytviken on South Georgia on 8 October 1913. Solveig Gunbjorg Jacobsen (daughter of Fridthjof Jacobsen, the assistant manager of the whaling station, and of Klara Olette Jacobsen), was born and registered by the Magistrate. There have been several more births, the most recent in 1979 aboard a private yacht. The earliest recorded grave on South Georgia is of Frank Cabriel, a sealer who died on 14 October 1820 and there are several other sealers' graves on the island. Hazards from the sea, weather, shipwreck, disease, accidents, the nature of employment, and other things have resulted in many deaths. All the whaling stations had cemeteries, a total of seven containing almost two hundred graves (1820–1983); and many men were lost at sea. Ages at death vary from 17 to 71 years and all are male. Several marriages have been celebrated on the island: the first, on 24 February 1932, between Mr A.G.N. Jones and Miss Vera Riches. Even divorce proceedings have taken place on the island, although initiated elsewhere.

The centre of administration was established at King Edward Point on 17 September 1912 after functioning at Grytviken from 2 December 1909. King Edward Point lies on a raised spur of morainic materials at 54°16'38·35″ S latitude by 36°29'16·74″ W longitude (the coordinates of a satellite survey station). It is the site of the British Antarctic Survey station named 'Grytviken' in 1978 (not without causing confusion with the abandoned whaling station 1 km away, to which this name is also commonly applied), and is now the only centre with a continual population. Another British Antarctic Survey station is presently established on Bird Island off the north-western extremity of South Georgia. The first building was erected there in 1958 and, over the years, several more have been constructed. The latest, and most substantial, was completed in 1982. In 1963, 1983 and 1984 it has been manned throughout the winter.

As well as these two stations, there are 14 field huts established by the British Antarctic Survey around the island for particular research work. Eight other forms of accommodation and six refuges are also used; they have very varied standards of

The sites used by whaling companies and sites where populations have lived over winter on South Georgia (other than sealer's encampments). (Author.)

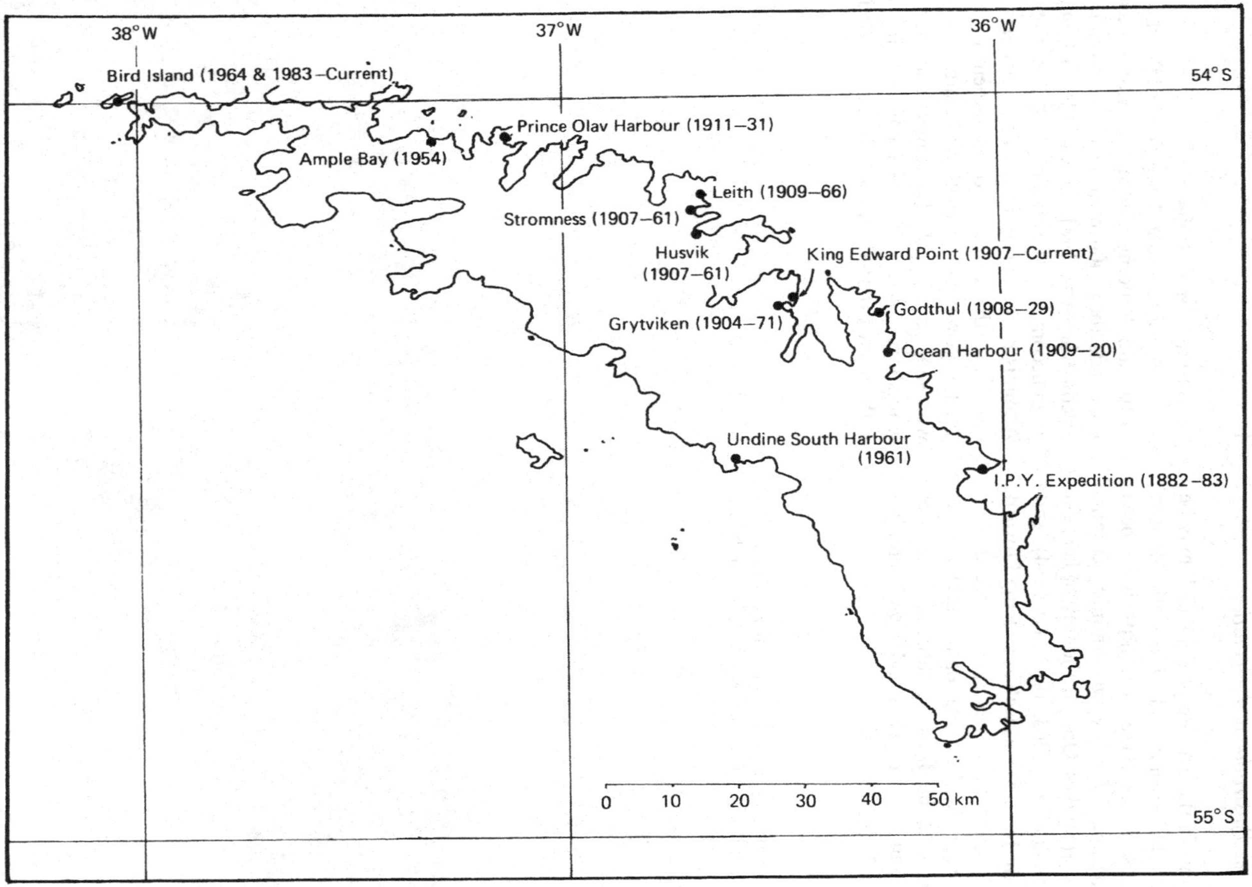

Bird Island (1964 & 1983—Current)

Ample Bay (1954)

Prince Olav Harbour (1911—31)

Leith (1909—66)

Stromness (1907—61)

Husvik (1907—61)

Grytviken (1904—71)

King Edward Point (1907—Current)

Godthul (1908—29)

Ocean Harbour (1909—20)

Undine South Harbour (1961)

I.P.Y. Expedition (1882—83)

38° W

37° W

36° W

54° S

55° S

0   10   20   30   40   50 km

comfort. Most of these are in places which may be reached by trekking after landing in a suitable area by launch. Some of these field huts have been regularly occupied for several summer seasons, mainly for seal studies. Only at the Bay of Isles (in 1954), Bird Island (in 1963, 1983 and 1984), and Royal Bay (in 1882–83), as well as at King Edward Point, have scientific parties spent more than a continuous year. Accommodation in other areas is generally in tents; the British Antarctic Survey type of pyramid tent is found to be good at withstanding most of South Georgia's weather.

King Edward Point

The settlement at King Edward Point consists of 10 major buildings and various stores, sheds, etc. It extends about 400 m along a track from the jetty (reconstructed in 1981) to Shackleton House, the main accommodation building. A Post Office and Administrative Office, near to the jetty has functioned since the early days of the settlement. Nearby are the power house (with associated stores and workshops); boat sheds, stores, and boatman's workshop, the radio and meteorological building with much external equipment; and a food store

The settlement at King Edward Point in February 1981. (Courtesy British Antarctic Survey, C.J. Gilbert.)

with freezing facilities. Discovery House, which stands further along the track, was built in 1925 as a laboratory and now houses construction and maintenance facilities together with the diving equipment and a store. A gaol, Customs House and two remaining former civil residences serve as stores and occasionally provide accommodation and laboratory facilities. At the end of the track stands Shackleton House, a large three-storey building constructed in the 1963/64 summer. It provides accommodation, dining facilities, laboratories, a surgery, some recreational facilities etc., as well as additional storage space. All the buildings and the majority of the other facilities on King Edward Point date from the days of the civil administration, when the whaling stations operated. The British Antarctic Survey has been pursuing a demolition policy for several years and had a long-term plan to build a new station near the jetty. Owing to the requirements of the garrison stationed at King Edward Point since April 1982, some changes in use of the buildings and other facilities have recently occurred.

The facilities established at King Edward Point, when it was manned by the British Antarctic Survey include well-equipped carpentry, engineering, electrical, and building workshops; as well as accommodation, laboratory, observatory, communications, administration, and stores buildings. Electric power was provided by diesel-driven dynamos which produce about 2000 kWh a day (440 and 240 V, 50 Hz). Running water, usually available throughout the year, was drawn from a dam in an adjacent valley about 2 km away (a new pipe was laid in 1980). A 25-line automatic telephone exchange operated internally. Heating of the major buildings was by electricity or oil-fired boilers fed from a main pipeline. Sea transport consisted of an 8.23 m launch *Albatros* (wrecked during the 1983 winter), a powered dinghy and inflatable rubber craft with outboard engines. Land transport, used around King Edward Point and Grytviken, includes tractors with trailers (and the author's bicycle) for use in summer; 'Snow-Mobiles', other tracked vehicles, and sledges for winter. A greenhouse allowed some salad vegetables to be grown throughout the year. Chickens are kept for the eggs. Medical facilities were available with a well-equipped surgery but rarely used (the standard of health was high

and British Antarctic Survey personnel are subject to a medical examination before arriving at the island). A medical officer was present during much of the summer and a dentist visits the station from one of the British Antarctic Survey's ships every year.

Sports and recreational facilities included football, darts, billiards, table-tennis, badminton and many others. It was usual for the King Edward Point team to defeat ships' companies on the football pitch at Grytviken whaling station, where up to a dozen 'international matches' are played annually. Excellent opportunities existed, (somewhat dependent on the weather) for outdoor pursuits such as trekking, diving, skiing skating, and sledging. Even swimming was not unknown in lakes at the height of summer. Persons who could afford the time were generally encouraged to spend up to three weeks on recreational excursions during the winter. The extreme beauty of the island made photography a very popular hobby, with a well-equipped darkroom to support it.

Indoor facilities included extensive libraries, films, tape and record collections which receive new and exchange material from ships of the British Antarctic Survey and Royal Navy. There was a well stocked bar with the most liberal licensing hours and many card, board and similar games. Hobbies (modelling, construction, electronics, painting, preparation of natural history specimens and much more) were generally abundantly catered for. Perhaps the greatest problem with all these possibilities was to get sufficient time to enjoy them, even during the long winter.

From 25 April 1982 a British garrison has been stationed at King Edward Point and the British Antarctic Survey has not been active there. The majority of the facilities of the Survey are in use by the garrison although various changes and additions have been made as required by the different circumstances.

The abandoned whaling stations

There are nine sites on South Georgia for which whaling leases were granted: Ocean Harbour, Godthul, Grytviken, Jason Harbour, Husvik, Stromness, Leith Harbour, Prince Olav Harbour, and Rosita Harbour. At Rosita Harbour nothing was built and at Jason Harbour only a small hut was constructed (it still stands and was recently

A plan of the scientific station at King Edward Point.
(Author.)

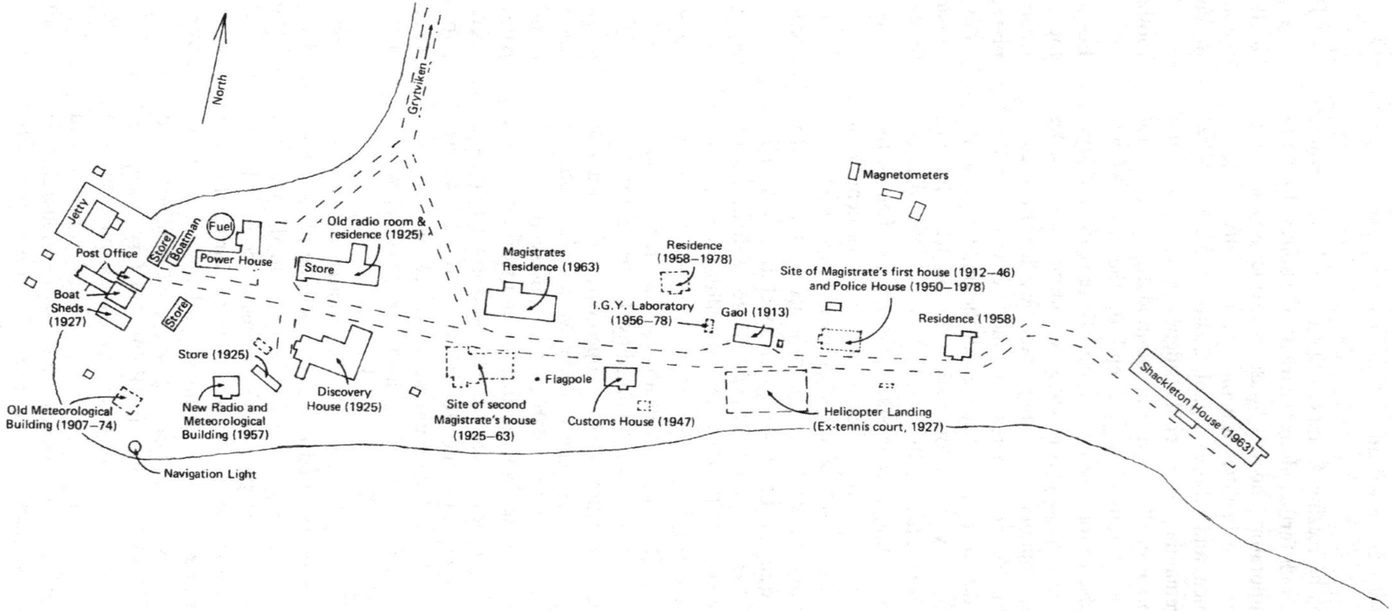

North

Grytviken

Magnetometers

Shackleton House (1963)

Site of Magistrate's first house (1912—46)
and Police House (1950—1978)

Residence (1958)

Helicopter Landing
(Ex tennis court, 1927)

Residence
(1958—1978)

Gaol (1913)

I.G.Y. Laboratory
(1956—78)

Magistrates
Residence (1963)

Flagpole

Customs House (1947)

Site of second
Magistrate's house
(1925—63)

Old radio room &
residence (1925)

Store

Discovery
House (1925)

Fuel

Power House

Store
Boatman

Store

Store (1925)

New Radio and
Meteorological
Building (1957)

Navigation Light

Post Office

Jetty

Boat
Sheds
(1927)

Old Meteorological
Building (1907—74)

refurbished). Ocean Harbour was closed in 1920 and now has only two ruined buildings, some machinery parts, and a small locomotive, as most of its equipment was taken to Stromness. Godthul has one large shed, three oil tanks, several wooden 'jolle' (the boats from which the whales were flensed), a dam and many barrels. The other five sites have far more substantial remains although, at Prince Olav Harbour, they are almost entirely ruinous (having been abandoned in 1931). Husvik, Grytviken, Strommess, and Leith Harbour were operating in 1960. When the latter three closed, by late 1965, they were left substantially intact and prepared so that they could be reopened should whaling resume. The effects of weather and human interference have caused many of their buildings to collapse and most others are very dilapidated. The remains of boilers, pressure cookers, rotary driers, power houses, and much else are there and attracted interest from a salvage collector (the consequences of which are related in Chapter 9). The abandoned whaling stations are strange places with many large factory buildings where the odour of whale lingers; dilapidated offices, laboratories, hospitals, accommodation, kitchens and much else; paths overgrown with plants; and a substantial population of rats. The whaling stations' history and operations are described in Chapter 5.

Survey, charts and place-names
The first chart of South Georgia was drawn by Captain Cook following his exploration of and landing on the island in January 1775. Cook named 18 features on this chart – as well as the island itself. With one exception, all remain current although one place-name has been moved as a result of later confusion. The next map, said to have been compiled in 1802 by a Captain Pendleton, was not published until 1906. The place-names it gave included almost all of Cook's, the exceptions being variants and one omission, together with 31 others derived from the sealers who were active on the

Grytviken whaling station in October 1982. (Author.)

island at the time. Thaddeus Bellingshausen prepared a chart of the island following his survey of the southern coast in 1819. This included five more place-names and renamed Cook's Pickersgill Island as Annenkov Island (Cook's name was subsequently transferred to a neighbouring group of islands). A passenger on a sealing ship in 1877, H.W. Klutschak, published the next known chart. This included about nine new names, as well as using (and to some extent confusing) translations of previous names.

The first detailed survey of a particular area of South Georgia was made by the German International Polar Year Expedition of 1882–83, of the region around their station at Royal Bay. The Swedish South Polar Expedition of 1901–03 made similar surveys of the Cumberland Bay region, the area around Grytviken, and elsewhere. These surveys were improved and added to rapidly after Grytviken and the other whaling stations were established and the resident Magistrate arrived early in the twentieth century. The first Royal Naval survey, after that of Captain Cook, was made by Captain Hodges aboard HMS *Sappho* in 1906 and, later that year, the first of the modern series of Royal Naval Hydrographic Charts for South Georgia was published. Subsequently the Royal Navy has continued to improve the charts on many occasions with data from its own surveys and other sources.

The first Magistrate, Mr James Innes Wilson, prepared several sketch maps of parts of the island which bore all Cook's names and a variety of others. A.E. Szielasko, D. Ferguson, R.C. Murphy and others contributed to charting local areas and added place-names where appropriate. The Filchner Expedition of 1911–12 prepared a map of the island, part of a blue print of which enabled Shackleton, Crean and Worsley to navigate on their epic trek from King Haakon Bay to Stromness Bay in 1916. Several other contributions to the charting and place-names of the island were made before 1928 when the Discovery Committee initiated a detailed survey of the coasts and anchorages. This resulted in a comprehensive series of charts being prepared. The first aerial surveys were made in 1938 from HMS *Exeter* using Walrus aircraft. Since the Second World War, many more hydrographic surveys (often using aircraft) have been made; these, and their predecessors, are related in Chapter 4.

The first extensive inland survey was made in 1928/29 by the Kohl-Larsen Expedition. Little more was done until 1951, when Duncan Carse led four expeditions of the South Georgia Survey from 1951 to 1957. These resulted in enormous improvements in the maps and are described in Chapter 4. The first Directorate of Colonial Surveys land map of South Georgia was published in 1950 and was deficient in many details; in 1958 Carse's map, published by the Director of Overseas Surveys, was a great contrast. Since 1960, helicopters have been used to take aerial photographs for surveying and, subsequently, satellites have been used for obtaining accurate positions and transmitting high-altitude photographs of the island.

The early naming of features of South Georgia had been a rather uncontrolled process without much coordination. Many features had several names in various languages applied to them. In 1945 the Antarctic Place-Names Committee was formed in London at the suggestion of Dr. B.B. Roberts. The Committee's terms of reference were 'to consider existing and proposed place-names in the Antarctic, and to make recommendations'. Dr Roberts was the first secretary of the committee and held the post to 1974. He compiled an index of all recorded place-names for South Georgia and other parts of the Falkland Islands Dependencies (as then defined) with details of their locations, references to maps or charts, origins, derivations, and other details. A series of general principles to ensure appropriateness, precision, stability, and correctness of place-names was also developed. From these, the first Gazetteer of the Dependencies was compiled in 1953 and regulations were promulgated in the Falkland Islands Gazette to make these official names and suppress variants. A Gazetteer for South Georgia, based on this with additions, was issued in 1954. It had 452 entries and included chart references. Several subsequent editions and amendments to these gazetteers have appeared.

In 1962 British Antarctic Territory was separated from the former Falkland Islands Dependencies which then came to include only South Georgia with Shag Rocks, Black Rock, and Clerke Rocks, and the South Sandwich Islands. Separate gazetteers were published for these places in 1977. The one for the newly defined Dependencies, with supplements in 1979 and 1983, includes 670 official place-names for South Georgia. The detailed historical records of the Committee have allowed appropriate and significant names to be allocated to features of South Georgia as required. Many commemorate persons, ships, and other items connected with the expeditions to the island. The index, started by Dr Roberts in 1945, has been maintained and now records nearly 3000 synonyms for place-names in the Dependencies from more than 400 published sources in eight main languages. Of these the majority are of English derivation and most of the remainder Norwegian; others have German, Spanish, Swedish, Russian, and Polish origins. In 1980 details of these were published with a discussion of the history and naming procedures in a most interesting work prepared by G. Hattersley-Smith, the second secretary of the Antarctic Place-Names Committee.

Shag Rocks, Black Rock, and Clerke Rocks

Shag Rocks lie at 53°32.5′ S latitude by 42°01.7′ W longitude and consist of two groups, each of three main rocks, about 200 m apart. They are sharply pointed and have an asymmetric profile: sheer north sides and less steep, foliation-controlled, south sides. The highest rock is about 71 m above sea level and the highest in the other group reaches 59 m. Black Rock lies at 53°38′ S latitude by 41°48.1′ W longitude about 18 km from Shag Rocks, and rises only 3 m above sea level. Another rock over which the sea breaks heavily lies 1 km to the east of Black Rock.

Geologically, these rocks consist of greenschists cut with quartzose veins. Directly comparable rock types are absent from South Georgia although they can be more closely matched with those of the South Orkney Islands and the South

Shag Rocks, from a photograph taken from RRS *Discovery* in 1930. (Courtesy Royal Navy.)

Shetland Islands. The rocks form the only exposed part of a small continental block, part of the Scotia Ridge, which is approximately one sixth the size of that of South Georgia. Biologically, as might be expected, their principal inhabitants are blue-eyed shags.

Shag Rocks were probably discovered in 1762 by the Spanish vessel *Aurora* but, if so, they were charted in the wrong position. Several other vessels subsequently sighted them and they were well known to the sealers and others in the eighteenth century, although their position remained equivocal. HMS *Dartmouth* made the first accurate determination of their position in 1920. The first landing was made in March 1956 by helicopter from the Argentine naval ship *Bahía Aguirre* by M.B. Giovinetto. In December 1972 HMS *Endurance* made a detailed survey of them during which two naval parties made geological and other collections. Black Rock was charted at this time but no landings have been made on it.

Clerke Rocks lie at 55°01′S latitude by 34°42′W longitude and also consist of two groups; the western one includes three large rocks the highest of which is 244 m above sea level, the eastern group includes Nobby and The Office Boys, the highest of which reaches 82 m. The rocks, which extend for 11 km, are of the same geological structure as the Drygalski Fjord Complex of the adjacent mainland and lie on the same continental block. They are frequented by numerous shags as well as other sea birds and have a sparse lichen flora. A patch of kelp indicating a shallow rock is reported about 7 km east south-east of Nobby.

Clerke Rocks were discovered in January 1775 by Captain Cook who named them after Lieutenant Charles Clerke aboard *Resolution*. The first landing was made by Norwegian whalers in 1928, when rock samples were collected. Only three other landings are recorded, on two of which biological and geological examinations were made.

Clerke Rocks, the north-east portion seen from the north. (Courtesy Trans-Antarctic Expedition.)

2

Discovery of and first landing on South Georgia

Prior to Captain Cook's exploration of South Georgia in 1775 the island had been seen twice and two other voyages had sailed close to it. Although details of the earliest two sightings are somewhat scarce, the exploration by Cook is well documented. Relevant passages of these early works are quoted at length in this chapter for they are difficult to find elsewhere and are of great interest. The discovery of South Georgia has sometimes previously been credited to Amerigo Vespucci in 1501 or 1502. Re-examination of the records has, however, shown this to be without foundation.

Discovery of South Georgia,
Antoine de la Roché (1675)
The first discovery of land in the Antarctic, south of the Antarctic Convergence, was most probably the sighting of South Georgia in 1675 by a London merchant, Antoine de la Roché. The original records of his voyage are no longer available and only a translated précis survives. la Roché was born in London, though his father was French. He had previously undertaken various mercantile voyages; in 1674 he sailed from Hamburg to Peru in a 350 ton vessel, the name of which is unfortunately not recorded. The return journey was made with a halt in Chiloé after which the ship endeavoured to pass the Le Maire Straits in April 1675. A translation of an account of the voyage describes this and subsequent events.

... being solicitous to pass by the said Strait Le Maire in April 1675, they could not, the Winds and Currents having carried them so far to the Eastward; and being unable to return towards the Land of the Strait of Magellan, nor to make Staten Land to sail into the No. Sea by Brower's Strait, and seeing that it was far advanced in April and beginning of Winter in that Climate, it would be much if they escaped with Life, particularly as they had no Knowledge or Intimation of the Land which they *now began to see toward the East* which making and using all endeavors to get near it, they found a Bay, in which they anchored close to a Point or Cape which stretches out to the SE with 28. 30. and 40. fath. Sand and Rock, in which situation they had sight of some Snow Mountains near the Coast, with much bad Weather; they continued there 14 days, at the end of which time having the Weather cleared up, they found that they were at that end of that Land, near which they had anchored, and looking to the SE and South, they saw another High Land covered with snow, leaving which, and the Wind setting in gently at SW and sailed out in sight of the said coast of the Island which they left to the Westward, seeing the said Southern land in the said Quarters, it appearing that from one to the other was about 10 lea. little more or less, and that there was a great Current to the NE, to which Point sailing, and steering ENE they found themselves in the No. Sea, in 3 Glasses disemboguing thro' the said passage, which they say is very short; for the land which appears to form the said New Island is small; which leaving and sailing one whole day to the NW, the Wind set in so strong and stormy at S that they sailed other 3 days to N till they were in 46″. ...

Thence they sailed to San Salvador in Brazil and onward to La Rochelle in France arriving on the 29 September 1675. Captain Francisco de Seixas y Lovera of Spain, an experienced mariner who had rounded Cape Horn several times, prepared the original of the extract quoted above for his *Description Geografica de la Region Magellanica*, published in 1690 (see Dalrymple, 1771).

The island la Roché discovered was almost certainly South Georgia. The discrepancy in geographical coordinates is reasonable considering the accuracy of navigation at that time – especially aboard a vessel suffering an Antarctic oceanic storm. Dr L. Harrison Matthews reviewed the description. His interpretation, based on an extensive knowledge of South Georgia, was that la Roché sailed past the southern side of the island, turned north near Cape Disappointment and anchored in Doubtful Bay or Drygalski Fjord for the 14 days, then continued north. The 'another High Land covered with snow', to the south-east would thus have been Clerke Rocks and the 'snow' may have been, in part, guano. During the period the ship was at anchor, no landing was described; presumably one was not made because the weather did not permit it. They departed as soon as conditions improved. In any case, the shores at the southern part of the island are largely composed of steep cliffs and thus the only descriptions of the land were of what was seen from the ship. The voyage was not one of exploration but a mercantile one blown far off course.

Sir Edmond Halley and *Paramour* (1700)

An interesting near miss of South Georgia was made in 1700 by Sir Edmond Halley aboard *Paramour*. He reached a position of 51°S by 43°W on a voyage to determine terrestrial magnetic variation in the Atlantic Ocean. At that position he found two species of penguins, other diving birds, seals and whales. He returned northwards after encountering a 'great fogg' and icebergs.

Second sighting of South Georgia; Jerez and *León* (1756)

The second sighting of the island was also made by a merchant vessel assailed by storms; the 468 ton *León* commanded by Gregorio Jerez of Spain. The vessel was in the service of French merchants and Nicolas-Pierre Guyot, Sieur Duclos of St Malo, France (the port from which most

French voyages to South American regions sailed), travelled aboard her. He kept a journal and an abstract, prepared by the geographer d'Apres Mannevillette, was published by Alexander Dalrymple in 1771 in the original language. Guyot later rose in naval rank and became the second in command of Bougainville's voyage circumnavigating the world.

León left Callao, Peru, on 8 April 1756 carrying a large quantity of bullion, cacao, alpaca wool, metals and pharmaceuticals. Additionally, she carried 50 passengers including the late Viceroy of Chile and his family. They were probably embarked on 30 April when the ship called at Valparaiso. By 26 June the ship had rounded Cape Horn under difficult circumstances in severe storms, her crew had been obliged to throw warm water on the freezing rigging to enable her sails to be manoeuvred. They were, by this time, well to the east of the Falkland Islands on a course that led to South Georgia. On 28 June increasing numbers of birds were noticed around the ship. At 09:00 on 29 June, land was believed to have been sighted but this could not be confirmed owing to the prevailing weather conditions. That evening, however, a small island was seen. This was probably Annenkov Island. Later the main island was sighted. A translation of the account made by James Burney in 1817 records:

At nine in the morning, we beheld a Continent of land extending about 25 leagues in length from NE to SW, full of sharp and craggy mountains of frightful aspect, and of such extraordinary height that scarcely could we discern the summits, although at a distance of more than six leagues. The quantity of snow on the land prevented our seeing if wood grew there. The observation on which we could most depend of any we were able to make, we being then three leagues distance from the small Island which was found to be at the like distance from the great land, is, of a very deep bay in this Continent, about eight leagues East and West with the said small isle. It is the only part which appeared to us fit to be inhabited; we might be distant from it 10 or 11 leagues: it appeared to us of large extent both in length and breadth.

There is to the left of its entrance, in the WNW from us a low point, which is the only one we could perceive. It appeared to us as if detached from the large land, and we are in doubt whether it is separate or joined by an isthmus.

Yesterday at four o'clock in the afternoon, died Don Domingo Dortez, Lieutenant General of the Armies of his Catholic Majesty, Count de Peuplades, and late President of *Chili*, aged 80 years. At ten this forenoon he was cast into the sea with the customary ceremonies. The Spanish crew saluted him with seven *Vive le Roys*, and respectfully wished him *Bon Voyage*. Latitude by account (at noon) 54°48′. Longitude 51°30′.

The 30th, from noon to four in the afternoon, the wind was from NW to SWbS, light, with fair weather, after which time it was calm, and we remained in this situation all the night. At break of day, the ship in perfect tranquillity, we tried for soundings, being then about ten leagues distant from the land; we found no bottom, nor was any current perceptible. We had constantly seen many birds and seals. At noon the land presented the same aspect, except that the summits of the mountains were covered with snow. By a good observation we found the latitude 54°50′ S: our longitude 51°32′.

Thursday July the 1st. The wind from the WNW, a light breeze. Steered to the SbW till sunset, to get to a distance from the land, and during the night our route was between SbE and SE. At daylight, the wind having shifted to NNW, with much fall of snow, estimating that we were at a sufficient distance from the land, we steered to the East, to see if this said land extended in that direction. At eight in the morning, the Easternmost point of the land bore N 5°, distant about 12 leagues. At noon, continuing on the same course, the latitude by account was 55°23′; longitude 51°. The 2d, light Westerly winds, the weather obscure with snow. Continuing our route to the ENE among much ice, we remarked much current, and more birds about us than usual, particularly of the white pigeons such as are seen near the Coast of *Patagonia*; many whales also: from all which we imagined we might be on a bank, but on sounding we found no bottom. We were then out of sight of land. Latitude by account 55°28′; Longitude 49°40′.

Thence they sailed towards the Cape Verde Islands which were sighted on 25 August. Guyot recorded the naming of the island after Saint Peter (Île de Ste Pierre) on whose feast day, 1 July, *León* passed the island to the south. This is the origin of the names St Peters Island, in English, and Isla de San Pedro, in Spanish, that have been applied to the South Georgia mainland. Guyot's calculation of longitude suffered from errors common at the time; some charts, published shortly afterwards, showed la Roché's land and Île de Ste Pierre separately.

Although neither of these voyages was intended to make discoveries, they were recorded in navigational descriptions and charts began to incorporate the island. Dalrymple's chart of 1771 shows

Île de St Pierre and another separate land mass with Straits of la Roché marked through it some distance farther east. Captain James Cook was familiar with these voyages prior to embarking on his second voyage when he circumnavigated the southern regions of the Earth.

Captain James Cook, the first landing and exploration (1775)

There was a notable expansion in British maritime exploration from the middle part of the eighteenth century. Commissioned voyages included those of Anson, Byron and Cook. An indication of the motivation behind this epoch of exploration is provided by a passage in the instructions to Commodore Byron dated 17 June 1764, during the reign of King George III.

Whereas nothing can redound more to the honour of this nation as maritime power, to the dignety of the Crown of Great Britain, and to the advancement of trade and navigation thereof than to make discoveries of countries hitherto unknown; and whereas there is reason to believe that there are lands and islands of great extent, hitherto unvisited by any European Power may be found, His Majesty, conceiving no conjecture so proper for an enterprise of this nature at a time of profound peace, which his kingdoms at present happily enjoy, has thought fit that it should now be undertaken.

Three of the most important of the voyages commissioned during this period were those commanded by Captain James Cook. During the second of these he reached South Georgia.

Cook was born in Yorkshire in 1728, the son of a farm labourer. He served an apprenticeship on a North Sea collier and continued sailing them until, at the age of 27, he joined the Royal Navy. There he pursued studies in surveying, mathematics and astronomy. Within 2 years he had been appointed the Master of a Royal Naval ship. In 1768 he was appointed to command HMS *Endeavour* on a voyage to observe (from Tahiti) the transit of Venus in 1769, and to make various explorations. The observations of transits of Venus across the sun's disc from different places at the same time allows accurate determination of the Earth's orbital radius. He returned to Great Britain, after circumnavigating the globe, in 1771. In 1772, he began his second great voyage of discovery and sailed from the River Thames on 22 June in command of

HMS *Resolution* of 462 tons, accompanied by HMS *Adventure*, of 336 tons, commanded by Captain Tobias Furneaux. The ships were built at Whitby in Yorkshire to a design similar to the northern English collier brigs on which Cook had spent his apprenticeship, one of many factors which contributed to the success of the voyages. One of the principal objectives of the second voyage was to investigate the 'Southern Continent', which was prominent on many maps of the period, but otherwise unknown. This was the first British Antarctic Expedition – indeed the first Antarctic Expedition of any nation.

Cook was accompanied by two naturalists on the voyage (as on his first); Johann Reinhold Forster and his son Johann Georg Forster (known as George). Among *Resolution's* officers were Lieutenants Charles Clerke and Richard Pickersgill who had served on Cook's first voyage, and Lieutenant

James Colnet who later commanded *Rattler* on a voyage of exploration near South Georgia. Lieutenant James Burney, aboard *Adventure*, later became an Admiral and compiled a list of early voyages which included those of La Roché and Jerez. Also aboard *Resolution* were William Wales and William Bayley, astronomers each with an assistant; William Anderson, a surgeon with a strong interest in ornithology; and William Hodges, an artist. A Swedish naturalist and geographer, Anders Sparrman was taken aboard at Cape Town early in the voyage and remained until the ship returned there in 1775. From the Cape of Good Hope they sailed eastwards to circumnavigate the Earth at high southern latitudes. On 17 January 1773 Cook and the expedition became the first to cross the Antarctic Circle. During the southern winter of 1773, the ships were separated by storms and *Resolution* continued alone. *Adventure* sailed for England after

The approximate courses of the first three voyages to South Georgia. (Author.)

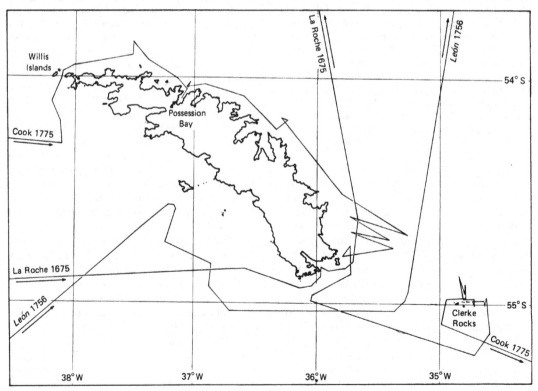

Captain James Cook from a contemporary etching.
(Courtesy Scott Polar Research Institute.)

the crew of one of her ship's boats was eaten by Maoris in New Zealand. Captain Furneaux sailed very close to and slightly south of South Georgia; without sighting it, during his return voyage from Cape Horn. He followed at first the 60° latitude to the Cape of Good Hope, where he arrived on 17 March 1774. He described the weather during the time he was, theoretically, in visual range of the island as hazy and cloudy.

Cook continued eastward across the South Pacific Ocean and passed Cape Horn. On Saturday 14 January 1775 Thomas Willis, a 'wild and drinking midshipman' saw land, which Cook later named after him. Cook's account continues:

(Saturday 14.) At 9 o'Clock the next morning saw an Island of ice, as we then thought, but at noon we were doubtful whether it was ice or land; at this time it bore E$\frac{3}{4}$S distant 13 Leagues, our Latitude was 53°56$\frac{1}{2}$′, Longitude 39°24′ West. Several Penguins, small divers, a snow Petrel and a vast number of blue Petrels about the ship. We had but little wind all the morning and at 2 pm it fell Calm. It was now no longer doubted that it was land and not ice which we had in sight: it was however in a manner wholly covered with snow. We were farther confirmed of its being land by finding Soundings at 175 fathoms, a muddy bottom, the land at this time bore EBS about 12 leagues distant. At 6 o'Clock the Calm was succeeded by breeze at NE with which we stood to SE. At first it blew a gentle gale, but afterwards increased, so as to bring us under double reefed Top-sails and was attended with snow and sleet.

Sunday 15th. We continued to stand to the SE till 7 in the morning when the Wind veering to the SE tacked and stood to the North. A little before we tacked we saw land bearing EBN. At Noon the mercury in the Thermometer was at 35$\frac{1}{4}$° [F]. The wind blew in Squals attended with Snow and Sleet and we had a great sea to encounter. At a Lee Lurch which the ship took Mr Wales observed her to lay down 42°. At half past 4 pm we took in the Top-sails, got down topgallant yards, wore the Ship and stood to the SW under two Courses. At Midnight the storm abated so that we could carry the Top-sails double reefed.

Monday 16th. At 4 in the Morning wore and stood to the East with the Wind at SSE a moderate breeze and fair. At 8 o'Clock saw the land extending from EBN to NEBN. Loosed a reef out of each Top-sail, got Topgallant yards across and set the Sails. At Noon observed in Latitude 54°25$\frac{1}{2}$′, Longitude in 38°18′ West, the land extending from N$\frac{1}{2}$W to East Six or Eight leagues distant. It appeared to be very mountainous and rocky and was allmost wholly covered with Snow. In this situation we

had 110 fathoms of water. The Northern extreme was the land which we first saw and proved to be an Island which obtained the name of *Willis's Island* after the person who first saw it.   (Beaglehole, 1961)

The Willis Islands are a small group at the western extremity of South Georgia. The main island is a prominent peak, 550 m high, clearly visible from a ship in the direction from which Cook was approaching.

Cook sailed through the strait separating the Willis Islands from the island immediately to the east in order to survey the northern side of his discovery. The island east of the strait he named Bird isle 'on account of the vast numbers that were upon it'. (Bird Island, as it was later renamed, has subsequently become a major study site for Antarctic ornithology; a scientific station was established on it in 1962.) The south coast was described by Cook as forming 'several Bays or inlets' with 'huge masses of snow or ice in the bottoms of them.' After passing through to the northern coast, Cook sailed eastwards passing and naming Cape North and Cape Buller before anchoring for the night.

On Tuesday, 17 January 1775 Cook continued sailing eastward along the northern coast passing and naming the Bay of Isles. At 07:00, an inlet was seen and a boat was hoisted out to make a landing. The landing party consisted of Cook, the Forsters, Sparrman and a midshipman. Cook's description of the bay and events continues:

The head of the Bay, as well as two places on each side, was terminated by a huge Mass of Snow and ice of vast extent, it shewed a perpendicular clift of considerable height, just like the side or face of an ice isle; pieces were continually breaking from them and floating out to sea. A great fall happened while we were in the Bay; it made a noise like Cannon. The inner parts of the Country was not less savage and horrible: the Wild rocks raised their lofty summits till they were lost in the Clouds and the Vallies laid buried in everlasting Snow. Not a tree or shrub was to be seen, no not even big enough to make a tooth-pick. I landed in three diffrent places, displayed our Colours and took possession of the Country in his Majestys name under a descharge of small Arms.   (Beaglehole, 1961)

This was the first national claim to Antarctic territory and is an important basis for British sovereignty over the island. The bay was named Possession Bay.

A brief examination of the natural history was recorded as:

Our Botanists found here only three plants, the one a coarse strong bladed grass which grows in tufts, Wild Burnet and a Plant like Moss which grows on the rocks.

Seals or Sea Bears were pretty numerous, they were smaller than those at Staten Island: perhaps the most we saw were females for the Shores swarm'd with young cubs...

...Here were several Flocks of Penguins, the largest I ever saw, we brought some on board which weighed from 29 to 38 pds..... The Oceanic birds were Albatross, Common Gulls and that sort which I call Port Egmont hens, Terns, Shags, Divers, the New White Bird and a small Duck such as are at the Cape of Good Hope and known by the name of Yellow-bills; we shot two and found them most delicate eating.

All the Land birds we saw consisted in a few small Larks, nor did we see any Quadrupedes. Mr Forster indeed saw some dung which he judged to have come from a Fox or some animal of that kind. The land or rather rocks bordering on the Sea Coast, was not covered with snow like the inland parts, but all the Vegetation we could see on the clear places was the grass above mentioned. The rocks seemed to contain Iron. After having made the above observations, we set out for the Ship and got on board a little after 12 o'Clock with a quantity of Seals and Penguins, an acceptable present to the Crew.   (Beaglehole, 1961)

Cook continued his survey of the northern coast passing, on that and the following day, places which he named: Cape Saunders, Cumberland Bay, Cape Charlotte (after Queen Charlotte), Royal Bay, Cape George, Cooper's Isle (later renamed Cooper Island) and Sandwich Bay.

On Friday, 20 January he found the coast, took a south-west direction and ended at a promontory which he later named Cape Disappointment. From the south of this he saw an island which he named after the Third Officer, Pickersgill Island. Cook then records:

Soon after a point of the main beyond this Island came in

'Possession Bay in the Island of South Georgia.' Copy of a woodcut in Captain Cook's description of the island published in 1777.

POSSESSION BAY IN THE ISLAND OF SOUTH GEORGIA

site in the direction of N 55° West which exactly united the coast at the very point we had seen and set the day we first came in with it and proved to a demonstration that this land which we had taken to be part of a great Continent was no more than an Island of 70 leagues in Circuit. Who would have thought that an Island of no greater extent than this, is situated between the Latitude of 54° and 55°, should in the very height of Summer be in a manner wholly covered many fathoms deep with frozen Snow, but more especially the SW Coast, the very sides and craggy summits of the lofty Mountains were cased with snow and ice, but the quantity which lay in the Valleys is incredible, before all of them the Coast was terminated by a wall of Ice of considerable height.

(Beaglehole, 1961)

Cook related the naming of the Island: 'This land I called the *Isle of Georgia* in honour of H. Majesty' and continued with his description of it. The name South Georgia was adopted generally by the early part of the nineteenth century.

The voyage continued and discovered some rocky islets to the south-east of South Georgia which were noticed just in time to avoid the vessel being wrecked on them. They were seen again the next day and named Clerke's Rocks (later renamed Clerke Rocks) after the Officer who first sighted them. Cook then departed from South Georgia and sailed on to discover the southernmost eight South Sandwich Islands.

In contrast to the earlier sightings of South Georgia, Cook's discovery was amply documented. Four other accounts were published and another log-book also described events. George Forster wrote *A voyage Round the World in His Britannic Majesty's Sloop Resolution Commanded by Capt. James Cook, during the Years 1772, 3, 4 and 5* published in 1778. Anders Sparrman wrote *Resa Till Goda-Hopps Udden, Södra Polkretsen och Omkring Jordklotet, samt till Hottentot og Caffer-Landen, Aren 1772–1776* published in 1783, which was translated into English in 1785 (*A Voyage Round the World with Captain James Cook in HMS Resolution*). The third account was published anonymously in 1775 as *Journal of the Resolution's Voyage in 1772, 1773, 1774 and 1775* and is believed to be the work of John Marra, a gunner's mate, from Ireland. It appeared two years before Cook's work was published. Following instructions, Cook had required all diaries to be surrendered and passed to the Admiralty, but Marra evidently had

not complied. He had been placed in irons earlier in the voyage and had a history of some indiscipline. A log-book, in manuscript form, kept by Lieutenant Clerke also amplifies some details. J.R. Forster also kept a diary of the voyage but, in contrast to the other published accounts which appeared soon after the voyage, it was delayed for over 200 years and published in 1982.

Some of these other accounts add interesting details to Cook's descriptions. George Forster, one of the naturalists, records that his father suggested the King's name for the island.

As it had been the main object of our voyage to explore the high southern latitude, my father suggested to Captain Cook, that it would be proper to name this land after the monarch who had set on foot our expedition, solely for the improvement of science, and whose name ought to be celebrated in both hemispheres.... It was accordingly honoured with the name of Southern Georgia, which will give it importance, and continue to spread a degree of lustre over it, which it cannot derive from its barrenness and dreary appearance.

On passing Bird Island he wrote that it

sloped gradually to the westward, being covered on that side by some grass, and with innumerable flocks of birds of all sorts, from the largest albatrosses down to the least petrels, for which reason it was named Bird Island. Great numbers of shags, pinguins, divers, and other birds played about, and settled in the water around us, this cold climate seemingly to be perfectly agreeable to them. Several porpesses were likewise noticed, and many seals were seen, which probably came to breed on these inhospitable shores.

On the landing in Possession Bay he wrote:

We landed in a spot which was perfectly sheltered from the swell, and where the land formed a long projecting point. Here we saw a number of seals assembled on a stony beach, and among them a huge animal, which we had taken to be a rock at a distance, but which proved to be exactly the same animal as lord Anson's sea-lion. The midshipman shot it through the head whilst it lay fast asleep, and we afterwards found a younger one of the same sort.... The animal we examined was about thirteen feet long, ... Here we likewise found a flock of about twenty pinguins, of a much greater size than we had hitherto seen; they were thirty-nine inches long, and weighed forty pounds.... These birds were so dull, as hardly to waddle from us; we easily overtook them by running, and knocked them down with sticks.

Of the taking possession of the Island, he related:

Here captain Cook displayed the British flag, and performed the ceremony of taking possession of these barren rocks, 'in the name of his Britannic Majesty, and his heirs for ever'. A volley of two or three muskets was fired into the air, to give greater weight to this assertion; and the barren rocks re-echoed to the sound, to the utter amazement of the seals and pinguins, the inhabitants of these newly discovered dominions.

At the conclusion of his description of South Georgia, George Forster made some interesting economic predictions:

But South Georgia, besides being uninhabitable, does not appear to contain any single article for which it might be visited occasionally by European ships. Seals and sea-lions, of which the blubber is accounted an article of commerce, are much more numerous on the desart coasts of South America, the Falkland, and the New Year Islands, where they may likewise be obtained at a much smaller risk. If the northern ocean should ever be cleared of whales, by our annual fisheries, we might then visit the other hemisphere, where those animals are known to be numerous. However, there seems little necessity to advance so far south as New Georgia in quest of them, since the Portuguese, and the North Americans, have of late years killed numbers of them on the coast of America, going no farther than the Falkland Islands. It should therefore seem probably, that though Southern Georgia may hereafter become important to mankind, that period is at present so far remote, and perhaps will not happen, till Patagonia and Tierra del Fuego are inhabited, and civilised like Scotland and Sweden.   (Forster, 1777)

Anders Sparrman's account noted some Patagonian Penguins, at Possession Bay which he claims were six feet in height. His opinion of the Island was:

It is thirty-one leagues in length, ten in bredth and of less value than the smallest farmstead in England.

The anonymous account attributed to John Marra is much briefer. He referred to the number of birds near Bird Island and described South Georgia as:

It appeared amazingly lofty, mountainous, craggy, and almost covered with snow.

The feelings of the mariners, believing at first that they had discovered the Antarctic continent, were:

While they continued sailing to the North Eastward, the land seemed in that direction to have no end; insomuch that all the mariners on board were overjoyed, imagining they had now found the Southern-Continent of which they came in search.   (Marra, 1775)

The elephant seals seen at Possession Bay were described as:

two monsters which lay on the beech were frightfully fierce... they killed one of them that measured 18 feet, and every way large in proportion; his head resembled the head of a shark, his eyes were fixed in the upper part of his head, and his phippers were armed with claws. He was shot with intent to make a drawing of him.... At half after one the boat returned with the monster on board, and at two they made sail steering ESE. (Marra, 1775)

Lieutenant Charles Clerke also provided further evidence of their disappointment on finding that South Georgia was an island and not part of the Southern Continent:

I did flatter myself from the distant soundings and the high Hills about it, we had got hold of the Southern Continent, but alas these pleasing dreams are reduc'd to a small Isle, and that a very poor one too – as to its appearance in general I think it exceed in wretchedness both Tierra del Fuego and Staten Land which places 'till I saw this, I thought might vie with any of the works of Providence in that particular'.   (Beaglehole, 1961)

J.R. Forster's account, published 1982, describes the navigation towards South Georgia and the earlier discoveries of La Roché and from *León*. His opinion of the region was not favourable:

The Storm increases, the Sea runs high, the Snow makes the Air thick, we cannot see ten yards before us, happily the wind is off shore. If a Capt, some Officers & a Crew were convicted of some heinous crimes, they ought to be sent by way of punishment to these inhospitable cursed Regions, for to explore & survey them. The very thought to live here a year fills the whole Soul with horror & despair. God! what miserable wretches must they be, that live here in these terrible Climates. Charity lets me hope, that human nature was never thought so low by his Maker, as to be doomed to lead or rather languish out so miserable a life.

On sailing past Bird Island he recorded:

We were in the Straights about 5 o'clock; on the Isle next to the shore innumerable Numbers of Birds of various kinds were seen, we observed Seals, common yellow-billed & Sooty Albatrosses, Shags, Port Egmont Hens & Quebranta-huessos, whiterumped & blue Petrels, Pintadas & black Shearwaters, small Divers, & larger ones with a reddish brown streak along both Sides of the Neck, Pinguins of two kinds & various other birds: the *Bird Isle* was lower than that on the West Side; & it was green, except were the steep rocks appeared nacked. Saw several Porpesses, with large white Spots or blotches.

After landing in Possession Bay he describes fur and elephant seals, one of the latter being recorded as thirteen feet long, then some of the birds encountered. A general description runs:

The whole Bay is included by high steep Rocks; wherever at the bottom of it a valley comes down, it is filled up with Snow changed into Ice, ending at the Water's Edge & breaking dayly off in pieces, with a great Noise. The rocks a blueish, heavy Iron-stone, or as they call it in Derbyshire *Dustone*. No other vegetables but a grass which we found on New Years Isle, & a common Burnet, of both very little; the rest either bare Rock or Snow & Ice. The Rock crumbles in Shingle. We saw Shags, Guls, Port Egmont-Hens, Quackerbirds or Sooty Albatrosses, whiterumped Petrels & several other birds. The Capt took a view of the harbour & then took *Possession in his Britannick Majesties Name & His Heirs for ever:* hoisted a Flag on the Land, & fired 3 Volleys, & then returned on board.'   (Hoare, 1982)

The account continues with many comments on the abundant birds and poor weather. Forster recorded temperatures daily and in the South Georgia region which gave an average of 2.4 °C on deck and 5.5 °C in his cabin.

The account of Cook's second voyage appeared in 1777 as *A Voyage Towards the South Pole and Round the World Performed in HM Ships Resolution and Adventure in the Years 1772–1775.* It included a chart and some engravings of South Georgia. An annotated edition, published in 1961 and edited by J.C. Beaglehole, has numerous additional details. Contemporary charts began to include South Georgia but confusion about Antarctic islands was far from resolved. In 1782, for instance, a French chart showed both Île de St. Pierre and Île de Roi George in different places.

Part of Captain Cook's chart of 1777 showing South Georgia. (South is at the top of the map.)

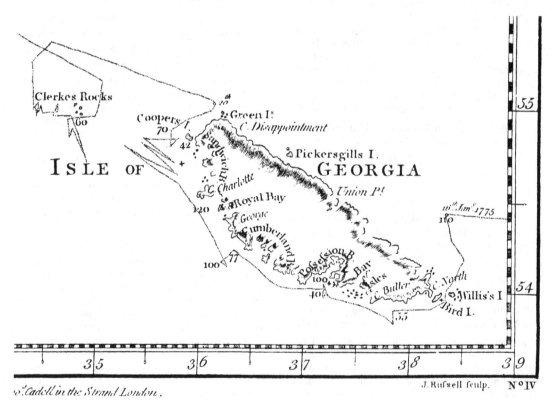

'. *Cadell in the Strand London.*   J. Rufsell fculp.   Nº IV

At the time Cook was circumnavigating the southern regions of the Earth global climate was being strongly influenced by the 'Little Ice Age'. During this period glaciers throughout the world were advanced well beyond their present limits, snow lay for longer and it fell in places where it is not now recorded. The 'Little Ice Age' reached a maximum at around 1700 when, throughout the world, temperatures were perhaps 1.5°C lower than at present. Consequently the accounts given by Cook and others on the voyage describe South Georgia as more ice-bound and glaciated than it is today. In discussing his discoveries of South Georgia and the South Sandwich Islands Cook described them as 'Lands doomed by Nature to perpetual frigidness: never to feel the warmth of the sun's rays; whose horrible and savage aspect I have not words to describe'.

3

Early history and the first epoch of sealing

Commercial exploitation of South Georgia began very soon after the reports of Captain Cook's discoveries were published. It became a major part of the island's history for most of the next two centuries.

Beginning of the first sealing epoch (1786)

The abundant elephant and fur seals described at South Georgia after Captain Cook's voyage rapidly came to the attention of the sealing industry. In the northern hemisphere both sealing and whaling, a closely related industry, had become well established centuries previously. During the latter part of the eighteenth century, in a constant search for new sealing areas, sealers had been extending their range farther south. Sealing was first recorded at the Falkland Islands in 1766 when Bougainville took a cargo of skins to France. Sealers were usually among the first arrivals in newly discovered areas where seals were reported. At the time of Cook's voyage, ships regularly passed either around Cape Horn or through the Straits of Magellan, on voyages which took them to Asia, Australia or the west coast of the Americas and to Europe and the east coast of the Americas. Navigation to areas not far distant from South Georgia was already well established.

There were several reasons for the movement of the sealers southwards including: an increasing demand for oil for lighting, textile and leather preparation and other uses prior to the widespread availability of mineral oils; and a demand for fur skins, for fur, felt and leather preparation. A more unfortunate reason for their spread was that these early sealing operations were of a basically self-limiting nature as virtually no attempt was made at conservation of the stocks by sealers or governments. Seals of all ages and both sexes were taken, removing complete populations from many areas. Seals rapidly became rare at known colonies and the sealers had to search elsewhere. One result of this was the near extinction of several species of seal and, eventually, a similar fate for this style of industry. Another consequence was that Masters of sealing ships maintained much secrecy about any discoveries of new places with large populations of seals. This, they hoped, would enable them to return to exploit the seals in subsequent seasons without competition. A sealer's success or failure depended largely upon whether he was the first in the field. Although there were perhaps hundreds of sealing voyages to the island, there is little information available about South Georgian sealing and the following description of voyages is therefore incomplete. Only half a dozen published accounts of sealing on South Georgia during the nineteenth century. are known.

One of the earliest references to sealing at South Georgia is a highly commendable note concerning seal conservation. This was written on 16 July 1788 by John Leard, a Master of the Royal Navy, to Lork Hawksbury, President of the Council for Trade and Foreign Plantations in London, as an appendix to a letter concerning sealing on the

Patagonian coast. Leard refers to the numbers of seals on the east coasts of Patagonia, Tierra del Fuego, Staten Island, the Falkland Islands and South Georgia, and makes suggestions for appropriate controls for maintaining a successful sealing industry such as limits on the ages, sex, and numbers of seals taken. He proposed that a 'South Sea Company' be established to regulate sealing and cites instances of previous depletions (and one of conservation) of seal populations in support of the argument. The subsequent history of the industry was unfortunately a discredit to all who participated. A course of action of unenlightened self interest was pursued over the years. This led to a rapid dwindling of the stocks of seals and to their virtual extermination over much of their range.

The fur seal was the species most profoundly affected on South Georgia. Elephant seals were also hunted but, although greatly reduced in numbers, they were never so close to extinction as the fur seal, and they have been relatively abundant for most of the twentieth century.

The beginning of sealing on South Georgia was probably 1786. Thomas Delano sailed from London in that year, in command of *Lord Hawkesbury*, after taking a full cargo of fur seal skins at Falkland Islands and South Georgia he returned in 1787. *Lucas*, from London, made a similar voyage in 1787. A Seal skull from 'New Georgia, near the ice towards the South Pole', was presented to the Museum of the Royal College of Surgeons, London, by a whaler, Mr Kearn. Allowing for time elapsed between its collection and publication of a brief account of it, the skull was probably collected during the 1789/90 season or possibly the 1788/89 one. The skull, which is accurately illustrated, is that of a leopard seal, a species often seen around the island.

More definite information is available about two vessels fitted out for sealing in the Falkland Islands by Elijah Austin, a merchant of New Haven, Connecticut. The brig *Nancy*, commanded by Daniel Greene, sailed on 15 May 1792, and the brigantine *Polly* commanded by Roswell Woodward sailed on 20 June 1792. Their voyages were very successful and obtained parts of their cargoes from South Georgia. *Polly* returned direct to the United States, reaching Connecticut in May 1794. *Nancy* sailed first to Canton, China, to sell the fur

skins for greater profit. Her voyage continued and circumnavigated the globe to return to the United States at New York in May 1795. A conflicting record of these voyages also exists although it has fewer supporting details. This record suggests the ships sailed from Connecticut in 1790. *Nancy* and *Polly* were the first United States vessels to sail in Antarctic waters.

The next known sealing voyage to South Georgia was made by Captain Pitman in *Ann*. She sailed from London on 18 June 1792 to arrive at South Georgia on 13 November 1792. Elephant sealing commenced and 50 tons of elephant seal oil was taken before fur sealing began on 3 December. Fifty thousand fur seal skins were taken before the ship departed on 2 March 1793. An unpublished letter concerning the voyage addressed to Francis Rotch, a well known sealing Master, in Rhode Island, United States of America, contains a note on South Georgia sealing taken from Captain Pitman's journal and is probably the earliest description of some aspects of this:

In running for the land of this Island be careful you are not deceived by the Islands of Ice that are frequent thereabouts. There are plenty of good Harbours from the NW to SE but on the eastern Coast there are sunken ledges that show themselves by Breakers & not by kelp. The NW is clear except outside of Willis's Island a good passage between this & the main island... The small skins are of little value. They are chiefly used here among Tanners and reconed excellent for Shoes, & the larger are the best. A vessel like the *Betsy* capable of bringing near 100 tons of oil may bring upwards of 30 thousand good sizeable skins... The skins packed in bulk have come out in much better order than those packed in Casks because they have an opportunity of examining them & salting them even if necessary. Other ships continued till the month of May & came away quite full. They found plenty of wild fowl but they chose the hearts of Elephants preferable to any other food. As the climate is cold, good cloathing & boots are necessary. These islands are about 5 days sail to the eastward of the Faulkland Islands. It is necessary to carry a good long boat with sails. In a small vessel this boat might be carried in frame & soon put together there.
(Letter-Book, W. Rotch Jr, Old Dartmouth Historical
Society Mss 2–3A1.)

In 1793 Captain James Colnet, in command of the *Rattler* on a voyage of exploration in the South Atlantic and Pacific Oceans to obtain information to extend sperm whaling, was in the vicinity of South Georgia, although he did not visit the island. He wrote, referring to a non-existent island reported to be west of South Georgia:

nor can I doubt, from the quantity of whales I have perceived near its supposed situation, that it would prove a much greater acquisition than the Island Georgia, to which many profitable voyages had been made for seal skins alone. (Colnet, 1798)

This is all he records concerning South Georgia, but it appears sealing had become well established by this time. Colnet had previously sailed in *Resolution* during Cook's second voyage and thus visited South Georgia in 1775.

First peak of sealing (1786–1802)

Some records are available for other voyages in the last years of the eighteenth and the first years of the nineteenth century. Two English ships were at South Georgia in the 1795/96 season: *Young William* commanded by Captain Henrie Mackie, and *Sally* commanded either by a Captain Farmer or a Captain Ellis. Both ships were from London. *Sally* was wrecked at South Georgia and her crew rescued by *Young William* which returned to London on 14 June 1796.

During the 1798/99 season, *Sybil* commanded by Captain Lewis Llewellin departed from England on 3 October and returned from South Georgia on 25 June 1799. This was a rather rapid voyage which implies it was successful in completing its cargo rapidly. In the same season, *Prince Edward* commanded by Captain Clark was wrecked on the return voyage.

The number of vessels recorded during the 1799/1800 season was greater: *Aurora, Lively, Earl Spencer*, and *Hercules* from the United Kingdom; and *Regulator* from the United States of America visited the island. *Regulator* was wrecked there, her crew built shelters near the wreck and were rescued at the end of that sealing season by a British sealer whose name is unrecorded.

Captain Edward Fanning from New York was the first sealer to leave a reasonably comprehensive published account of a visit to South Georgia, made in the 1800/01 season. This did not appear in print, however, until 1834 when any details given about the seal locations would have been well out of date. He sailed from New York in the *Aspasia* on 11 May 1800. She carried 22 guns

and a 'Letter of Marque' from the Government of the United States of America, which provided his voyage with some official character. *Aspasia* called at Brazil and Tristan da Cunha before reaching South Georgia in September. It had been expected that she would find the crew of *Regulator*, a vessel owned by the same company and wrecked in the north-west of the island during the previous season. Captain Fanning found their abandoned shelters and the wreck, probably in what is now named Right Whale Bay. He later made contact with one of *Regulator's* crew on the British sealing vessel *Morse*. Fanning was informed that the 14 000 fur skins she had taken had been sold to another ship in which the rest of the crew took passage home.

Fanning relates how he deployed three shallops (10 m cutters for inshore work), one converted from a launch, one purchased from another ship, and the third made partly from the wreck of *Regulator*. He was able to take full advantage of that season and obtained 57 000 fur seal – probably the greatest number taken from South Georgia by one ship. He also reported the arrival of 17 other sealing vessels at the island as the sealing season commenced, and gave the total number of skins taken during that season as 112 000. The commencement of the sealing season was described by Captain Fanning:

The first appearance of a change from the winter to the summer season at South Georgia, is discoverable in November; the ice then begins to break away, and the seals to come up; this is followed with an immediate destruction of their numbers by the sealers, with as much briskness as due regard to the skins being kept well conditioned, will admit of.

Another note described the notorious winds of South Georgia.

Some idea of the same may be had by the fact, that the light cedar whale-boat moored at the stern of this ship, and held by the warp at her bows, had been taken up by these violent gusts, and turned over and over, before again striking the water, the same as a feather attached to a thread, and blowing in the wind.' (Fanning, 1833)

Fanning also described how astonished the crew was to discover the abundance and quality of 'fine cod, some eighteen inches or so in length...some were taken weighing thirty and forty pounds'.

Fanning departed from South Georgia on 8 February 1801 and experienced difficulties sailing against strong winds, in the presence of many icebergs and ice floes, in thick fog and misty weather as he sailed around Cape Horn. The voyage continued to China to sell the fur skins. Fanning records meeting 31 other sealing vessels along the coast of Chile with similar cargoes destined for China.

Apparently only North American sealers enjoyed the China Market for fur skins and they often circumnavigated the world in the course of trading at it. The skins they sold were taken from many southern islands and parts of South America. The ships involved in this trade were commonly armed as a precaution against pirates. While there was little attraction for pirates in taking cargoes of seal skins and oil, after the ships left China they carried far more valuable items, making them much more attractive and therefore highly vulnerable. The voyage round the world to return to New England passed some of Asia's most notorious pirate coasts. The China market was abandoned after the first peak of sealing and all skins were subsequently taken to North America or Europe.

As well as those of *Aspasia* and *Morse* several other sealing voyages are recorded in the 1800/01 season (*Earl Spencer, Duke of Kent, Eliza,* and another *Sally*). Several other voyages *Earl Spencer* made at this time were to the 'South Seas' for sealing. *Duke of Kent* was despatched by D. Starbuck, a member of a New England family involved in the sealing and whaling trade, who had established himself in London. *Eliza* sailed from London in August 1800 and returned from South Georgia in April 1801. This rapid voyage might have indicated a successful one with a full cargo easily obtained; however she was despatched elsewhere with a different master for her next voyage, thus it is also possible she was unsuccessful at South Georgia and another voyage there was considered as not worth the effort. Captain Nathanial Storer of New Haven, Connecticut, sailed aboard *Sally*, a 240 ton ship with a crew of 45, on 22 May 1800 carrying a 'Letter of Marque'. His 9 year old son, Peter, was also aboard. They first landed on the coast of Patagonia where a 28 ton shallop was built, and then sailed to the Falkland Islands in December 1800 where a few seals were taken. The greater part of their cargo was obtained at South Georgia where they spent the next two seasons. The voyage then

went round Cape Horn to China, stopping for small numbers of seal skins at some Pacific Ocean islands. At Canton 45 000 fur seal skins, mainly from South Georgia, were sold and, with a cargo of 'tea, silk, namkeens, &c', they returned to New Haven on 2 June 1803 after an absence of 3 years and 10 days. Three men died on the voyage. Captain Storer later made another sealing voyage aboard *Huntress* never to be seen again. His son, who became a sealing master, recounted the voyage of *Sally* in his ninetieth year.

The 1801/02 season has three recorded voyages. *Earl Spencer* made another visit and was lost on South Georgia although her crew were saved. *Sprightly* and *Dragon*, commanded by Captain H. Barton and Francis Todrig respectively were also there. They returned to England on 1 April and 2 July 1802 and were the last British sealers to sail to the island for 12 years.

Dr R.C. Murphy, who made a voyage in a New Bedford vessel in 1912/13 (described later in this chapter), and became well acquainted with New England sealers in consequence, wrote that in 1802 one of them, Captain Pendleton aboard *Union*, visited the southern side and compiled a map of South Georgia. This was not published until over a century later, and then by an Italian geographer (Faustini, 1906). The text accompanying the map does not give conclusive information about its origin but the map has several names in common use during the early sealing period and is the second map produced of the island (if the date 1802 is correct). There is no supporting record of the 1802 voyage, however Captain Isaac Pendleton sailed in the *Union* from New York late in 1803 on a sealing voyage to Îles Crozet (which he did not find), and then continued to Australia. He may have made an unrecorded voyage to South Georgia the previous season.

South Georgian sealing appears to have suffered its first decline at about this period, with a lull in recorded voyages of almost a dozen years. It is reasonable to assume that the large numbers of skins taken until 1801/02 had so seriously depleted seal stocks that it became uneconomic to make voyages to the island. Much of the evidence for this is from recorded voyages, and one should bear in mind that these probably represent only a minority of voyages. The Napoleonic Wars between the

United Kingdom and France together with the 1812–15 war between the United Kingdom and United States of America both inhibited voyages to such remote places as well as causing great losses of shipping.

Second peak of sealing (1814–23)

Following this lull, the next recorded sealing season on South Georgia was in 1814/15. William Beacon, formerly Master of the ill-fated *Earl Spencer*, sailed to South Georgia aboard *Recovery* during that season. The catch was presumably poor as she did not return for the next season.

For each of the next three seasons several voyages are recorded with much detail. These provide some of the best descriptions of South Georgia sealing and appeared in a very interesting autobiography by Thomas W. Smith, a Londoner, who recorded an eventful life as a seaman. Following an unfortunate childhood he went to sea as an

apprentice and made 18 voyages. These are recounted in his autobiography published in Boston, Massachusetts, in 1844 when he was ashore in New Bedford following invalidity from a series of accidents. Some of the dates he records are probably mistaken and have been corrected by comparison with other records in this account. In two of his three voyages to South Georgia he was shipwrecked (on one of them, twice).

On the first voyage he sailed on *Norfolk* from London in 1815. She was a French built ship of 650 tons, probably captured during the Napoleonic Wars, and carried two shallops of 24 and 35 tons. Her crew consisted of 52 men, 16 of whom were apprentices. Smith records calling at the Cape Verde Islands, to take a cargo of salt and believing that the voyage was bound for Africa to secure elephants of the terrestrial variety. Look-outs were stationed after the ship sailed past 45° S latitude as a precaution against colliding with icebergs. They

Chart of South Georgia drawn during the first epoch of sealing. Published by A. Faustini in 1906.

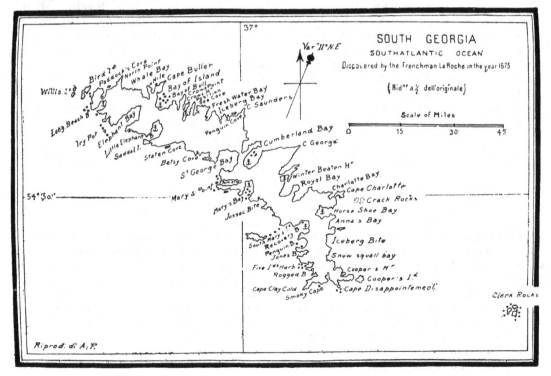

arrived at South Georgia in December and first sighted Bird Island. The ship sailed to Royal Bay and Smith commented favourably on the verdure of the island. At Royal Bay, *Norfolk* was secured for summer and the subsequent winter with four anchors and all her upper masts and arms were taken down. While the shallops were being prepared, sealing began in the vicinity using six boats. Elephant seals were killed using clubs and lances, their blubber was conveyed to the ship.

Smith describes South Georgia as:

... about 120 miles in length while its extreme bredth does not exceed 12 miles ... It abounds with large bays, which are 22 in number; 14 on the north side, which are remarkable for their safe harbours. The south side has eight bays which possess very unsafe and dangerous harbours for vessels to ride in. Out of 22 bays there are fifteen which contain icebergs [i.e., glaciers] of the bluest kind at their heads. The average sizes of which can be no less than a mile in bredth and four miles in length in proportion as the creeks, in which they are formed extend in-land between the mountains.

He relates how the Captain, 'an old voyager to this island', gave warning of the dangers of ice and crevasses, partly by describing how a seaman had died after falling in one while trying to cross the island to collect mail from the first ship to arrive in spring. The spectacular calving of glaciers is also described, sounding 'like the loudest thunder', and sizes and shapes of icebergs are also referred to. The natural history of elephant seals as related to sealing is also summarised. He advised that sealers should arrive before the 'pupping-cow season',

which commences in October. The elephants which get on shore during this season are the pupping-cows and bulls. They come up on sandy beaches which are made suitable to their condition at this time of the year. Immediately after their landing on shore, by peculiar instinct they form themselves into pods along the beaches. This arrangement is made by the bulls, which keep a constant look out for those which come out of the water, and immediately drive them to their respective pods and keep a constant watch to keep them in this place. Sometimes he takes his position in the middle of his pod from which he keeps rising now and then to watch, at others on the outside, and when this is the case they generally go round the pod once in about two hours. The largest of the bulls, which are from 20 to 22 feet in length and from four to five in height, will fight for hours most furiously for mastership of the pods.

The next season in the life of the seals was described as the 'brown-cow season'.

These are large barren cows and come up any-where wherever they can find a chance.

This is followed by the 'young bull season':

These come up and lay among the bogs to shed their skins, and after two months on shore, they go off with scarcely any fat on them.

The 'March bull season' is the last in the sealers' calendar when bulls that have survived the hunting season haul out to moult.

These bulls are very large and fat and three of them generally make two tuns of oil ...

and are thus highly sought after by the sealers.

Smith describes sealing operations on South Georgia very well:

As to the manner of making up the voyage I remark in the first place, it is essential to choose a good ship-harbour in a central part of the island, to divide the distance fairly for the shallops and boats from each end of the island to the ship. In the second place the ship should be provided with two good shallops to take the blubber on board of the ship. These shallops are sent, one to the east and the other to the west of the ship. There are two or three boats to attend each shallop to kill the elephants and take the blubber on board of them. The boats' crews are so situated at times, as to endure the severest hardships, cold, starvation and hunger. The boats have to pass through breakers, over sunken rocks and bars and to land on dangerous open, sandy beaches, among ice and tremendous surfs, which often upsets the boats, notwithstanding the care and caution of the crews. They leave the ship and seldom return under three or four months, during which the boat is the house and home of the crew. They always sleep under the boat, which is turned bottom-up on a sandy or strong beach or rocks, which is frequently the case; but in the winter and spring, the ground being deeply covered with snow, the boat's crew are under the necessity of sleeping on it.

Fur seals and elephant seals are described as seals and elephants respectively in Smith's account. In 1816 he describes finding few of the former and these were secured 'by examining every creek and corner'. When found, they were immediately killed and skinned. However, by the time Smith arrived at South Georgia most sealing was for elephant oil.

Life on the island was particularly rigorous for the sealers. Food shortages and losses with consequent endeavours to survive on seals and penguins; boats, shallops and ships wrecked; death by drowning, freezing, falling down crevasses, etc.; the strength of the winds and storms; sleeping on rocks, snow and ice are all described regularly in this account. Finally Thomas Smith states:

I suffered much with the severity of the cold, which affected my feet to a great degree. I had neither shoes nor stockings, having previously worn them out, and it was impossible to obtain any from the slop-chest, as it was empty, or from any of the crew.   (Smith, 1844)

During the first winter of Smith's voyages all men returned to the ship for the first part of it. They were chiefly employed in cutting ice and shovelling snow to preserve the ship.

Smith's first voyage gained 3500 barrels of oil and 5000 fur seal skins which sold in London for £50 a ton and £2 each. The crew, with the exception of the apprentices, made a profitable voyage.

Despite the hardships, Smith made a second voyage aboard *Norfolk* which sailed in June for the next season accompanied by the schooner *Ann* and a shallop. The shallop was wrecked during a storm in Cooper Bay with Smith aboard; he survived but two other men died. *Ann* then made a voyage to the South Sandwich Islands but found neither fur nor elephant seals there. The captain of *Norfolk* had purchased a replacement shallop for £700 from another sealer working on the south side of South Georgia, the 80 ton *Lovely Nancy*. Unfortunately she too was lost, on rocks in the Bay of Isles, with Smith and a full cargo of elephant seal blubber aboard. *Norfolk* returned to London with 2400 barrels of oil and 10 000 fur skins but the crew received no payment owing to the bankruptcy of the vessel's owner.

Thus Smith had to make a third voyage in 1817. The ship *Admiral Colpoys*, was wrecked on 28 November 1817 at South Georgia with a large cargo of oil aboard. Her cable was severed by an iceberg, despite attempts to demolish this by cannon fire, and the vessel drifted on to rocks. Smith obtained a passage back to England, perhaps on *Norfolk*, and arrived as poor as when he left. Again he was obliged to go sealing and voyaged to the newly discovered South Shetland Islands sealing grounds aboard *Hetty*. There he suffered further misfortunes and thenceforth undertook no more sealing voyages.

Of the several personal accounts of Antarctic sealing which exist, Thomas Smith's is one of the more detailed and particularly relevant to South Georgia. It is easy to understand, however unfortunate, that men engaged in such arduous work as that of the sealers in such wild and inhospitable places, with the types of equipment, clothing and foodstuffs available at the time, must be expected to make all they could, and care for none who came after them. They therefore killed all old and young, male and female seals they came across, for an immediate return.

Russian voyage of exploration (1819)

A voyage of exploration, comparable with Captain Cook's second expedition, reached South Georgia on 27 December 1819. This was the circumnavigation of the southern regions of the globe commissioned by the Tsar Alexander I, who had also ordered a similar northern expedition. It was led by Captain Thaddeus Bellingshausen of the Imperial Russian Navy, aboard *Mirnyi* (Pacific), accompanied by Captain Mikhail Lazarev in *Vostok* (Orient). Bellingshausen was of Estonian origin, born on the island of Sarremaa in the same year as Captain Cook was murdered. The commander originally selected to lead the voyage, Commodore Tashmanoff, had the misfortune to lose his ship. He declined to command the expedition and recommended Bellingshausen for the post. Bellingshausen consequently had only about six weeks notice prior to the sailing of the voyage.

Special attention was given to supplies of appropriate foodstuffs, clothing, equipment, etc. and the health of the expedition remained relatively good. Wages paid to the seamen were to be eight times the normal rates with a bonus of one year extra at the conclusion of the voyage. Bellingshausen had several improvements made to the ships, at short notice, just prior to their departure. On sailing, *Vostok* carried 117 men and *Mirnyi* 73, including a priest.

The expedition had intended to carry two naturalists. Two from Germany were recruited but, unfortunately, they declined to go at short notice and replacements could not be recruited in the time available. The Tsar visited the expedition on 5 July; on 15 July they sailed from Kronstadt. They proceeded to Spithead from where the two Captains took a coach to London to purchase charts and navigational instruments. They sailed south, calling at Tenerife and Rio de Janeiro, and then on to South Georgia. Very early on 15 December 1819 (on the Julian Calendar which Bellingshausen kept, on 27 December on the Gregorian Calendar), South Georgia was sighted. The first view was of the Willis Islands and it was described with comments about numbers of whales, many species of birds and of giant kelps. Bellingshausen encountered two sealing vessels at Undine Harbour:

An extremely heavy surf from the west-south-west thundered against the cliffs. I kept along the coast of South Georgia at a distance of $1\frac{1}{2}$ to 2 miles. Going at the rate of 7 knots, we noted several bays in which there was probably good anchorage to be found. From one of these bays a sailing boat, flying the British colours, came out. On their approach, we asked them to come up to us and hove to.

In the boat was a steersman and two sailors. The former told us that they had not recognised us from a distance, and supposing that we also had come for whaling, they had intended to pilot our vessels into the bay, hoping that they would receive payment for their trouble. Two three-masted vessels, belonging to a British Whaling Company, one the *Indespensable*, the other the *Mary-Ann*, under the command of Captains Brown and Short, were lying in the bay whence the boat had come. The depth of the anchorage was 18 sazhen, bottom mud; a large brook of fresh water runs into the bay which is called Port Mary. The vessels had been there already for four months; the whalers extract the blubber from the seals; their work takes them into all the bays, and, as shelter for the night, they turn their boats keel upwards and light fires underneath them. For melting the blubber of these animals they use as fuel the skins of penguins which frequent this part in very great numbers at this time of year. They also saw very frequently albatrosses and other sea birds but of land birds only larks and a kind of pigeon, also called a sheath-

Chart of South Georgia from Bellingshausen's voyage in 1819. Drawn by F. Debenham 1945. (Courtesy of the Hakluyt Society.)

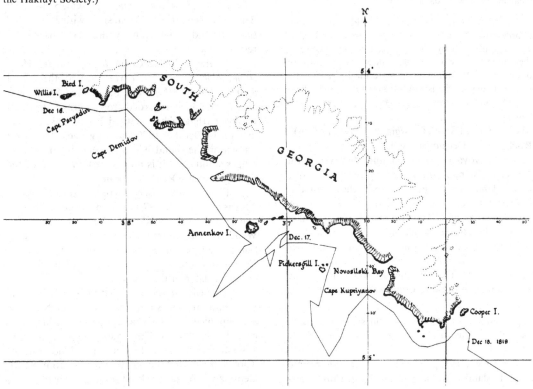

bill. The only plant life there was a moss. In return for this information, I ordered our guests to be given grog, sugar and butter.

Bellingshausen also met a Russian seaman who had deserted his ship in England and was among the crew of one of the sealers.

Bellingshausen continues to describe the southern coast, which was examined with a telescope to detect vegetation. 'Excepting here and there a yellowish green moss we saw none'. A running survey was conducted and several place-names bestowed on the island. This covered the south-west part and is an important, although somewhat briefer, complement to that of Captain Cook's north-eastern survey made 44 years earlier. Cook's 'Pickersgill Island' was renamed Annenkov Island and the original name later devolved onto a much smaller group of islands. *Vostok* was delayed and separated from *Mirnyi* after encountering the sealers and *Mirnyi* commenced regular cannon fire to indicate her position. They rejoined that evening.

On departing from South Georgia in poor weather Bellingshausen wrote;

I had little hope that the weather would improve soon in order to look for a safe anchorage; moreover the land was inhabited only by penguins, sea elephants and seals; there were but few of the latter, since they are killed by the whalers. On sailing along more than a half of the southern coast of the island, we saw not a single shrub nor any vegetation; everything was covered with snow and ice.

The last part of South Georgia they saw was Clerke Rocks on 29 December 1819.

The voyage continued and discovered the Traversay Islands – the northern three South Sandwich Islands; on 27 January 1820 they probably sighted the continent of Antarctica without recognising it as such. Thus the expedition may have been the first to have done so. They continued the circumnavigation to Sydney, Australia, and then wintered in the South Pacific. After returning to Sydney, the voyage continued to Macquarie Island where Bellingshausen commented, after meeting sealers there:

The conditions of life here were in some ways more tolerable than that of the sealers we had seen in South Georgia.

Thence they sailed again to the ice front, discovered Peter I Island and Alexander I Land (now Alex-ander Island), investigated the South Shetland Islands, then returned to Kronstadt (passing Shag Rocks and calling at Rio de Janeiro) on 4 August 1821.

Bellingshausen wrote a detailed account of his voyage which was edited and published in Russian in 1831. A German précis appeared in 1902 but it was not until 1945 that the whole work became available in English (see Debenham, 1945).

Second peak of sealing continued

The number of vessels at South Georgia increased greatly in the 1818–19 season. *Norfolk* made another voyage while *Ann, Arab, Echo, Grand Sachem, Indespensable, King George*, and *Mary Anne* are also recorded. Several of these vessels were prize ships, from the United States or France and owned by British companies. For the 1819/20 season many of them returned; *Echo, Indespensable*, and *Mary Anne* are recorded while *Arab* and *Norfolk* were probably there also as was *Recovery* which had been present in 1814/15. *Indespensable* and *Mary Anne*, the sealers met by Bellingshausen, took 736 seal skins with 350 tons of seal oil and 9000 skins with 100 tons of oil respectively.

In the 1820/21 season *Indespensable, Mary Anne, Norfolk*, and *Recovery* visited the island again. *Francis Allen* from the United States was also there and is recorded from a grave at Ocean Harbour (that of her steward Frank Cabrail). He died on 14 October 1820 and his grave is the oldest known on the island. Unfortunately its marker no longer stands and it is now uncertain which of the eight graves at Ocean Harbour it is.

Five British vessels are recorded in the 1821/22 season: *Ann, King George* and *Tartar*, with *Jane* and *Beaufoy* which made a combined voyage. James Weddell commanded *Jane* on his first voyage to South Georgia, he made another aboard her in 1823 and wrote a detailed account of it which is described later in this chapter.

King George was recorded in June 1823 at Ile Saint-Paul in the south Indian Ocean She was accompanied by the shallop *Success* which displaced only 28 tons. Three men were rescued from a sealing party stranded on the island, one of whom stated that he knew *Success* from South Georgia where she was built some years before, and that he had helped rig her out. *Success* continued to Hobart, Tasmania, arriving on 11 July 1823, an amazing voyage for such a small craft. William Goodridge, from whose account these details are taken, also described a method sealers used to catch small birds on South Georgia for food:

by lighting a blazing fire after dark with Sea Elephant's blubber and dried grass, at the foot of some tolerably high cliffs; this would attract them in quantities, and they would fly with such force against the rock as to stun themselves, and falling down we obtained a large supply.

Benjamin Morrell and *Wasp* (1822)

A description of South Georgia sealing and a voyage to the island in 1822 occurs in a work published in 1832, written by Benjamin Morrell of New York. The account has some factual foundation although Morrell draws strongly on his imagination for various descriptions in the work. It has been claimed that, in his day, he was known as the 'greatest liar in the Pacific'. Perhaps one of the better examples of his powers of imagination is the finding of a very colourful bird 'resembling the bird of paradise' in the South Sandwich Islands. The dates and positions he records are frequently rather vague.

Morrell sailed from New York on 30 June 1822 in *Wasp* in company with Robert Johnson, Master of *Henry*, about whom there is little more recorded. The voyage was to be one of sealing and exploration. The latter aspect was becoming more important to the sealing industry as seals were, by that time, becoming far less abundant in the known sealing areas. Morrell called at Rio de Janeiro, several places on the coast of Patagonia, the Falkland Islands and, after making a search for the non-existent Aurora Islands (probably confused with Shag Rocks), continued to South Georgia where he arrived on 20 November 1822. The brief account he gives of this visit is:

...steered for the island of South Georgia, where we safely arrived on Wednesday, the 20th, and came to anchor in Wasp's Harbour, on the north side of the island, at one o'clock, P.M. At two, P.M., I sent the boats in search of seal; but after an absence of three days they returned unsuccessful, on Sunday, the twenty-fourth, at ten, A.M., having Circumnavigated the whole island without discovering a single seal.

South Georgia is an island in the Southern Ocean, bearing E. by S. from the Falklands, distant about 260 leagues...its whole circumference being about seventy leagues. Wasp's Harbour, where we now lay at anchor, is in lat. 54°58′ S, long. 38°25′ W. The sides of the island are deeply indented by bays, some of them so deep on opposite sides as almost to meet in the centre. The mountains are lofty, and the tops perpetually covered with snow; but in the valleys there grows a strong-bladed grass in great plenty.

November 24th. – The sole object of our visit to this cheerless port being frustrated by the absence of seal we weighed anchor on Sunday, the 24th, and proceeded to sea.   (Morrell, 1832)

Several things are of interest in the account; that he reports a search for seals that found none; that the bays almost bisect the island which was believed by several sealers and is clearly shown in the 'Pendleton' chart (described in Chapters 1 and 3), and that the ship's boats circumnavigated the island in three days. This last point is highly unlikely: the shortest distance involved (without entering any bays) is about 420 km. It is also of interest that Morrell was not successful in finding any new sealing areas during the rest of his voyage. There is internal evidence that Morrell's work owes much to that of James Weddell, published in 1824.

James Weddell and *Jane* (1823)

James Weddell undertook a second voyage for both sealing and exploration from 1822 to 1824. The account of this (Weddell, 1825) is one of the greater works of Antarctic exploration. Weddell was engaged by a James Strahan of Leith, Scotland to make the voyage with the ship he previously used, *Jane*, and a crew of 22. The cutter *Beaufoy* of London again provided an escort, she was commanded by Matthew Brisbane and had a crew of 13. Captain Brisbane made a subsequent voyage to South Georgia in 1827 which is described below.

The vessels sailed from Britain in September 1822 and called at Madeira, the Cape Verde Islands for the usual cargo of salt, thence to the coasts of Patagonia where Weddell made some coastal surveys. In January 1823 he called at the South Orkney Islands, their position was surveyed and it was reported that only a few seals were there. He continued sailing south and reached 74°15′ S at 34°16′45″ W on 20 February 1823, the southernmost point reached by a ship at that longitude, a record which lasted for almost 150 years. He then sailed north, as the state of health of his crews was deteriorating (despite a great increase in their rum ration) although there was still open water further south. *Jane* and *Beaufoy* became separated on the northward course but were reunited at South Georgia. Weddell arrived at the island on 12 March 1823 and was very pleased to reach it. He described his crew eating various green plants to prevent scurvy – possibly the leaves of *Acaena* or conceivably *Ranunculus* species (there is nothing else reasonably likely available). The first record of seismic activity was described, it was detected by the behaviour of a dish of mercury being used as a horizontal mirror for surveying purposes. The equinoctial gales are also recorded.

Weddell's account of his visit continues:

With the wind blowing strong at west, we steered to the northward in company; and at 10 A.M. we saw the island of South Georgia, bearing N. by W. distant about 9 or 10 miles. Notwithstanding the forbidding appearance of this land, every one, I believe, in the two vessels, feasted his eyes upon it; and at 3 in the afternoon both ships came to anchor in Adventure Bay, (S.W. part of Georgia,) in 7 fathoms water, over a bottom of strong clay.

Our arrival here, though it was not a country the most indulgent, we considered to be a very happy event. Our sailors had suffered much from cold fogs and wet during the two months they had been navigating to the south; and as we had been nearly 5 months under sail, the appearance of scurvy (that disease so fatally attendant on long voyages) was to be dreaded. Our vessels, too, were so much weather-beaten, that they greatly needed refitting; so that taking into account our many pressing wants, this island, though inhospitable, was capable of affording us great relief.

Our crews here fed plenteously on greens which, although bitter, are very salutary, being an excellent antiscorbutic: with regard to meat, we were supplied with young albatrosses that is to say, about a year old: the flesh of these is sweet, but not sufficiently firm to be compared with that of any domestic fowl.

Our harbour duties, and a search upon the island for animals for our cargo, were immediately commenced and carried on with zeal, although we experienced frequent interruptions from the heavy gales which were now prevalent: it being near the time of the autumnal equinox of the hemisphere.

I took opportunities of making various observations on shore, and found the head of the bay to lie in latitude 54°2′48″, and in longitude by the mean of two of

the best of my chronometers, 38°8′14″. The variation of the compass at the same place by azimuth, was 11°15′ east. The head of this bay being surrounded with mountains, I ascended the top of one of them for the purpose of taking the altitude of the sun when at some distance from the meridian, but after placing my artificial horizon, I was surprised to find, that although there was not a breath of wind, and everything around perfectly still, yet the mecury had so tremulous a motion, that I could not get an observation. The ground was evidently agitated internally; though it was only by means of the quicksilver that I could detect it.

On the 17 April, our harbour business being completed, both vessels put to sea, and with the wind at east, we directed our course towards the Falkland Islands.

Weddell continues with a good account of the biology of the king penguin, about which he wrote:

In pride, these birds are perhaps not surpassed even by the peacock, to which in beauty of plumage they are indeed very little inferior...

Three other species of penguin are mentioned: macaroni; jack-ass (gentoo); and stone-cracker (chinstrap). The derivation of the first name is given as:

The macaroni is so called from its having been likened to a fop or macaroni, though I must confess I do not see the similitude.

A description of the wandering albatross (*Diomedea exulans*) follows:

A full grown albatross sometimes measures 16 or 17 feet from the tip of one wing to the tip of the other when expanded; but more commonly they average about 12 feet. These birds are so abundantly covered with feathers that, when plucked, they appear not above one half the original size, and our astonishment at their apparent magnitude immediately vanishes. I have found them when cleaned to weigh from 12 to 25 lb. ... They have great power in their beaks, and, when on the nest, I have observed them defend themselves for half an hour against an active dog.

This is the first reference to a non-indigenous animal on the island. The giant petrel (*Macronectes* species) was regarded as a great nuisance to the sealers by Thomas Smith. Weddell commented:

Their fondness for blubber often induces them to eat so much that they are unable to fly. A flock of perhaps five or six hundred has been known to devour 10 tons of sea-elephant fat in six or eight hours.

Weddell's account contains descriptions of

the natural history and exploitation of fur and elephant seals, he also discusses the sealing industry and recounts aspects of its unfortunate history:

These animals are now almost extinct; but I have been credibly informed that, since the year in which they were known to be so abundant, not less than 20,000 tons of the sea-elephant oil has been procured for the London market. A quantity of fur seal skins were usually brought along with a cargo of oil; but formerly the furriers in England had not the method of dressing them, on which account they were of so little value, as to be almost neglected.

At the same time, however, the Americans were carrying from Georgia cargoes of these skins to China, where they frequently obtained a price of from 5 to 6 dollars a-piece. It is generally known that the English did not enjoy the same privilege; by which means the Americans took entirely out of our hands this valuable article of trade.

The number of skins brought from off Georgia by ourselves and foreigners cannot be estimated at fewer than 1,200,000.   (Weddell, 1825)

Some additional information about the island's geography is provided and he mentions that boats were frequently transported over the neck of land which separates Undine Harbour from Elsehul. This is about 1 km wide and 50 m high, its use avoids sailing small boats around the unsheltered, precipitous western extremity of the island when proceeding with them between the southern and northern sides at that end. The mountainous topography of the island is also described. Weddell gives some credence to the island being divided by a strait filled by glaciers. He also postulated a geological connection between the Americas and Antarctica through South Georgia and other islands.

Spread of sealing to other Antarctic regions

Exploration of the regions farther south than South Georgia and Tierra del Fuego, and indeed generally in the far southern hemisphere, had continued since Cook's second voyage. Many voyages which rounded Cape Horn reported icebergs, low temperatures, tempests, etc. which gave an idea about conditions farther south. The next discovery of land in that direction was made by William Smith, an Englishman, Master of the brig *Williams*. This was found on 19 February 1819 while he was sailing from Montevideo to Valparaiso on a mercantile voyage. He took a more

southerly course than usual, owing to weather conditions, and encountered what are now known as the South Shetland Islands. Smith reported his findings and, shortly afterwards, they came to the attention of British and United States sealers. Further explorations revealed enormous quantities of fur seals on the islands.

It is not unreasonable to consider, however, that there may have been earlier voyages to the South Shetland Islands, and even to South Georgia prior to those recorded. In 1820 Jeremiah Reynolds, a strong proponent of Antarctic exploration, assessed the situation well when he wrote to the Secretary of the United States Navy about geographical discoveries incidental to sealing voyages:

I regret that I am not at liberty to communicate in writing all the interesting facts . . . In the seal trade secrecy in what they know has been deemed a part of their capital.

For the sealers, the discovery of the South Shetland Islands was very timely. Fur seal populations were suffering their second drastic reduction on South Georgia and at many other established sealing areas. In consequence, very intensive sealing began on the South Shetlands, sufficiently intense to be commonly referred to as like a 'gold-rush'. As might have been anticipated, the prospects for these seals were as bleak as elsewhere.

The South Orkney Islands were discovered on 6 December 1821 by Captain George Powell, an Englishman, in *Dove* accompanied by Captain Nathanial Palmer, from the United States, in *James Monroe*. These, and the South Sandwich Islands discovered by Captain Cook in 1775, and Bellingshausen in 1819, were also visited by sealers but proved not to have large seal populations.

One result of these discoveries of new areas combined with the virtual extinction of the fur seal and great reduction of the elephant seal on South Georgia was a great decline both in the number of voyages to the island, and a reduction of the takings of those that did.

Voyages from 1829 to 1846

Matthew Brisbane made a sealing voyage from London to South Georgia in 1829. The ship he sailed was *Hope* but, despite her name, she was wrecked on 23 April on the island. Brisbane and his crew were able to reach Montevideo using a shallop and reported to the British Consul on 20 May. His subsequent history was most unfortunate as, while he was in charge of the settlement at Port Louis in the Falkland Islands, he was murdered by some gauchos and Indians on 26 August 1833.

Captain James Brown from New England made a voyage to South Georgia in *Pacific* in the 1829/30 season. He left Portsmouth, Rhode Island, on 1 October 1829, called at the Cape Verde Islands for salt and arrived at South Georgia on 29 December. He departed on 5 March 1830 having secured only 256 fur seal skins and 45 barrels of elephant seal oil, an extremely small cargo by previous standards.

A gap of nine years occurred before the next

known visitors arrived at the island. Two vessels, *Mary Jane* and *Medina*, are recorded at Prince Olav Harbour on a grave marker – a rectangular plate of copper with an inscription made by perforations. Some other graves are near it which R.C. Murphy, who saw them in 1913, believed were of seamen from one of them. The inscription reads:

Iohn Anderson
Mate of Schr Mary
Jane of N York
Capt Joseph E Parsons
Died Nov 23rd AD 1838
in this port on board
of the Brig Medina

Sealers' graves at Prince Olav Harbour, December 1979. One of Iohn Anderson the others unknown. (D. Matthews.)

of N York Capt
Elijah Hallett
aged 36 Years
An honest man

These vessels sailed in company and also visited Shag Rocks, which were sketched. *Mary Jane* took 600 fur skins.

The next vessel recorded at the island was there in July 1846. Another series of graves indicate her presence. At the Grytviken cemetery a repaired wooden marker is inscribed: 'In memory of W.H. Dyre, Surgeon of the Esther of London. Jas Carrick, Master. July 1846.' and four of the nine unmarked graves in that cemetery are said to be of her crew, all having reportedly died of typhus. It is uncertain if this was a sealing voyage; a surgeon was not usually carried on such a voyage and the ship probably arrived at South Georgia about the middle of winter. If it were it would have been the last British voyage in the first epoch of South Georgia sealing. There was a gap of over twenty years before the next known voyage to the island.

First Letters Patent (1843)

On 23 June 1843 the first British Letters Patent to provide for the government of the Falkland Islands and Dependencies were promulgated. These were revised and consolidated by Letters Patent in 1876, 1892, 1908, and 1917. They provided that the Governor of the Falkland Islands should also be Governor of the Dependencies and empowered him to make legislation for their administration. The extent of the Dependencies remained undefined, however, until 1908.

Third peak of sealing (1869–1913)

The resumption of records of voyages was quite rapid from 1869 and indicated a third peak of sealing which also extended to several other Antarctic regions. Numbers of fur seal skins and quantities of elephant seal oil obtained were poor compared with the earlier records, even though the seal populations had had 20 years in which to recover. All the voyages recorded in this period were from the New England region of the United States of

America. Most were in newly constructed vessels, as the majority of the New England whaling and sealing fleet had been destroyed during the United States Civil War (1861–65).

The only record from 1869 is of 800 fur skins being taken from South Georgia, probably by Captain James W. Budington of Groton, Connecticut (who later made other voyages to the island); there are no further details of this available. Three voyages are known from 1870 to 1871; one from an inscription, another from a grave, and the third from literature. The German contingent of the International Polar Year found a piece of wood near the Nachtigal Glacier, Doris Bay, in 1882. On it was inscribed; 'T.K. Purdy 1870 Bk Peru, SA Norwalk Ct' and '1877 Sch Eothen' thus recording the visits of these vessels. On 10 January another burial was made at the Grytviken cemetery, the inscription on the grave marker is: 'In memory of H. Brockloe, Cooper of Bark Trinity, New London, Conn. Aged 35. Died Jan the 10th 1871. D. Rogers, Master'. The death of the Cooper was particularly unfortunate for a sealing vessel as he was essential for preparing the barrels for elephant seal oil. *Trinity* was known to have been accompanied by *Flying Fish*, commanded by Alfred Turner, on this voyage. She had left New London on 23 July 1870 and returned on 21 April 1871 with 250 fur skins and 210 barrels of oil. Brockloe's grave marker is in place at Grytviken and was repaired in 1919.

J.W. Budington was in South Georgia again in 1874 and secured 1450 fur skins. Of these, only 86 were taken from the place which yielded a few in 1869 and the remainder from a rookery he found elsewhere. In the next year (1875) he recorded that five vessels visited the island and took 600 skins, and in the year after that (1876) four vessels took only 110 skins. These voyages must have been mainly for elephant seal oil which would have made up the majority of the cargo, the voyages would otherwise have been financially disastrous for their principals. *Golden West*, from New London, visited in the 1875/76 and 1876/77 seasons with *Flying Fish* on the first and *Trinity* on the second of these.

There were three voyages recorded in 1877 (*Eothen, Trinity* and *Flying Fish*). The first is known from a brief inscription found at Royal Bay, previously referred to, and the second from a note that she accompanied the third, as on a previous voyage. Of the voyage of the *Flying Fish* much more detail is available from the account of a passenger aboard her.

Heinrich Klutschak (1877/78)

In 1877/78 Heinrich W. Klutschak, an Austrian traveller, visited South Georgia as a passenger aboard the New England sealer *Flying Fish*, the vessel which first visited the island in 1871. He published an interesting account of the voyage in a German geographical journal which included two illustrations and a map of the island. The map, prepared from the work of United States sealers, was probably the fourth of the island after those of Cook, Pendleton, and Bellingshausen.

Klutschak joined the voyage in the Cape Verde Islands. He described a rather irregular procedure used for recruiting part of the crew of

Flying Fish. The vessel sailed close to and stood off the Islands' capital during daylight hours, and then a boat was lowered that night. This went to within swimming distance of the shore and recruits swam out to join it thus starting a sea-going career and avoiding military service. This situation was somewhat regularised by the consul of the United States of America who later came aboard to sign and notarise the vessel's muster roll.

Flying Fish passed Shag Rocks on 22 September 1877 and Klutschak gave a description of them. He suggested a geological relationship between South America and Antarctica as part of a submarine ridge of high mountains joining them and showing itself as Shag Rocks, South Georgia and various other islands (some no longer considered as part of the Scotia Ridge). He also postulated that the mountains of South Georgia were extinct vol-

Chart of South Georgia from Klutschak's visit in 1877.

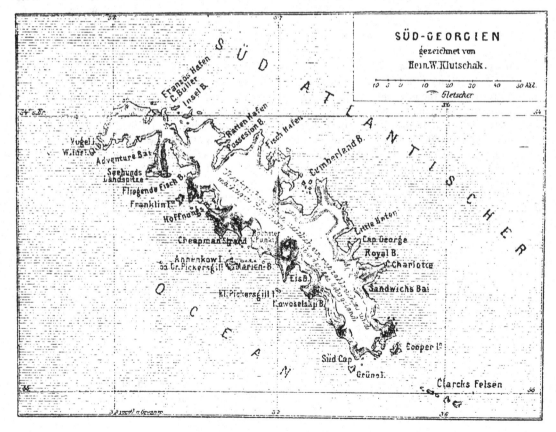

canoes and referred to great beds of lava he claimed were on the island. His descriptions of the meteorology were more accurate and mentioned the differences between the north-east and south-west sides of the island, correctly imputing this to the shelter afforded by the Allardyce Range to the north-eastern coast.

The ship anchored in Wilson Harbour, referred to as Fliegende Fisch Bai after her previous visit, and a description of the bay is given. A comment on the weather experienced was that for one reasonable day they had to suffer one inclement week. Sealing is described: elephant seals were killed with bullets through the heart or brain, rather than by the earlier procedures with clubs and lances. Klutschak is very critical of the industry and pessimistic about its future. He commented about its cruel, wasteful and unnecessary aspects, the destruction of females and, in consequence, the young and considers extinction of the species certain. The leopard seal is described and reported as of little interest to the sealers.

In December, after taking shelter from storms in a bay at the southern end of the island for 3 weeks, *Flying Fish* sailed to the north-east side of the island. Klutschak refers to the vegetation and bird life with enthusiasm after the bleak south-west side. Rats are reported for the first time, at Prince Olav Harbour. They may have been present on the island for almost a century by then. The lack of fear of man by albatrosses and penguins is described. Of the latter he reports three species and refers to accounts of a penguin oil industry but is unable to confirm the truth of this. Sealers' try-pots and the wreck of a French ship on the north-western coast are also referred to (the ship was probably a prize taken during the Napoleonic wars, which then became a British sealer).

Fur seals were found on the north-western parts of the island. They were described as congregating on low rocks washed by surf, rendering it difficult for sealers to get close enough to kill them; perhaps this was a behavioural adaption to promote their survival. The killing was accomplished by

'The glacier front in Ice Bay', a copy of a woodcut by Klutschak.

clubbing or shooting, sometimes after the seals had been driven inland. Several hundred skins were taken in January and February 1878 and salted down in the hold. The ship departed on 28 February and endeavoured to make the return voyage as fast as possible to keep the skins in good enough condition to retain their value. Klutschak left *Flying Fish* in Pernambuco, Brazil, early in April. The skins were resalted at that port before *Flying Fish* continued her northward voyage.

Trinity and *Flying Fish* visited South Georgia again in 1878; commanded by B.N. Rogers and Simeon Church respectively. *Flying Fish* sank off Cape Horn, and it was also *Trinity's* last voyage to the island as she was wrecked on Heard Island in 1850. She secured only 250 barrels of seal oil.

Sealing Ordinances (1881)

The first of a series of Ordinances to regulate sealing and protect seals was enacted in 1881. It

'The entrance to Cumberland Bay,' a copy of a woodcut by Klutschak.

provided for a closed season between 1 October and 1 April inclusive for sealing in the Falkland Islands and their Dependencies and in the adjacent seas. At South Georgia this was without much effect, except to mark the beginning of regulation of the industry.

International Polar Year
Expedition (1882–83)

From August 1882 to September 1883, a major scientific expedition visited South Georgia: the German contingent of the first International Polar Year. This is described in Chapter 4 – the first land-based scientific expedition.

Third peak of sealing continued

George Comer of East Haddam, Connecticut had made voyages to many sub-Antarctic and other islands for sealing since 1879. Two voyages to South Georgia are described from his notes. On the first, he was mate of *Express* from Stonington and was on

the island from 14 October 1885 to 11 February 1886. He records: 'We had heard reports of the numbers of seals formerly found there, but we did not get a seal and only saw one'. Comer restricts the word seal to mean fur seal only. During the voyage, 123 leopard seals and only two elephant seals were killed which made about 60 barrels of oil. This is the first reference to the taking of any numbers of leopard seals and, presumably, was due to the exceptional dearth of elephant seals. Whilst the leopard seal yields oil, it provides much less than the elephant seal as well as being a more dangerous animal to hunt. Comer also recorded that: 'The coal is all gone and the cook burns the leopard skins'.

The visit was also notable in that Comer collected skins and eggs of birds from South Georgia for G. Verrill, an ornithologist who published an account of them and some of Comer's voyages. Comer also recorded descriptions of the birds, their behaviour, nests, etc. and of egg collecting to supplement rations. He provided a list of 17 bird species from the island, with their vernacular names as used by the sealers.

George Comer's second voyage involving South Georgia was aboard *Francis Allyn* of New London commanded by Joseph Fuller. The ship had visited several southern sealing islands rather unsuccessfully, and made an excursion from Gough Island to South Georgia between 22 August 1888 and 23 January 1889. Comer remained at Gough Island during this period with a gang of 12 sealers but was able to record that the visit to South Georgia yielded only three fur seal skins. The voyage as a whole obtained 850 barrels of seal oil but it is uncertain what proportion of this was from South Georgia.

Another voyage is recorded by an inscription on a grave at the Grytviken cemetery: 'Joseph H. Montaro. Bravo C.D. Vera, A.B. Sch. S.W. Hunt. Died Feb 28th, 1891. Aged 19 years. R.I.P.'. It is probable that he was a young Cape Verde Islander recruited during a call to purchase salt and may have been attempting to gain experience as a sailor and avoid military conscription, as related by Klutschak.

Captain Budington was again on South Georgia in January 1892 when he secured 135 fur skins, none of which, however, came from rookeries he had previously known. He summarised the situation then prevailing: 'The seals on South Georgia are practically extinct'.

Following these rather unsuccessful sealing voyages there are no more recorded until early in the next century. In 1899, the Government of the Falkland Islands and Dependencies repealed the Sealing Ordinance of 1881 and replaced it with an improved one. The new Ordinance maintained the closed season and introduced requirements for licensing sealers; these required them to return an accurate account of the number of seals taken. The Ordinance allowed the making of regulations to protect females, young and certain species of seal, to amend the closed season and to declare sealing reserves. Further Ordinances and regulations followed as the industry was led to a more rational exploitation of the resources. This is described in Chapters 5 and 8, as it mainly concerns what might be called the second epoch of sealing.

Captain C. A. Larsen, who is closley associated with South Georgian history, established a whaling station at King Edward Cove in 1904. He also began sealing and took 80 elephant seals in 1905. Ernest Swinhoe visited the island in *Consort* leading the South Georgia Exploration Company from August 1905. Part of the expeditionary expenses were to be met by profit from sealing; after leaving a camp and some domestic animals at King Edward Cove, the ship went sealing. Swinhoe's report suggests that they did not secure any seals however; other aspects of the visit are described in Chapter 4.

Several sealing voyages were recorded shortly after the beginning of the new century, in 1905, 1906, 1907, 1908, 1911, 1912, and 1913. They principally took elephant seal oil but secured fur skins whenever possible. In 1918 C. A. Larsen described two catches of 170 and 270 fur seal skins taken during those years and stated that, on several occasions, he had met United States sealers who reported having caught fur seals at the northwestern end of South Georgia in the period. As well as those of *Daisy*, described below, some details of four of these are recorded. One arrived at Grytviken about 1906 from the South Sandwich Islands and a circumnavigation of South Georgia; on 10 March 1907 *Stranger* called there and again on 1 April 1908 with a full cargo bound for Buenos Aires and thence back to the United States. Another sealer was seen in the Bay of Isles in February 1909 but sailed away rapidly when detected. A fourth, *Agnes G. Donahoe*, was seen off the north-eastern coast on 21 December 1911 and was believed to have sailed from the South Sandwich Islands. Living with few expenses, some of these sealers could afford to visit South Georgia for only a comparatively small amount of elephant seal oil, the occasional fur seal skin and whatever amounts of whale oil they obtained during the voyage to and from the island.

Benjamin Cleveland and *Daisy* (1912/13)

The last voyage of the first epoch of South Georgia sealing was made by Captain Benjamin Cleveland of the United States of America. He had been sealing from New Bedford, Massachusetts, for many voyages (two of which were to South Georgia). The second of these, during which he passed his 69th birthday, was the last of the epoch. Little information is available about the first of these two voyages but an abundant amount (including photographs), is available for the second one.

Captain C.A. Larsen of Grytviken was exploring South Georgia and the South Sandwich Islands in the steam yacht *Undine* to discover sites for the expansion of the whaling industry. He encountered Cleveland, sealing near Undine Harbour in *Daisy*, in November 1908. In early 1909, *Daisy* was seen in the Bay of Isles. Larsen may have informed Cleveland about recent developments on South Georgia and the Sealing Ordinances. In any event, someone informed him; a letter requesting permission to undertake sealing at South Georgia was forwarded to the Colonial Office, on his behalf, by the United States Consul in London. The inquiry was referred to the Governor of the Falkland Islands and Dependencies in Stanley who replied with details of the permission required under the Seal Fisheries Conservation Ordinance. A letter to this effect was forwarded through the United States Consulate in London. Cleveland, however, later strongly denied he was aware of any legal impediments to his subsequent voyage – and subsequently disregarded provisions made to conserve seals on South Georgia. Undoubtedly he believed that the spread of law, government and consequent bureaucracy had surpassed acceptable proportions when it arrived on South Georgia and involved itself with him. However, it remains that he had

been made aware of the requirements several years prior to his last voyage, and had later even written to the Magistrate at King Edward Cove in 1911, about a sealing licence.

Captain Cleveland's last South Georgia voyage was remarkable for one most fortunate matter. It carried Robert Cushman Murphy, a naturalist from the Museum of the Brooklyn Institute of Arts and Sciences. Apart from gathering a large number of natural history specimens and making many scientific observations, Murphy kept a comprehensive diary which provided excellent details of the voyage, sealing, South Georgia and many related aspects. An edited version of his journal and letters was published in 1947. In 1967 a book of his photographs was published to illustrate the voyage and supplement the edited diary. Most of the details about Murphy's involvement in the voyage are more appropriately included in Chapter 4, with aspects of the island's history outside the first sealing epoch. The sealing aspects of the voyage are, however, outlined below.

Captain Cleveland sailed from New York to Barbados to join *Daisy*, where she lay following a previous voyage. Her crew consisted of 20 men of very mixed background, several were from the Cape Verde Islands. Portuguese may have been more widely spoken than English on board. She sailed rather indirectly to the Cape Verde Islands, and Brazil to arrive at South Georgia on 23 November 1912.

On the next day, she reached Cumberland Bay and was towed by the whale catcher *Fortuna* into King Edward Cove, where a whaling station and administrative outpost had been established in the previous decade. *Daisy* remained there to 12 December awaiting mail and this permitted Murphy to investigate the area and make some collections. The Magistrate (Mr J.I. Wilson) and Captain Cleveland did not get on very well together and Murphy records Captain Cleveland saying: 'What is the World coming to, when a peaceful whaler and sealer can't go about its business without being pestered and bled white by a gang of ...,...,..., lime-juicers!'. Cleveland and Murphy were welcomed by Captain Larsen who invited them to dinner. Murphy posted several items from King Edward Point to his wife and advised her: 'The stamped covers should be a philatelist's delight. Sell 'em;!' (Murphy, 1947).

Some elephant seals were taken while *Daisy* was in Cumberland Bay. On 13 December she was towed out of the bay and set sail for the Bay of Isles, where she arrived and anchored on 15 December. Murphy engaged himself in natural history, mainly from a shore base, while the crew went sealing. They combined forces in one aspect – the preparation of two elephant seal skulls for the Museum. Later, the ship moved into Possession Bay where Prince Olav Harbour whaling station was visited and a few more seals secured. Murphy mentioned some graves from the schooner *Elizabeth Jane* of New York dated 1835 there; he probably saw those from *Mary Jane* of New York dated 1838, described previously. After securing several more elephant seals, to make a total of 1641 for the voyage, *Daisy* sailed on 15 March 1913 – the last day of the first epoch of South Georgia sealing.

Daisy was equipped with two try-pots on her deck just abaft the main mast. These were surrounded by brickwork and were where the elephant seal blubber was processed into seal oil. She also took whales on the way to and from South Georgia, and treated their blubber similarly. Murphy gave an interesting description of the treatment of elephant seals and was rather critical of the waste involved:

The old American method of utilizing the blubber is wasteful at every stage. After the slain 'elephant' has been allowed to bleed thoroughly, the hide is slit lengthwise down the back, and then transversely in several places from the dorsal incision to the ground. The flaps of hide are next skinned off, and the remaining investment of white blubber, which may have a maximum thickness of about eight inches, is dissected away from the underlying muscle and cut into squarish blanket pieces. The animal is then rolled over and the same process repeated on the ventral side. Thus the hide, and the considerable amount of blubber which clings to it are lost at the start.

The blanket pieces of blubber are hauled to the water's edge to be strung on short ropes called 'raft-tails'. These are towed to the anchored ship where each laden raft-tail is looped about a hawser which extends from bow to stern, and the blubber is permitted to soak for forty-eight hours, or thereabouts, until the red blood corpuscles have been practically all washed away. During the soaking process a certain proportion of oil is lost, and moreover, flocks of ravenous 'Cape Pigeons' (*Petrella*) and other ubiquitous sea birds feed upon the floating fat with an interminable hubbub, both night and day. When the blubber is hauled on board it is cut into narrow strips

called 'horse pieces', and is afterwards 'minced'. The mincing differs from the same process in sperm whaling only in that the fat is cut very finely with hand knives. At this stage an additional loss of oil occurs particularly if the temperature of the air chances to be well above the freezing point. Finally the minced blubber is 'tried out' in the familiar deck try-works of the old whaling type. There is so little residue or 'scrap' from boiled sea-elephant blubber that the Heard Island sealers of the last century used to calculate 'a cask of oil from a cask of blubber'.
(Murphy, 1918)

Only one of the photographs of the accompanying book illustrates anything of South Georgia, showing *Daisy* at anchor in King Edward Cove. The rest show many scenes on the ship of blubber processing and other activities. However, Murphy illustrated the large number of papers he wrote later with many fine plates of the island.

Summary and methods of the first sealing epoch

The epoch of the first type of sealing on South Georgia started shortly after Captain Cook's visit in 1775 and lasted about 130 years, ending in 1913 with the departure of Captain Cleveland. At first only fur seals were taken for their skins. These proved so valuable, especially when the trade for them was developed with China, that the animals were hunted almost to extinction. Any remnants of the population which survived the early period were always taken when found, as their value remained. The very long period the population took to recover is very largely a consequence of this. Elephant seals were exploited only for their oil, which was extracted from the blubber. Profit could be derived from them only when they were reasonably abundant; thus, although greatly depleted in numbers, their survival was at no time so precarious as that of the fur seal. Their populations re-established much more rapidly after the end of the first sealing epoch.

Since about 1970, the fur seal has begun to recolonise beaches where it was once abundant on South Georgia (and elsewhere). Prior to this only small isolated populations, mainly at or near Bird Island, were known. (Until 1977 a diary was maintained at the scientific station at King Edward Point to record fur seals if they were sighted; it has since fallen out of use, as too many were seen too often for records to be so precisely kept.) Population pressure at the established seal rookeries is probably the factor most important in promoting the fur seal's spread to unoccupied areas. Although a marked Antarctic fur seal from South Georgia has been recorded in Tierra del Fuego and some other long distance travels are known, it appears that, for breeding purposes, fur seals tend to return to the region where they were born. The present rate of growth of their populations has been estimated as about 10% a year.

Elephant seals remained distributed throughout their range although greatly reduced in numbers. This species did, therefore, not have any vast recolonisation to accomplish before re-establishing its populations. The numbers had sufficiently recovered to permit a successful controlled exploitation of the species to commence from 1909 – the second sealing epoch described in Chapter 5.

The first epoch of sealing on South Georgia had three peaks, a major one around 1800, a smaller one around 1820 and a third, even less pronounced, around 1870. The first peak was mainly of fur sealing; the subsequent peaks were principally elephant sealing, but fur skins were also taken whenever possible. Several isolated voyages are also recorded, most apparently not profitable. These interpretations are based on voyages for which there are some records, however tenuous. The records are far from perfect because of the secrecy of voyages as previously discussed; it is not, however, unreasonable to deem their imperfections random and thus to deduce that the postulated three peaks are probably real.

Sealer's operations were fairly constant during the whole epoch. Although only two species of seal were principally taken and the methods of dealing with them were different, it is reasonable to regard it as one industry since most of the vessels involved were equipped to exploit both species. Occasionally other seal species and sometimes whales were also taken for their blubber.

The extracts from the literature of sealing quoted above describe many aspects of it on South Georgia. There was a general similarity in operations throughout the period and these are outlined below. On arrival many of the sealing vessels used to anchor with several cables in the various bays of South Georgia and prepare themselves against storms etc. Occasionally they would overwinter, in order to be ready at the start of the next sealing season. As an alternative, vessels staying for a shorter period merely spent their time sailing between the various harbours, taking seals at every opportunity. Anchored sealing vessels were equipped with 'shallops', a type of cutter generally about 10 m long and displacing less than 50 tons. These were occasionally left on the island over winter for use during subsequent seasons and were often brought down on the sealing vessels in sections for local assembly. There is one record of a sale of a shallop on the island. In 1902, two members of the Swedish South Polar Expedition (Andersson and Duse), found a shallop left at King Edward Point:

Close under the mountains crept a little bay I had not seen before, with a low point shooting out between it and the main fjord – and now comes the strange part of the story – on this point, and drawn up some distance from the shore, lay a *large green-painted boat*. The boat had evidently lain there for many years, for the tussac grass grew high and close around it. It was a large undecked centre-board boat, thirty feet long and eleven feet wide, almost too large to have been brought here as a deck boat on board a vessel, but too small to have sailed alone here to this stormy coast.

(Nordenskjöld, Andersson, and Larsen, 1905)

This shallop was later incorporated into the jetty foundations at King Edward Point and last revealed herself during the subsidence of part of the approach to the present jetty in the mid 1970s. Shallops were able to penetrate many beaches on the island which were inaccessible to larger vessels. They were also used for transporting blubber or processed oil to the sealing vessel and supplies to the sealers established in various places on the island. For reaching the beaches and landing, both sealing vessels and shallops carried ship's boats of a type that could negotiate surf and land on very rough shores.

There are several records and relics of sealers staying on land for extended periods and occasionally overwintering. Some of the island's caves provided shelter and a number of ruined huts are known. The supplies issued were minimal and, in as far as practicable, the sealers lived off the land. Cases of sealers being marooned owing to the wreck of a ship or other reasons are recorded. As fur seals became rarer and elephant oil production increased as a proportion of the industry, more operations became shore-based. Temporary camps, and possibly some longer term ones too, traditionally used the upturned boat as a shelter.

Fur seals, the bulls of which were known as wigs and the cows clapmatches, were killed with clubs and then stabbed through the heart to bleed. Skinning was done directly; it was important that this be completed while the carcasses were fresh, as skinning a stiffened seal is very difficult and any decomposition may cause the fur to spoil. The method of skinning depended on the intended market for the skins, those for China being prepared with less blubber on them than those for Europe and North America. The prepared skins were washed in sea water to remove blood and dirt, then laid or pegged out on shingle, turf or the like to drain and dry as far as possible. A competent man could prepare about 30 skins a day – and up to 30 men were deployed from a sealing vessel.

Preservation was done with salt, usually obtained at the Cape Verde Islands on the outward voyage. About 2 kg were required for each skin. It was important that they were adequately salted as the return journey crossed the Equator and opportunities for putrefaction in the hold of a sailing ship in the tropics were enormous. One cargo of rotten seal skins arriving in London was recorded as eventually being dug out of the ship's hold and sold as fertiliser.

The salted skins were arranged in 'books' or 'kenches'. The former method involved stowage in layers, interspersed with salt. The latter method required the skins to be built into a circular pile, with raised edges, so that the salt (converted into a pickle by extraction of fluid from the skins and attached blubber) remained in contact with them. When necessary, resalting and restowing were done later on board.

The preparation of fur seal skins is somewhat more complex than with other fur skins. The seal's coat has two types of hair: guard hairs and underfur. For a commercial fur skin it is necessary for the guard hairs to be removed. The Chinese had probably developed a process for this before 1750. In the United Kingdom processes were patented for it in 1799 by T. Chapman and, in 1801, by J. Brunswick. These rendered the skin of even greater

value in Europe. Fur skins were also used for making leather, when all fur was removed. In China, the fur resulting from this was used for felt making. Seal leather was used for many purposes, particularly for making boots, shoes and gloves. Seal furs were principally used in Europe and North America for garment manufacture.

Elephant seal operations were broadly similar in many aspects. The seals were driven to a suitable place (the water's edge if the blubber was to be taken aboard the sealing vessel), and killed with lances, clubs or – with larger bulls, and later in the industry as the animals became rarer – rifles. They were left to bleed from lanced incisions and many reports describe the enormous amount of blood they had. The blubber was stripped off by making various incisions. This was described in detail by Murphy, quoted earlier in this chapter.

The blubber was soaked in the sea to remove the blood and was then transported to a 'try-works' which was either on land or on the sealing vessel. The 'try-works' generally consisted of two 'try-pots', usually cauldron-shaped vessels, with a capacity of about 400 litres. A common variety had flat surfaces on opposite sides, to permit them to be joined in pairs and sit efficiently over the same fire. A hearth, generally built of bricks, supported them over a fire. On the sealing vessels, this arrangement was often just abaft the main mast and must have ensured that the masts, spars, rigging, and sails became very sooty and oily. Elephant seal blubber, in suitably sized pieces, was cut into thin sections and boiled in a try-pot. The oil separated and was ladled into barrels. The remains of the skin, crisp (virtually fried) pieces, were hooked out and used to fuel the fire which made the process at least partly self-sustaining. Fuel (mainly wood and coal) was provided to start the process. (Reports of the discovery of coal on the island in the late 1800s were probably the discovery of a sealer's fuel depot.) Penguin skins were also reported to be used as fuel on South Georgia. They were certainly used at other elephant sealing areas and, with the feathers acting as a wick for the fat, may have been an unfortunate necessity. Preparation of penguin oil by a process very similar to that for seal oil has also been reported from South Georgia. There is very little evidence for this, although at the Falkland Islands, Macquarie Island and other similar islands it formed the basis for an industry which followed the decline of sealing.

Elephant seal oil was used for many purposes, among which were lighting and, to a lesser extent, lubrication (before mineral oils replaced it in the late 1800s). Seal oil was also, and remained, an item used in textile processing to replace natural oils and in preparing leather to render it supple and water-proof. The latter use was one of the things which ensured a continuing demand for it during the second sealing epoch.

The sealers were paid according to the success of the voyage, generally by a series of shares on the profits realised. These were referred to as 'lays'. The lay a man received depended on his rank, experience and other factors. A Captain might receive $\frac{1}{11} - \frac{1}{15}$ of the nett profit, the mate $\frac{1}{20} - \frac{1}{30}$, and so on in proportion to perhaps $\frac{1}{80} - \frac{1}{125}$ for seamen and $\frac{1}{200}$ for new recruits. Some captains and others did well from the industry but most profit chiefly went to the owners of the vessels – and often the Captain had shares in the ownership.

Many relics of this period may be found on the island. These are presently becoming more apparent owing to some effects of the increase in numbers of fur seals. Fur seals tend to kill and erode tussac grass near the beaches by 'nesting' in large tussocks. Thus as the margin of tussac communities retreats with increasing fur seal populations, items buried beneath it are being revealed for perhaps the first

A sealer's try-pot at Grytviken made by 'Johnson and Son, Wapping Dock, London,' photographed in Autumn 1979. (Author.)

time in a century or more. Try-pots are the most widespread and conspicuous relics. They are one of the best known pieces of sealers' equipment and several have been transported from South Georgia to museums around the world. On the island they may still be found at Grytviken, King Edward Point, Elsehul, Trollhul, Nilse Hullet, Elephant Cove, Wilson Harbour, Undine South Harbour and elsewhere. A maker's name may still be determined from one of those at Grytviken whaling station: Johnson and Son, Wapping Dock, London'. This company and 'Mount and Johnson' of the same address, probably made most of the South Georgia try-pots.

The three graves with markers and at least six unmarked ones from the sealing era have been described above. Some voyages are recorded only by information on these and it is possible there are others as yet unknown. A skeleton with evidence of a bullet through the skull was disinterred at

Ocean Harbour in 1910 and might be regarded as evidence of some sealer's rivalry a long time ago. Another skeleton was found in a cave near Prince Olav Harbour early in 1913 which was presumably that of a sealer. Several caves show signs of habitation, including those at Maiviken, Fortuna Bay, Will Point in Royal Bay, a cleft on Dartmouth Point and a cave near Carlita Bay. Several expeditions have reported finding things within them and most show evidence of some walling and of a hearth. The sealers' cave near Carlita Bay is particularly well preserved, with a low entrance wall and remains of burnt seal blubber still discernible on the central hearth. As well as the artifacts found within them, these places occasionally have legible inscriptions on their walls.

Huts or other habitations were also constructed at several places and their ruins are known from Hestesletten, near Cape Vakop, Hope River and at Diaz Cove. Most are small structures and that near

A hearth in a sealers' cave near Carlita Bay, February 1982. (Author.)

Cape Vakop is only about 2 m². The floor is wooden with 2 cm thick planks and walls, built either of stone and turf to about 1 m high or cut into a bank behind. The upper portions of them and the roof were of galvanised iron with small corrugations. It is a tribute to the quality of the galvanising of the period that, after about a century close to the sea, most of it is still not seriously corroded. A wrought iron blubber-hook was also found in the vicinity.

Wooden spars and similar items are commonly found on the beaches and elsewhere on the island. Many of them appear to be of considerable antiquity. The 'eight foot high pile of wreckage and relics from ships', described by Shackleton at King Haakon Bay in 1916, was reduced to nothing other than the occasional weather-worn spar – to the disappointment of the Combined Services Expedition who visited the site in 1966. This type of relic, together with brick works (from try-pot bases and the like) are the main things turning up as a result of fur seal activity. Unfortunately, at present, age,

origin and use of these sites can rarely be positively determined. It is likely that an archaeological investigation of the places referred to above and such as Rosita Harbour, Prince Olav Harbour, Doris Bay, Trollhul, Koppervik, Holmstrand, Shallop Cove, as well as many others, would yield much interesting information.

In 1929 a harbour at Diaz Cove was entered by the Kohl-Larsen expedition. This area had not been visited since the previous century. There, the wreck of a wooden ship was found with the remains of a dwelling place. The Magistrate requested that a further inspection be made and Dr L. Kohl-Larsen retrieved parts of a flintlock gun, parts of a telescope, clothing remains and various parts of the wreck. One may conjecture the eventual fate of these sealers when such relics were left behind.

To conclude the discussion of this part of the history of South Georgia, and the regrettable example of uncontrolled exploitation of the two seal species, it is appropriate to consider Antarctic and

Remains of a sealers' hut near Cape Vakop, January 1982. Note the stove and tripod. (A. Pinhiero-Torres.)

other southern sealing generally. The first sealing epoch was by no means a specialised South Georgian phenomenon but an important part in the development of the whole southern region, and many places elsewhere. Many geographical discoveries arose from sealers seeking to extend their operations to unexploited areas. Other peri-Antarctic islands with seal populations; Gough Island, Tristan da Cunha, Bouvetøya, Prince Edward Islands, Îles Crozet, Kerguelen, McDonald Islands, Heard Island, Îles Amsterdam and Saint-Paul, Macquarie Island, the sub-Antarctic islands of New Zealand, as well as the South Sandwich, South Orkney, and South Shetland Islands, together with the Falkland Islands, Tierra del Fuego, Patagonia, and some other places suffered a similar exploitation. The course of history and effects on both seals and the early sealing industry were very similar in most of these places. Comparable examples exist for exploitation of many other animal species throughout the world, some of which are quite recent. Modern sealing operations at South Georgia and several other places were very different, once stable, controlled industries under scientific management were developed. The recovery of the South Georgian and other fur seal populations from near extinction is also remarkable evidence of the resiliency of animals to the effects of such exploitation.

4

Expeditions, visits and other events on South Georgia from 1882

The account of the discovery, early history and first epoch of sealing on South Georgia, given in Chapter 3, briefly described the International Polar Year Expedition at South Georgia in 1882–83. This was the first of many scientific and other expeditions which visited the island. These may be conveniently grouped, together with some other aspects, and treated as a separate chapter of the island's history. Other historical divisions concern the exploitative aspects from the beginning of the whaling era and the development of the administration, which are outlined in Chapter 5. As well as the expeditions described in this chapter, several others to South Georgia have been proposed which, for a variety of reasons, were abandoned, some at comparatively advanced stages.

The sections in this chapter reflect the amount of material available concerning expeditions etc., as well as their relative significance for South Georgia (many also operated elsewhere). The size of the sections should not be regarded as closely correlated with the latter aspect and the comments about research conducted by the Falkland Islands Dependencies Survey and British Antarctic Survey are also relevant in this matter (Chapter 5).

International Polar Year Expedition (1882–83)

The German South Georgia expedition of 1882–83 was the first land-based scientific exploration of the island and was the only Antarctic expedition of the first International Polar Year, although others had been planned. The International Polar Year was a world-wide cooperative programme initiated by Lieutenant Karl Weyprecht of the Austro-Hungarian Navy and was comparable with the International Geophysical Year of 1957–58 which also established a station on South Georgia. One of the principal proponents of the International Polar Year was Dr Georg von Neumayer who coordinated the German contributions. He had been deeply involved with the geomagnetic observations made at Göttingen, Germany, where Gauss had founded the first major geomagnetic laboratory. Subsequently he was responsible for the establishment of a geomagnetic observatory at Flagstaff, Melbourne, Australia, and became its first director. Eleven nations participated in the observations of the first International Polar Year and 14 observatories were established. Germany operated two stations, one at Cumberland Sound, Baffin Island and the other at Royal Bay, South Georgia. The latter was selected because it had a longitude almost diametrically opposed to that of the Flagstaff observatory.

The expedition left Hamburg on 3 June 1882 to arrive in Montevideo on 4 July. There they transhipped to *Moltke*, a three masted steam corvette commanded by Captain Pirner, and sailed for

South Georgia on 23 July. *Moltke* was the first powered vessel to reach South Georgia. On 12 August they sighted the island near Possession Bay and sailed south-eastwards. The ship entered Cumberland Bay and anchored for the night of 16 August. An attempt to land on the 17th was unsuccessful owing to severe storms, and they considered abandoning South Georgia for a station on the Falkland Islands as the ship was running short of coal. Further eastwards they entered another bay but the weather was still too severe to permit landing. However, in Royal Bay a landing was effected on 20 August at a harbour later named after the ship. Between 1 and 2½ metres of snow were shovelled away to clear a site on shore before supplies were unladen from the ship with the assistance of 100 seamen. The erection of the station, much of which was pre-fabricated, commenced. The living quarters, together with some other buildings, were completed before *Moltke* sailed on the 3 September for the Falkland Islands. When she arrived at Port Stanley Captain I.H.M.D. Seaman of the *Kosmos* shipping company of Hamburg, was provided with instruments by the expedition which he used to operate a second order meteorological observatory there for the International Polar Year.

The 11 members of the expedition, led by Dr K. Schrader, included physicists, a meteorologist, a botanist, a zoologist and medical officer, an engineer and artist, a mechanic, a cook, a carpenter, a sailmaker, and a seaman. The station had a living quarters, two magnetic observatories, two astronomical observatories, a darkroom, a zoological laboratory, sheds, stables, and gardens; several fixed survey marks were also established.

The scientific programme was determined mainly by the requirements of the International Polar Year. These included astronomical, meteorological, geomagnetic, gravimetric and tidal observations. The opportunity was also taken to undertake much biological, geological and other research. A transit of Venus was observed on 6 December 1882 with a telescope deployed in a building with a revolving cupola. This involved timing the passage of the planet Venus as it moved across the sun's disc during one of the rare times when she is between a particular part of the Earth and the Sun. The data obtained allow accurate calculation of terrestrial and planetary positions and distances. The observ-

ation was successful – the expedition evidently having had much luck with South Georgian weather. The telescope had also been used to observe a comet in September 1882. Tidal observations were recorded and the data from them indicated on 27 August 1883 the effect of the eruption of the volcano Krakatoa on the previous day.

Meteorological and geophysical observations were made hourly from early in September. On 'term days', twice a month, geomagnetic observations were made every minute and, for a one hour period, every twenty seconds. These were made synchronously by all the stations of the International Polar Year throughout the world.

The mountings of several of the instruments are still at the site. There are over 24 of these – 60 cm cubes of granite, which must have presented great difficulties during unloading. During the spring thaw some of them sank into the peaty soil, which necessitated the recalibration of several instruments. An electric time synchronisation system operated throughout the station and was connected to a master clock, the first telegraph system to be used in the Antarctic.

Other observations conducted included a comprehensive investigation of the animals, plants, glaciers, geology, and the preparation of maps of the Royal Bay area. The published results of this research described over 20 aspects of the island's natural history; many of the species collected were new to science. Photographs were taken on the island; unfortunately none were published. An interesting description of the numerous whales seen in Royal Bay was given.

The expedition was also responsible for some of the earliest deliberate introductions of animals and plants to the island. A hound 'Banquo', 3 oxen, 17 sheep, 6 goats with 3 kids, and 2 geese were brought and the expedition's botanist, H. Will, attempted to grow potatoes, Swedish varieties of rye, barley, and wheat together with some salad vegetables. Some of the animals were allowed to roam freely, grazing on tussac during the summer. Stables were built to house them through the winter. As well as fresh meat, the goats provided milk for much of the time the expedition was on the island. When released at Royal Bay, the two geese flew some distance to a grassy ledge on a cliff where they remained until they died. Rats were not reported.

Much local exploration was undertaken in the area from St Andrews Bay in the north, to the southern side of Royal Bay (which was reached by boat). Several peaks were climbed and an excellent map of the locality prepared. Parts of this map still form the basis of Admiralty charts of the region. Measurements and surveys of the Ross Glacier and some other glaciers were made. The recorded positions, rates of movements, and dimensions of these glaciers are of great interest as the expedition was on the island during a period of maximum glacial advance and when a large section of the Ross Glacier separated.

The expedition was relieved on 1 September 1883 by the corvette *Marie*, commanded by Captain Krokisius, which arrived from the Falkland Islands. She departed on 6 September for Montevideo when the station was closed. The expedition left maximum and minimum self-registering thermometers and many bottles with enclosed messages at South Georgia. Several visits were later made to the site; one of these was that of the Filchner Expedition in 1911, when the station was reopened and a short

series of meteorological and other observations were made. By 1931 it was in ruins and an attempt to inhabit one of the rooms resulted in a description that as much snow fell within it as without.

In January 1982 there was little of the walls standing above ground level, apart from earth banks and some posts. The author was easily able to determine what the various buildings were from the original plan, as much of the floor materials remain. The granite instrument mountings are also a distinct feature, as are the remains of the transit of Venus cupola. A view of the station was published in the descriptive account of the expedition and the author was able to obtain a similar photograph 100 years later. Many interesting artifacts remain on the site including porcelain insulators from the time line, cast iron stoves from the dwelling house and zoology laboratory, a large number of glass and earthenware bottles (originally left in a pyramid), many metal castings for various purposes, and other similar items. Unfortunately a careful search made from HMS *Endurance* in 1984 found no sign of the self-registering thermometers.

The German International Polar Year Expedition station at Royal Bay in 1882–83. Copy of a woodcut by E. Mosthaff.

Jason, Hertha, **and** *Castor* **(1894)**

Towards the end of the nineteenth century stocks of whales in many parts of the northern hemisphere were decreasing and more legislation restricting whaling was being enacted. This resulted in the industry becoming active in seeking new grounds to exploit. Several expeditions were despatched to investigate the Antarctic regions (an analogous situation to that with seals a century earlier). Two of these left Europe in 1892 and returned in 1893, one from the United Kingdom and the other from Norway. Each included a man who was to become prominent in Antarctic exploration; the Scotsman W.S. Bruce on the former and the Norwegian C.A. Larsen on the latter. These expeditions did not visit South Georgia although the Norwegian one passed close to the island on 7 November 1892. Their objectives were very similar and they cooperated substantially. The expeditions met each other during December 1892 in the Weddell Sea. In 1894/95, another expedition aboard *Antarctic* was engaged in a similar exploration of the Ross Sea; in 1902 she carried an expedition to South Georgia.

A second Norwegian expedition departed barely 2 months after the first returned. This consisted of three vessels; *Jason*, just back from the previous voyage, commanded by the same Master, C.A. Larsen who lead the expedition, together with *Hertha* commanded by C.J. Evensen and *Castor* commanded by M. Pedersen with complements of 36, 25 and 30 men respectively. *Jason* sailed from Sandefjord on 12 August 1893. The ships investigated parts of the Antarctic Peninsula, South Shetland Islands and Weddell Sea. Some sealing was undertaken on Seymour Island. Further north, the Falkland Islands were visited and landings made near Cape Horn. *Jason* separated from *Hertha* and *Castor* early in 1894, after they had arranged to meet later at South Georgia before

The site of the German International Polar Year station one century later, January 1982. (Author.)

starting their return voyage to Norway. *Castor* and *Hertha* arrived in Moltke Harbour, Royal Bay on 20 April 1894. The German International Polar Year station there was visited and a note to that effect left on a wall. The first recorded whale harpooned at South Georgia was killed in Moltke Harbour on that day from *Hertha*. It was a Southern Right whale which was unfortunately lost when the line which secured the harpoon broke. Meanwhile *Jason* had arrived in Cumberland West Bay and anchored in a harbour later named after her. *Hertha* and *Castor* reunited with her next day and they sailed north to return to Sandefjord on 5 July 1894. The ships had secured 13 223 seal skins (which must have been of species other than fur seals) and 1100 tons of seal and whale oil, believed to be mainly from Seymour Island. None was recorded as being taken at South Georgia. Little information was published concerning these pioneer voyages to Antarctica though much improvement to charts resulted. It was C.A. Larsen's first encounter with South Georgia and the beginning of his long and important association with the island. The expeditions brought back reports of abundant whales in the region and C.A. Larsen made inquiries at the Royal Geographical Society, London, in February 1896 about establishing a whaling station on South Georgia.

Economic enquiries and the lease of South Georgia (1900)

Some enquiries about South Georgia had been received by the Government in Stanley from late in the nineteenth century. These resulted in enquires and correspondence only until 1900, when a lease of the island was advertised in the Falkland Islands Gazette.

The first hint of commercial operations on South Georgia, other than whaling and sealing, came in 1872. A Buenos Aires merchant applied to the Falkland Islands Government to rent Beauchêne Island and other islands in the Falkland Islands together with South Georgia in order to obtain guano. The exploitation of deposits of penguin guano in the Falkland Islands had been under consideration by the Governor, Colonel G.A.C. D'Arcy, at the time following a cessation of the demand for penguin oil and a desire that the birds be protected. Guano was collected, as a result of

the application, from the Falkland Islands but the operation did not extend to South Georgia.

Another commercial enquiry came after the International Polar Year Expedition, probably as a result of attention being drawn to South Georgia in consequence. The Governor, Mr T. Kerr, received an inquiry from a retired Royal Naval officer, Captain R.D. Inglis, in 1887 about purchasing or leasing the island with the object of farming sheep on it. The Governor's reply described South Georgia as 'covered with snow to great depths, surrounded by icebergs, and fringed with glaciers' and the matter was pursued no further.

In 1898 the Governor, Mr W. Grey-Wilson, sought information about South Georgia when it became known that the Government of Argentina had offered a reward for the discovery of coal on it. Captain F.M. West who had visited South Georgia twice on sealing voyages described, to the Governor, several deposits of coal, claimed to have used it on a ship and gave some other information about the island. There is no naturally occurring coal although a few small deposits of low-grade lignite are known. West may have imaginatively reported depots left by earlier sealers.

In 1900 a private application to lease South Georgia for farming purposes was received by the Falkland Island Executive Council. After consideration the Council decided that a lease of South Georgia should be advertised to give all colonists an opportunity to submit applications. Accordingly, an advertisement was promulgated in the Falkland Islands Gazette on 2 October 1900. It stated that the Government was prepared to grant a mining and general lease of the island for 21 years (renewable) to any responsible individual or company and advised that applications should be submitted prior to the end of March 1901. Some of the conditions for this prospective lease were: an annual fee for a sealing licence was to be paid together with a percentage of the profits from all undertakings; the lessee was to undertake to render and declare such returns as the Government might require; and a deposit, as surety for compliance with the terms of the lease, was to be lodged with the Colonial Treasury. This advertisement apparently yielded no results beyond enquiries. However, in 1903 a group of farmers in Chilean Patagonia approached the British Vice-Consul in Punta Arenas to try to negotiate a grazing lease of the island and the Tierra del Fuego Sheep Farming Company Limited contacted the Governor in 1904 but the matter was not pursued. These contacts, however, eventually led to an expedition to South Georgia by the South Georgia Exploration Company from Chile, which arrived in 1905. This is described later in this chapter.

Swedish South Polar Expedition (1902)

The Swedish South Polar Expedition of 1901–03, led by O. Nordenskjöld, sailed to the Antarctic Peninsula and South Shetland Islands where, at Snow Hill Island on the east of the Peninsula, a station was established. The leader, with five colleagues, intended to spend only the 1902 winter there and accidentally spent the subsequent one also. After disembarking the land party the ship (*Antarctic*, commanded by C.A. Larsen) sailed to Tierra del Fuego, the Falkland Islands and South Georgia to continue explorations during the winter. On 22 April 1902 she arrived at South Georgia and anchored in Jason Harbour where her Master had been about 8 years previously, when in command of *Jason*. Aboard were J.G. Andersson, C.J. Skottsberg and S. Duse who made zoological, botanical, and geological collections and observations, prepared charts, and accomplished a large amount of other investigation during the visit.

Collections were first made near Jason Harbour; these included an elephant seal skin and skeleton. A programme of scientific work was planned with a shore camp to be made at Cumberland Bay while dredging and hydrography were undertaken from the ship. Moltke Harbour and the site of the German expedition were also to be visited at the request of Professor Neumayer of Hamburg. This was achieved on 27 April, after a delay owing to storms. The note left by *Castor* and *Hertha* in 1894 was found in the main building, which remained in reasonable condition. The other buildings, however, had deteriorated markedly. While in Royal Bay, the opportunity was taken to make another survey of the front of the Ross Glacier.

The *Antarctic* returned to Cumberland Bay and a party was landed in a small cove on 1 May. The group comprised Andersson, Duse and Skottsberg with a Falkland Island lad engaged for the South Georgia voyage. The beauty of the landing place, which was named Maiviken (May Cove) was commented upon. They commenced biological investigation and surveying as soon as they had established a tent encampment. Local exploration led to a number of discoveries of artifacts from sealers who had previously visited the area. A cave, which had the remains of two hearths and various sundries from human occupation, was found close to the camp site. In the course of exploration, Andersson crossed the pass in Bore Valley (which he named on account of its glacial origin), after having followed the stream through the lake system in the northern part of the valley. On crossing a pass at about 200 m and descending to the shore on the other side, he discovered the sealers' shallop, described at the end of Chapter 3, together with seven 'try-pots' and associated brick works of a hearth. One of the try-pots he records as having the inscription 'Johnson and Sons, Wapping Dock London' (one still there has this inscription and may well be the same). He named the bay where these were found Grytviken (Pot Cove). Several days later, Andersson and Skottsberg found the graves on the southern side of Grytviken and recorded the three inscriptions which were then legible. Skottsberg also discovered a grass later determined to be *Poa annua* in the area, the first recorded non-indigenous plant on the island other than the Royal Bay introductions of 1882.

On 11 May they moved camp and sailed the ship's boat along the south side of Cumberland West Bay. An uncomfortable camp was established at Papua Beach from which further exploration and biological observations were made. On 12 May, *Antarctic* returned following a journey to Possession Bay and the Bay of Isles while conducting hydrographic work and making collections of marine organisms. After one more day of surveying, the land party rejoined the ship after a perilous crossing of Cumberland West Bay in a small boat during a storm. After collecting the remains of the camp at Maiviken, *Antarctic* sailed into Grytviken on 14 May, where she remained for the rest of the duration of the expedition's visit. During this time geological, zoological and botanical studies were continued as well as hydrographic and land surveying. The first fossil from South Georgia was collected near Moraine Fjord. This required Andersson, Duse and two seamen (equipped with rock drills, blasting powder, a tent and provisions for several days), two days of boring and blasting to extract the mollusc.

During the expedition's last week on the island, advantage was taken of the marine zoological investigations; large numbers of excellent fish were obtained which were a welcome addition to the diet. The weather deteriorated during the mid part of June and almost 1 m of snow covered the local terrain, effectively terminating further scientific work. The ship sailed on 15 June 1902 for the Falkland Islands.

The results of the visit to South Georgia were published in many monographs and included a series of charts prepared by Duse. The expedition had been equipped with cameras and took many photographs of the island. C.A. Larsen learnt much more about the region and its commercial potential on his second visit.

The *Antarctic* and the expedition then became involved in one of the most interesting Antarctic rescues. The ship sailed south to embark the party at Snow Hill Island late in 1902, but she could not reach them owing to the amount of sea ice that summer. Three men were landed at Hope Bay to sledge to Snow Hill Island but they too were unable to reach it. Finally the *Antarctic* was trapped by ice on 9 January after which she was crushed and sank on 13 February 1903. Her complement were able to reach Paulet Island with some supplies and part of the scientific collections. Thus, the expedition was stranded in three separate groups for the 1903 Antarctic winter. They were able to survive with much extemporisation and lived principally on locally available food (mainly penguins and seals). Several attempts to effect rescue were made. The Argentine ship *Uruguay*, commanded by Captain Irizar, eventually reached Snow Hill Island on 8 November 1903, just as a party from the survivors of *Antarctic* arrived after trekking south from Paulet Island. The sledging party had reached there about a month earlier. A member of the British National Antarctic Expedition (1901–04), Mr Ernest Shackleton, had provided assistance to the Argentine Government in the equipping of *Uruguay*. The members of the Swedish expedition were rescued together with much of their scientific collections, notes, and other expeditionary records. C.A. Larsen and the crew of the *Antarctic*, all very

experienced in navigating the Antarctic Ocean, were in turn involved in saving *Uruguay* after Captain Irizar agreed to surrender command for the duration of a very severe storm and the Norwegians saved the ship by partly dismasting her on 15 November. The expedition and complement of the ship were received in Buenos Aires with a heroic welcome before returning to Scandinavia (Nordenskjöld, Andersson & Larsen, 1905).

C.A. Larsen and the beginning of whaling (1904)

The next expedition to visit South Georgia was one of the most significant in the history of the island and resulted in the establishment of a permanent population. C. A. Larsen, who was the instigator of the expedition, had founded a company in Buenos Aires, the Compañia Argentina de Pesca, to commence whaling on the island. The details of this are described in the commercial section of South Georgia history in Chapter 5. Regarding expeditions, however, some details are appropriate here. C.A. Larsen arrived at South Georgia on 16 November 1904 with about 60 other Norwegians aboard three ships; *Louise, Rolf* and *Fortuna*. A whaling station was established at Grytviken which yielded the first whale oil on 24 December. In February 1905 *Rolf* sailed to Buenos Aires with the first cargo of barrels of whale oil. The Argentine naval transport ship *Guárdia Nacionál*, on charter to the company, arrived at the whaling station on 16 June where she unloaded cargo, including 1000 tons of coal, ordered by Larsen, and remained for about 2 weeks, during which her Master conducted a survey of the harbour for the company. She then sailed for Punta Arenas, Chile, with C.A. Larsen on board. Regular voyages were made to supply the Grytviken whaling station thenceforth to 1965, when it closed. Similarly, with communications to the other whaling stations and government settlement at King Edward Point, South Georgia was frequently connected with the rest of the world.

C.A. Larsen had been approached by Einar Lönnberg of the Museum of Natural History in Stockholm. He had heard of the intended expedition to South Georgia and had asked if a biologist might accompany it. Larsen consented and Eric Sörling was appointed. He sailed with Larsen's fleet to arrive in November 1904 and remained until

Captain Carl Anton Larsen photographed just prior
to his departure for South Georgia in 1904.
(Courtesy Hans-Kjell Larsen)

September 1905. During this time he was able to prepare three whale skeletons, several skins and skeletons of elephant, Weddell and leopard seals together with those of many species of birds; preserve collections of fish, various embryos and some marine invertebrata. He also made notes about the whales, seals, fish, birds and the introduced rats. Sörling was partly responsible for collecting meteorological data from January to August 1905. A monograph on the fauna of South Georgia was published by Lönnberg from the data Sörling collected; this contains some remarkable early photographs of the island. Much concern about the preservation of the fauna was expressed by Lönnberg in this and other publications. He recommended:

Above all, wanton destruction should be strictly forbidden and heavily punished. For it has been witnessed how, by the crew of an Argentine vessel, merely for "fun's" sake elephant-seals have been shot and killed only to be left to rot on the beach, or wounded taken their refuge to the sea only to miserably die afterwards. And likewise it has been witnessed how a crowd of ruffians have broken off the wings of penguins and let them loose to see how they behaved. To such barbarisms there ought to be put an end, not only in the name of science but in the name of humanity. (Lönnberg, 1906.)

South Georgia Exploration Company (1905)

An interesting series of events occurred later in 1905 at Grytviken. Following developments from the advertised lease of South Georgia (described earlier in this chapter), and subsequent negotiations, the South Georgia Exploration Company was formed and registered in Punta Arenas, Chile. The Falkland Islands Government was somewhat cautious about the motives of the company as connections with seal poachers and other irregularities were suspected. However their proposals were accepted and a visit to the island was arranged. The Company's manager, Ernest Swinhoe, was to lead the first expedition aboard *Consort*, chartered in Punta Arenas. Part of the expenditure required was to be met by the proceeds from sealing.

Consort, carrying livestock and other requirements to commence farming, sailed for Port Stanley where she arrived in July 1905. The appropriate contract was completed and from 1 July 1905 South Georgia was leased for £1 a year. *Consort* then sailed to the island, arriving on 9 August. She stayed about 3 months and returned to Punta Arenas after calling again at Port Stanley. Swinhoe prepared a report on the expedition which was presented to the Governor. This gave a description of the discovery of, and some difficulties experienced with, the whaling station (which had been established on the island without appropriate authority), as well as giving a general account of the expedition. The report first describes the arrival off Bird Island and the inspection of the Bay of Isles, Possession Bay and surrounding areas during which the opinion was reached that South Georgia was unsuitable for sheep farming. On 14 August the whaling station was contacted and *Consort* sailed into King Edward Cove. The animals (24 sheep and 4 horses) were landed and were able to feed on tussac grass.

Shortly after establishing camp and despatching *Consort* on sealing operations, Swinhoe visited L.E. Larsen (brother of C.A. Larsen), who was acting as manager of the Grytviken whaling station. Discussion of lease requirements and authorities ensued and Swinhoe left the acting manager a written statement of his claims. When *Consort* returned to King Edward Cove, after failing to secure any seals, the expedition prepared to depart as they were relying on sealing profits. Eighteen animals, all the sheep with lambs and the horses, were left behind on *Consort*'s departure. Swinhoe discussed in his report the possibility of exploiting iron or other minerals of commercial value on the island but believed that whaling was the only business which could satisfactorily yield commercial results there. The visit of the *Guárdia Nacionál* and her survey was also reported.

The statement of claims given to L.E. Larsen by Swinhoe was despatched to the head office of Compañia Argentina de Pesca in Buenos Aires. As a result, the president of the company, Pedro Christophersen, with Captain Gullermo Núñez (the Director of Armaments of the Argentine Navy, who was also a principal shareholder in and technical adviser to the company) visited the British Legation in Buenos Aires. They produced the letter from Swinhoe and claimed to be unaware of British sovereignty over South Georgia. They also raised the matter of Compañia Argentina de Pesca applying for a lease for the site of the whaling station, through the British Legation. A report was despatched to the Governor of the Falkland Islands, Mr

W.L. Allardyce. This he received shortly before *Consort* reached the Falkland Islands from South Georgia. The lease of South Georgia held by the South Georgia Exploration Company was renewed, with somewhat different terms, in 1906. It was later purchased by Messrs Bryde and Dahl of Norway, who used its whaling provision.

HMS *Sappho*, visit of inspection (1906)

The Governor, after receiving these reports, informed the Colonial Office in London. He indicated that he was suspicious of the protestations of the management of Compañia Argentina de Pesca about their being unaware of the political status of South Georgia and that he was concerned about the consequent loss of revenue. Further discussions followed, involving the Falkland Islands Government, British Government and the companies concerned. In order to clarify the matter HMS *Sappho*, commanded by Captain M. Hodges, was despatched to investigate and report on the situation at South Georgia. She sailed directly from Montevideo to arrive at the island on 31 January 1906. Captain Hodges subsequently reported that:

The Officers of the ship commenced a survey of this bay which we have called King Edward Cove, the first thing next morning. Captain Larsen who is in charge of the station took me on shore and shewed me over it. The station consists of Captain Larsen's house which was brought from Norway in sections and put together here, some huts for the men, the Boiling factory, a ship fitted with powerful winches for heaving the whales on shore, blacksmith's shop, etc., in fact everything is most complete and Captain Larsen has now brought down a dynamo, and the factory will soon be lit with Electric light. A wooden pier has also been run out which allows a ship drawing sixteen feet of water to come alongside to discharge or take cargo.

In the bay we found the 'Louise', a dismantled barque which is used as a general store ship and has about a thousand tons of welsh coal on board, two whaling steamers, the 'Rossita' and 'Fortuna', of 200 tons, and alongside the pier the 'Cachalot', a steamer of 900 tons, purchased by the company to take stores to the station from Buenos Aires, and to take back the oil. She is commanded by Captain Larsen's younger brother, while an elder brother is in charge of the factory. Both Captain Larsen and his elder brother have brought their wives with them, and Captain Larsen has brought his whole family consisting of five daughters and two sons down for the summer.

The unpublished report continues to discuss whaling and gives monthly catch data from December 1904. The total catch to January 1906 had been 236 whales comprising 189 humpback, 22 fin, 18 blue and 7 right whales; these yielded a total of 7434 tons of oil. Whaling had continued through the winter but with comparatively small catches. The amount of oil the company could produce was limited by the number of barrels the coopers could make Captain Larsen admitted taking 80 elephant seals which yielded 168 barrels of oil, and that he was considering fishing for 'cod' (*Notothenia rossii*) to sell in South America. Four horses and some sheep left behind by the South Georgia Exploration Company were reported as well as pigs, poultry, and some rabbits. The latter had been released near the whaling station and were apparently thriving. A description of the weather then followed.

On 3 February Captain Hodges accompanied Captain Larsen on a whaling voyage to Fortuna Bay and Antarctic Bay, during which they discussed the siting of beacons on the island. After making a landing in Antarctic Bay to visit a king penguin colony, they were stranded when the weather deteriorated rapidly. They suffered a long delay with much difficulty before reboarding – the possibility of spending the night there had been contemplated. Officers of HMS *Sappho* prepared a chart of Cumberland Bay and sailing directions for the north-east coast of the island during her visit. Collections of plants and rocks were also made which were despatched to the Imperial Institute, London, for examination. *Sappho* departed for Montevideo on 5 February 1906 at 06:00.

Discussion certainly took place between Captains Hodges and Larsen concerning the licensing of Compañia Argentina de Pesca's whaling station, which was then being arranged through the British Legation in Buenos Aires and the Government at Stanley. Some conflicting accounts describe this as difficult. These suggest that there was either a Norwegian or Argentine flag flying over Grytviken to which Hodges objected, that Hodges gave Larsen a time (said to be either 15 or 30 minutes) to remove the flag before *Sappho*'s guns, trained on the flag pole, would open fire to the same effect, and that Larsen lowered the offending flag. Neither the official account prepared by Captain Hodges nor the Norwegian histories refer to this and no supporting contemporary reports include it. Although the supposed action is sometimes regarded as part of the traditional history of South Georgia, it is undoubtedly apocryphal. All other accounts indicate that relations between Captains Hodges and Larsen were amicable and cooperative.

Development of whaling

Interest in southern whaling received an enormous stimulus as it became generally known that a land station had been successfully established in Antarctic regions. Applications were received by the Government of the Falkland Islands and Dependencies in Stanley for licences to establish land stations and to operate floating factories in many areas under their jurisdiction, in particular at South Georgia. Another 'gold-rush' epoch in the island's history resulted which is discussed in greater detail in Chapter 5. A further consequence of this was an enormous increase in the number of voyages to and from South Georgia. It is impracticable to catalogue these beyond 1905 other than on a selective basis. Descriptions of voyages which, although basically for whaling purposes, are of importance in the scientific investigation of the island are given below.

A. Szielasko and *Fridtjof Nansen* (1906)

One of the earliest whaling voyages, that of *Fridtjof Nansen*, was most unfortunate. She had been specially equipped as a floating factory and arrived at South Georgia in November 1906, accompanied by two whale catchers, only to be wrecked on a previously undiscovered reef which was subsequently named after her (until 1979 when it was renamed 'Nansen' reef). Nine of her crew were drowned. Dr A.E.A. Szielasko, her physician, remained at Grytviken after the other survivors of the wreck had been repatriated. While he was on the island he completed some investigations of several bird species as well as making geographical observations and compiling charts. An ice cap on the Barff Peninsula was named after him.

Voyage of *Undine* (1908)

The voyage made by C.A. Larsen aboard *Undine*, a steam yacht, was referred to briefly when he encountered Captain Cleveland sealing in November 1908 at the southwestern part of South Georgia. The voyage was to make a reconnaissance survey of parts of South Georgia and the South Sandwich Islands, in search of harbours or anchorages useful for the whaling industry. Little was found although many areas (particularly on the southwest coast of the island) were examined and roughly charted. During the visit to the South Sandwich Islands, C.A. Larsen was seriously affected by the volcanic gases of Zavodovski Island. He returned to Grytviken in a debilitated condition.

Visits of *Uruguay* (1909 etc.)

In February 1909 the Argentine Naval ship *Uruguay*, previously involved in the rescue of the Swedish Antarctic Expedition, arrived in South Georgia. She was on a voyage to relieve the meteorological station at Laurie Island (in the South Orkney Islands) which had been established by the Scottish National Antarctic Expedition in April 1903. The station had continued functioning, supported by the Argentine Meteorological Office at the invitation of the British Legation in Buenos Aires. She was commanded by Captain C.S. Somosa, who made brief surveys of parts of Cumberland Bay and Moltke Harbour while at South Georgia. She called again on 12 February 1910, 27 February 1911, 28 February 1915, 11 March 1918, and 6 March 1919 on the same voyage. The annual relief of the meteorological station was thenceforth usually made from South Georgia by a chartered sealing vessel or whale catcher of the Compañia Argentina de Pesca.

Carl Skottsberg (1909)

Carl Skottsberg first visited South Georgia in 1902 aboard *Antarctic* with the Swedish South Polar Expedition of 1901–03. In 1907–09 he undertook another expedition, to Patagonia and thence to South Georgia where he arrived, with P.D. Quensel, aboard *Cachalot* on 15 April 1909 and remained to 4 May as a guest of C.A. Larsen. Owing to the time of year and brevity of his visit, the only scientific observations he was able to accomplish were on marine algae. During his visit, however, he was able to make voyages on *Undine* and *Carl*. He recounted these experiences, which included a description of whaling from the latter vessel. A reunion of three members of the 1901–03 expedition occurred as the cook (Axel Andersson) was then serving at the whaling station.

Appointment of the Magistrate (1909)

Owing to the vast increase in activity on South Georgia following the establishment of the whaling industry, the Government considered it timely to establish a local administration. Mr J.I. Wilson was appointed Stipendiary Magistrate on 20 November 1909 and arrived at South Georgia on 30 November. He was accommodated at Grytviken until a residence was established for him at King Edward Point in September 1912. The administrative headquarters of South Georgia have subsequently remained there. The Magistrate had a great variety of duties to perform which are described in Chapters 1 and 5.

In the same year, 1909, further legislation was passed to regulate sealing in the Falkland Islands and their Dependencies. This marked the beginning of the second epoch of sealing in South Georgia which was in great contrast to the first. It was for elephant seal oil only and was regulated by licences so that exploitation was controlled to ensure the success of the industry and consequent survival of the seal species.

Wilhelm Filchner and *Deutschland* (1911 & 1912)

A German expedition led by Dr Wilhelm Filchner conducted geological, geophysical, meteorological and some other scientific investigations on South Georgia in 1911 and 1912, making three visits in the process. The expedition also contributed much to the charts and hydrographical observations of the island. The original concept of the expedition involved two ships, in mutual radio contact, sailing in the Ross and Weddell Seas and a land transverse being made between them. However, finance was restricted and only one ship was available to the expedition. She was a Norwegian sailing vessel, *Bjørn*, with auxillary engines. *Bjørn* was renamed *Deutschland* and commanded by Captain Vahsel. Her complement was 35, which included 5 scientists and 2 medical officers.

The expedition sailed from Hamburg on 3 May 1911 and conducted extensive oceanographical investigations on the way to Buenos Aires, which was reached on 7 September. At Buenos Aires they took on board some Manchurian ponies and Greenland dogs which had arrived by steamship. Incidentally, *Deutschland* was moored next to Amundsen's ship (*Fram*) in Buenos Aires. The latter was overwintering while Amundsen was preparing for his race to the South Pole on his 'northern' expedition. *Deutschland* sailed for South Georgia on 4 October 1911 to arrive at King Edward Cove on the 31st after a very rough passage.

On arrival, the animals were unloaded to recuperate from the voyage and scientific investigation and exploration began. This was greatly aided by Captain C.A. Larsen who made available his steam yacht *Undine*. One of the effects of the expedition's visit was to cause the recently established Post Office to run out of some denominations of stamps, giving rise to an improvisation (see Chapter 6).

For the first voyage of the expedition, *Undine* went to Royal Bay where the station of the German contingent of the International Polar Year from 1882–83 was reopened and occupied by three men, including Filchner, with much assistance from the crew of *Undine*. The main building of the station was described as being in tolerable condition but most of the rooms were full of snow and the roof somewhat defective. The other buildings were reported as being almost entirely in ruins. Scientific investigations undertaken included: geomagnetic and meteorological observations; another survey of the Ross Glacier; and, mainly from *Undine*, a hydrographic and coastal survey of Royal Bay. The meteorological observations were made at the same times as those of the observatory at King Edward Point and included large numbers of balloon flights to determine direction and speed of air currents at various altitudes. Two of the staff were replaced by one man on 30 October, just before *Deutschland* sailed to visit the South Sandwich Islands. The station remained open for almost 4 weeks, until the expedition left the island.

After her work at Royal Bay *Undine* sailed south and a landing was made at Cooper Bay on 25 November. She then sailed past Cape Disappointment; those on board were impressed by this brief view of the magnificently wild scenery of the south-east coast. The voyage returned along the former route and continued northwest. Landings and collections were made at Fortuna Bay, Antarctic Bay and Possession Bay before deteriorating weather forced them to take shelter in Stromness Bay. Here they visited and described the two shore stations and the floating factory of the whaling companies established there. After a brief call at Royal Bay, for the staff change mentioned above, they returned to King Edward Cove on 30 November 1911.

The rest of the expedition, again on board *Deutschland*, made an excursion to the South Sandwich Islands departing from South Georgia on 1 November 1911 and returning to Husvik, on the 14th when they took coal on board before sailing to King Edward Cove to arrive on the 19th. The South Sandwich Islands voyage suffered very severe weather with, by Filchner's estimate, 20 m-high waves. The ship had demonstrated her seaworthiness, but the storms had prevented much scientific work being accomplished.

An unfortunate accident occurred shortly after the return of *Deutschland*. The third officer, responsible for wireless operations, Walter Slossarczyk, went out in a small boat in King Edward Cove on 26 November. He was never seen again although the boat was later found, floating upright well outside Cumberland Bay, by the whale catcher *Fortuna*. A memorial cross was erected for him which still stands slightly uphill from the Grytviken cemetery.

The second cruise of *Undine* was from 22 to 29 November. She sailed to Bird Island and a landing was made; that night she anchored in Elsehul and a landing for a photogrammetric survey was made there the next day during which a hydrographic survey of the vicinity was also made. On the 24th, she sailed through Stewart Strait and down the south-west coast. This was investigated from Undine harbour to Schlieper Bay and Wilson Harbour, where another photogrammetric survey was made. A landing was made at Ice Fjord followed by a trek over Tawny Gap, which allowed the ocean to the north of the island to be seen, but the weather was unsuitable to permit a survey. On 26 November they anchored in King Haakon Bay where 12 glaciers all flowing independently into the sea were described. On the 27th they continued sailing southeast and, despite a low cloud ceiling, commented on the magnificent scenery, with its alternation between mountain and glacier. The night was spent at Cooper Bay. Next day Drygalski Fjord and Larsen Harbour were explored charted and named. The geologist on board, Dr F. Heim, found the first evidence of the different structure of that end of the island. Past and present effects of the glaciation in the area were also commented upon. Another fjord was entered in the vicinity which was later named Slossarczyk Fjord after their dead colleague. On the 29th they collected the two persons at Royal Bay and closed the station. King Edward Point was reached the next day. There one of the expedition's medical officers, Dr Ludwig Kohl, suffered an attack of appendicitis and was operated on by his colleague at Grytviken. He remained there to recover after *Deutschland* departed, fortunately the expedition was able to recruit a replacement for him from the *Harpon* which arrived from South America. Dr Kohl subsequently married one of C.A. Larsen's daughters and returned to South Georgia with her in 1928 on the Kohl-Larsen expedition, described later in this chapter.

The results of the voyages aboard *Undine* allowed much improvement of the charts of South Georgia. Astronomical positions and photogrammetric survey details were available as well as hydrographic survey in several areas. Effects of a former, much greater, glaciation were described and a large number of photographs of glaciers and various other land forms were produced.

On 11 December 1911 the expedition left South Georgia aboard *Deutschland* for the second time, carrying gifts of oxen, sheep and pigs presented to them by the whaling station, as well as the ponies and dogs. On 2 February 1912 they reached 77°45′S, their closest to the Pole. An attempt to establish a base on 'Stationseisberg' was only temporarily successful, being abandoned when the ice broke up. A second attempt also failed. As winter approached, the sea began to freeze and by mid March *Deutschland* became trapped in the ice. She drifted until 26 November, covering about 1000 km. The expedition had been well organised during this time; the animals were accommodated on the ice, local excursions were arranged, the ship was lit electrically and conditions were, under the circumstances, good. They were fortunate with the ice conditions and their course, as distinct from the experience of *Antarctic* crushed in 1903 and *Endurance* to be crushed in 1915. The most unfortunate event was the death of Captain Vahsel whose health had been poor from early in the voyage.

After reaching open water, *Deutschland* sailed north to South Georgia to arrive in King Edward Cove on 19 December 1912. There she left much of her unused supplies of polar clothing, food, equipment as well as the ponies – which ran wild on what was known as Hestesletten (Horse plain), after their predecessors in 1905 from the South Georgia Exploration Company. The dogs were also released to be maintained by the whalers. *Deutschland* then sailed for Buenos Aires, where she was commissioned by the Argentine Government to relieve the meteorological station at the South Orkney Islands. On the return voyage she called at King Edward Cove in March 1913 (Filchner, 1922).

David Ferguson's Geological Survey (1912)

Further geological investigations of the island were made at about this time when one of South Georgia's whaling companies, Christian Salvesen's, arranged for a mineralogical exploration of it. The company had been granted mineral and pasturage rights on the island for five years by the Governor. David Ferguson of Edinburgh was contracted to prospect for economic minerals; he made a geological survey of the island from 7 January to 19 April 1912. No mineral deposits of economic significance were discovered but the investigation contributed to an understanding of the geology and physiography of the island. Several scientific accounts of this resulted from Ferguson's work and he made a similar survey for the same company in the South Shetland Islands during the next summer.

R.C. Murphy, Captain Cleveland, and *Daisy* (1912/13)

The voyage of Captain Benjamin Cleveland aboard *Daisy* in 1912/13 was mentioned earlier as the last voyage of the first epoch of sealing. Robert Cushman Murphy accompanied the voyage as a naturalist and wrote an excellent description of it. Murphy also made collections of the flora and fauna while on South Georgia. The project began when the Director of the American Museum of Natural History contracted with the owners of a New Bedford ship, which was going to South Georgia, to carry a naturalist. The Brooklyn In-

stitute of Arts and Sciences was a co-sponsor of the project and R.C. Murphy, of their staff, was appointed to the post. Murphy had married Miss Grace Barstow just prior to sailing and kept a detailed diary of the voyage for her, as well as his own comprehensive notebooks. These were eventually published as *Logbook for Grace*, an excellent account of the voyage, sealing and South Georgia. In the small luggage space he had available Murphy was able to take with him for reference copies of everything which had been published about the island.

Daisy arrived at South Georgia on 23 November 1912 and remained at King Edward Cove until 13 December. During this time Murphy secured a collection of over 100 birds skins, seal skulls, whale embryos, fish, invertebrates, plants, and rocks from the area. Some of the crew of *Daisy* saw the reindeer introduced by the Norwegians only a year previously. They reported them to Murphy as big, long-horned, mountain goats and he too was perplexed until informed about their origin. Murphy also made a voyage aboard the whale catcher *Fortuna*, travelled around much of the district on foot and in a boat, *Grace Emmiline*, from *Daisy*. From Cumberland Bay, *Daisy* sailed to the Bay of Isles where Murphy visited Albatross Island with a sealing gang. The next day, while out in the ship's boat, he found a small cove near Salisbury Plain where, with the assistance of some of the crew of *Daisy*, he established a camp. *Daisy* remained anchored close to the area and Murphy returned to her to spend the nights, after working on shore during the daylight hours. He experienced great difficulty with his tented encampment and the weather. The tent was blown over several times and, even with an oil stove, it proved difficult to keep sufficiently warm to allow him to complete his work in anything other than marked discomfort. During this period he was able to prepare more bird skins, secure seal skeletons, and make observations of several bird species in the area. He also compiled a chart of the Bay of Isles which formed the basis of the Admiralty Charts of the area until 1929. Of the several place-names he recorded on it, one (Grace Glacier) was in honour of his wife.

After leaving the Bay of Isles *Daisy*, with Murphy aboard, visited the whaling station at Prince Olav Harbour. While there he made more biological observations in Possession Bay and was able to trek to Antarctic Bay accompanied by the mate. On the evening of 15 March 1913 they sailed north and departed from the island. Murphy exposed many photographic plates while on South Georgia which he used to illustrate the large number of papers he wrote about the voyage. Most of these plates were developed in sea water aboard *Daisy* and include excellent scenes of South Georgia. José Correira, the cooper of *Daisy*, assisted Murphy and was trained in the preparation of bird skins during the voyage. While they were at King Edward Cove arrangements were made with C.A. Larsen for

Correira to return in the 1913/14 season. This he was able to do and he remained at Grytviken, employed in the whaling station for several months of this season, using his free time to collect and prepare a series of specimens which Murphy had been unable to obtain. These were forwarded to the American Museum of Natural History where Murphy's original collections were lodged. Correira later took part in many other ornithological expeditions initiated by Murphy.

In December 1970 Murphy made a second visit to South Georgia, accompanied by Grace, his wife, while aboard the tourist ship *Lindblad Ex-*

The sealer *Daisy* in Cumberland Bay photographed from King Edward Point by R.C. Murphy in 1912.

plorer which called at King Edward Cove for part of a day. He delivered an interesting lecture to the British Antarctic Survey personnel and many of the ship's complement at Grytviken which he had visited 58 years previously.

Major Barrett-Hamilton (1913/14)

A scientific investigation of whales was made during the 1913/14 summer by Major Barrett-Hamilton assisted by P. Stammwitz, both of the British Museum (Natural History). The first prisoner (who, as described in Chapter 1, had previously escaped from gaol at King Edward Point) was later also recruited to assist the investigation. Many measurements, descriptions and samples were taken from 294 whales at Leith Harbour whaling station. Some of the material collected by Major Barrett-Hamilton is on display in the whale hall of the British Museum (Natural History), London. The research came to an abrupt end when Major Barrett-Hamilton died suddenly on 17 January 1914. His assistant transferred the data and specimens to the Museum and a detailed report on them was prepared by James Hinton of the Colonial Office.

Sir Ernest Shackleton, *Endurance* and *James Caird* (1914 & 1915)

After sailing from Buenos Aires on 26 October 1914 the British Imperial Trans-Antarctic Expedition, led by Sir Ernest Shackleton aboard *Endurance* (built in Sandefjord as *Polaris* in 1913), arrived at South Georgia on 5 November 1914. Shackleton had been a member of Captain Scott's British National Antarctic Expedition of 1901–04 aboard *Discovery*. He had also led the British Antarctic Expedition of 1907–09 aboard *Nimrod*. The 1914 expedition had planned to cross the Antarctic continent from the Weddell Sea to the Ross Sea over the South Pole and to embark on *Aurora* which was to await them. The plan was similar to that of Filchner which, because of financial difficulties, had been abandoned. The expedition was at South Georgia and remained for a month, arranging for stores and coal as well as generally making final preparations for the overland part of the journey. Much information for the voyage was obtained from the Masters of the whale catchers and other persons familiar with

navigation in ice. The expedition's dogs also benefited from some shore leave and a diet of whale meat, supplies of which were taken on board for the voyage south. The tethering points Shackleton used for the dog lines remained near Grytviken until a few years ago. Surveying was done at King Edward Cove and marks were fixed before *Endurance* sailed for the Antarctic continent on 5 December 1914. One of these marks was in the form of a cross which was erected above Hope Point for navigational purposes.

Coats Land was sighted on 10 January 1915 but, owing to ice conditions, the expedition was unable to land. On 19 January 1915, *Endurance* became beset by ice at 76°34′S latitude by 31°30′W longitude from where she drifted over 2500 km to 69°05′S latitude by 51°30′W longitude near the Antarctic Peninsula where, on 27 October 1915, she was crushed and sank. All aboard survived. By sledging to open water and thence in *Endurance's* boats they reached Elephant Island (one of the easternmost of the South Shetland Islands), on 15 April 1916, in difficult circumstances at the beginning of the southern winter. The urgency of effecting rescue at the earliest opportunity resulted in Shackleton and five colleagues making a most perilous, albeit successful and famous voyage. One of *Endurance's* boats, *James Caird*, 6.9 m long, was decked, covered with canvas and otherwise modified for a voyage to South Georgia. It was decided to sail there, over 1500 km, across one of the most tempestuous oceans of the world, as wind and current favoured that direction rather than a geographically shorter alternative (to Tierra del Fuego or the Falkland Islands). She sailed on 24 April 1916.

After a very difficult voyage of 16 days during which severe storms were experienced and they had run short of food and water, *James Caird* arrived at the southern side of South Georgia on 9 May 1916. A landing was made at the entrance to King Haakon Bay from where, after a short halt to recover from the immediate effects of the voyage, Shackleton and the others sailed to the head of the bay to establish camp beneath the upturned *James Caird* in the manner of the former sealers. This was named 'Peggotty Camp' after a not dissimilar dwelling described in a work of Dickens. On 17 May a reconnaissance was made of the pass, later named Shackleton Gap, leading towards Possession Bay. This provided an idea of the difficult travelling conditions on the island. The trek to Stromness Bay, the nearest place of habitation, was made by three of the party (Shackleton, F.A. Worsley and T. Crean; the first two later wrote accounts of it). Three men (T. McCarthy, W. McNeish and J. Vincent) remained at Peggotty Camp as two were not sufficiently fit to make the journey and one remained to attend them. Departure was at 03:00 on 19 May, in good weather and by the light of a full moon. The equipment they carried was very basic: a small compass, a primus stove full of fuel, a pot, 16 m of knotted rope, a pair of binoculars, a modified adze, and a piece of blueprint of a chart of South Georgia from the Filchner Expedition. Boots had been modified by the addition of brass screws from *James Caird*, which passed through the soles from within to improve the grip on the ice. These were completely worn out at the end of the journey.

The trek was exceedingly difficult over much

James Caird landing on South Georgia at King Haakon Bay with Shackleton and colleagues. From a composite drawing and photograph by G. Marston.

soft snow, crevasses and very steep areas; it required several detours and some retracing of the route. Early in the morning of 19 May the northern side of the island was visible at Possession Bay. It was seen to be impossible to continue around the coast, so they turned inland. During the rest of that day they passed many obstacles in generally bright, sunny conditions. The journey would have been vastly different without their good fortune with the weather. Halts were made only for meals, five in total. As evening approached it was decided to continue through the night with the aid of moonlight. At 06:30 on 20 May 1916 a steam whistle from one of the whaling stations was heard, the first contact with the rest of the world since 5 December 1914. They descended into Fortuna Bay and trekked to the plateau behind the Stromness Bay whaling stations from where buildings, men and ships were seen. The last part of the trek required an unexpected descent through a waterfall. The three, with the log of the journey, arrived at Stromness whaling station after 36 hours of very difficult trekking from Peggotty Camp. Worsley described the party as:

Ragged, filthy and evil-smelling; hair and beards long and matted with soot and blubber; unwashed for three months, and no bath nor change of clothing for seven months.   (Worsley, 1940)

Shackleton recounts that:

Down we hurried, and when quite close to the station we met two small boys about ten or twelve years of age. I asked these lads where the manager's house was situated. They did not answer. They gave us one look—a comprehensive look that did not need to be repeated. Then they ran from us as fast as their legs would carry them. We reached the outskirts of the station and passed through the 'digesting house', which was dark inside. Emerging at the other end, we met an old man, who started as if he had seen the Devil himself and gave us no time to ask any question. He hurried away. This greeting was not friendly. Then we came to the wharf, where the man in charge stuck to his station. I asked him if Mr Sørlle (the manager) was in the house.

> 'Yes,' he said as he stared at us.
> 'We would like to see him.' said I.
> 'Who are you?' he asked.
> 'We have lost our ship and come over the island,' I replied.
> 'You have come over the island?' he said in a tone of entire disbelief.

The man went towards the manager's house and we followed him.'
I learned afterwards that he had said to Mr Sørlle:

> 'There are three funny-looking men outside, who say they have come over the island and they know you. I have left them outside.'

A very necessary precaution from his point of view.

> Mr. Sørlle came out to the door and said, 'Well?'
> 'Don't you know me?' I said.
> 'I know your voice,' he replied doubtfully. 'You're the mate of the *Daisy*'.
> 'My name is Shackleton,' I said.

Immediately he put out his hand and said, 'Come in, Come in'.   (Shackleton, 1919)

Despite protestations that they were in no fit condition to enter the house Mr Sørlle immediately took them in, had them seated, provided them with coffee and food, then baths and changes of clothing were arranged. He gave orders for *Samson*, a whale catcher, to leave that night for King Haakon Bay. Worsley accompanied her and the three men at Peggotty Camp reached Stromness the next day. HMS *Kent* visited South Georgia in search of the expedition and had left Stromness only a day or two before Shackleton arrived there.

After one more night to recover, Shackleton reported to the Magistrate and made arrangements to depart on the whale catcher *Southern Sky* for Elephant Island on the 23 May to rescue the rest of the expedition at the earliest opportunity. Shackleton was accompanied by Crean and Worsley on this voyage while the others sailed directly to the United Kingdom. *Southern Sky* could not get closer than 100 km to Elephant Island owing to the condition of the pack-ice, thus they sailed to the Falkland Islands to use the telegraph station there. A second attempt with *Instituto de Pesca No. 1*, lent by the Government of Uruguay, got to within 30 km of Elephant Island but no farther. A third attempt with *Emma* from Chile and arranged privately was also unsuccessful. A fourth on *Yelcho*, lent by the Government of Chile, reached Elephant Island on 30 August 1916, after the ice had fortunately briefly dispersed. The rest of the expedition were rescued and arrived in Punta Arenas on 3 September. After a welcome there they proceeded to Valparaiso for Santiago to receive a tumultuous reception.

Most of the intended objects of the expedition were not attained, principally owing to the extremely unfavourable weather, and ice conditions in the Weddell Sea in the summer of 1914/15. Virtually nothing went as anticipated after *Endurance* left 'the gateway to the Antarctic', as Shackleton called South Georgia. However, the achievements of the expedition in surviving what transpired and its eventual successful conclusion are among the greatest in polar exploration. The significance of these events was not fully realised at the time as the First World War occupied thoughts and activities throughout the world. Shackleton's leadership was very highly regarded by his colleagues and its efficiency is demonstrated by all involved returning alive after enduring such an adventure.

In March and April 1917 Frank Hurley, the expedition's photographer, visited South Georgia again to obtain film and photographs of the island's wildlife to replace those lost with *Endurance*.

The First World War (1914–18)

The most significant effect on South Georgia of the First World War was the enormous increase in demand for (and price of) whale oil. It was required in the war effort for the manufacture of glycerol to make nitroglycerine, the basis of many explosives, as well as for its use in edible fat production. Soap and other materials resulting from glycerol production became, for the duration of the war, virtually by-products. Throughout the War, seven bays of South Georgia had whaling factories operating in them.

In order to increase oil production, the restrictions on numbers of whale catchers permitted and requirements for full utilisation of whale carcasses were repealed for the 1916/17, 1917/18, and 1918/19 seasons. The number of whale catchers deployed from South Georgia reached its maximum, 32, for the first two of these seasons. Although the incomplete use of whale carcasses resulted in greater production of the highest grades of oil, the consequent wastage and pollution from discarded skrotts in the vicinity of the whaling stations was enormous. Some of this is described in Chapter 5.

Towards the end of the war, regulations controlling export of whale oil were introduced in order to reduce the chance of it becoming directly or indirectly available to the enemy. These had several complex effects on the pricing and markets of the whaling industry which persisted for many years.

Visit of *Carnegie* (1916)

A United States research vessel, *Carnegie*, under the command of Captain W.J. Peters, conducted geomagnetic observations on a peri-Antarctic voyage from and to New Zealand from December 1915 to April 1916. The Department of Terrestrial Magnetism of the Carnegie Institute of Washington had organised this as part of a series of geomagnetic studies which started in 1905. *Carnegie* called at King Edward Cove from 12 to 18 January 1916 where she took on board fresh water and provisions. This was the last call of a United States vessel at South Georgia for almost 50 years.

HMS *Dartmouth* and *Weymouth* (1920 & 1921)

A Royal Naval voyage reached South Georgia from Port Stanley in 1920 at a very convenient time for the Magistrate. He and a whaling company were having great difficulties with a number of whalers who were refusing to comply with their contracts and had come out on strike at Grytviken. Things were becoming serious as the Magistrate and three other Government Officers could do nothing against 50 rioting men who had threatened to attack him and the settlement at King Edward Point. At a critical moment, on 17 January 1920,

HMS *Dartmouth*, a four funneled light cruiser commanded by Captain H.W.W. Hope, entered King Edward Cove. As soon as her commander learnt of the state of things he sent a squad of men ashore to assist the Magistrate to restore law and order. The principal malefactors were arrested, disarmed and deported to their place of origin in South America. This had a salutory effect on those who had offered them support. It was reported that some whalers were convinced the Magistrate summoned help in some mysterious way though there was no wireless station on the island until 1925. The strikers were mostly labourers of various nationalities recruited at the docks of Buenos Aires and not regular Norwegian whaling employees.

For much of the rest of her time at South Georgia HMS *Dartmouth* continued hydrographic survey operations from positions both ashore and afloat, principally of Cumberland Bay. She departed for the Falkland Islands on 28 January 1920.

HMS *Weymouth*, commanded by Captain Beil, called at King Edward Cove almost a year later, from 8 to 14 January 1921. She arrived from and returned to Port Stanley and found the situation peaceful on South Georgia.

Sir Ernest Shackleton and *Quest* (1922)

After the First World War, Sir Ernest Shackleton planned another polar expedition, to the Arctic. He had obtained financial support from the Government of Canada as well as some smaller amounts from elsewhere. A vessel, *Foca I*, renamed *Quest*, was purchased in Norway for transport and several members of previous expeditions led by Shackleton joined this one. Rather late in the proceedings Canadian support was withdrawn, leaving Shackleton with a partly arranged expedition, little finance, and doubtful objectives. At this critical time Mr John Rowett came forward to take an active part in the work; he provided practically the whole finances for the expedition. Thus it became known as the Shackleton-Rowett Expedition. The new objectives included an examination of several South Atlantic and Antarctic islands together with hydrographical survey and oceanographical work being conducted while at sea. Comprehensive meteorological observations were to be recorded throughout the voyage.

The ship was modified and equipped in South-

Sir Ernest Shackleton from a portrait by R.G. Eves
painted in 1921. (National Portrait Gallery copyright.)

ampton and London. She was provided with radio apparatus and a gyroscopic compass. On 17 September 1921 she left London. She called at Plymouth, Lisbon and Madeira where the expedition obtained supplies from the same firm as had provisioned James Weddell in 1822 on his voyage to South Georgia. The first scientific work of the expedition was performed at Saint Paul's Rocks, in the equatorial Atlantic. This included ornithological and meteorological observations, biological and geological collections together with a series of soundings around the rocks. The ship continued to Rio de Janiero where she remained for several weeks while repairs were made and her engines overhauled. The voyage continued directly to South Georgia, which was reached on 4 January after a period of very severe weather during the latter part of the journey.

Shackleton and several other men aboard were familiar with King Edward Cove and Grytviken where *Quest* anchored and they remembered much about it from their earlier visits. Two members of the expedition were already on South Georgia, having arrived by a transport vessel of a whaling company earlier in the season. They were making geological investigations at different ends of the island.

Early in the morning of 5 January 1922 Shackleton died from a heart attack while aboard *Quest* moored in King Edward Cove. Command of the expedition devolved upon Frank Wild, formerly of expeditions led by Scott, Shackleton and Mawson, and a descendant of Captain Cook, who ordered that it should proceed – a decision which he was sure would have met with Shackleton's approval. Shackleton's body was embalmed at Grytviken hospital and transferred to the church there. It proved impossible to contact the outside world with the radio transmitter on *Quest* so a message was arranged to be transmitted from the steam ship *Albuera* as soon as she came within range of a coast station. Shackleton's body was conveyed by *Professor Gruvel* to Montevideo, accompanied by Frank Hussey who had been on the *Endurance* expedition. A Uruguayan guard of honour received the coffin to convey it to the Military Hospital in Montevideo. The next day, a message was received from John Rowett that Lady Shackleton had expressed her wish that Sir Ernest be buried at South Georgia. His

body was returned aboard *Woodville*, after a service at the English Church in Montevideo. She sailed with a ceremonial escort provided by the Uruguayan Government. *Woodville* reached South Georgia on 27 February 1922 in a blinding snow storm and the coffin was taken ashore to the Grytviken Church. On Sunday 5 March a service was held there with the Magistrate officiating and Sir Ernest Shackleton was buried in the Grytviken cemetery. A granite grave stone, sent from Britain, was erected in 1928, and unveiled by the Governor.

Preparations for the southern voyage continued. These involved some more repairs to *Quest*, the purchase of equipment from the whaling stations and the taking aboard of coal. Wild was disappointed to find that the depot of polar food, clothing and equipment left by Filchner in 1912 and inspected by Shackleton in 1914 had been dispersed.

Sir Ernest Shackleton's grave at the Grytviken Cemetery. The stone was unveiled in 1928 by Governor Hodson. (Author.)

Sledge dogs, also left behind at that time, were expected to be available, but they had proved such a nuisance and so voracious that they had been shot some years previously. *Quest* sailed to Cooper Island where two geologists were collected from their camp where they had been for some days. She surveyed Larsen Harbour and spent the night in it. On 18 January the voyage continued south by way of Clerke Rocks and the South Sandwich Islands. A search for a reported island was made, which confirmed it did not exist. By the end of February they were well within the pack-ice. The southernmost point reached by the expedition (22 February 1922) was 69°17′ S latitude by 17°09′ E longitude. They continued westwards across the Weddell Sea towards Elephant Island where elephant seal blubber was obtained in order to supplement dwindling coal supplies. The camp site from Shackleton's ill-fated British Imperial Trans-Antarctic Expedition was inspected as they passed. The voyage continued directly to South Georgia as winds and currents propelled them faster than the engines could have done. The route followed was very similar to that of *James Caird* 8 years previously and, like that voyage, they first arrived near Annenkov Island.

   Quest reached Leith Harbour on 6 April 1922 where Hussey was able to give a full account of Shackleton's burial. She underwent more overhaul but was no longer considered fit to make a voyage through pack-ice, so preparations began for a

Shackleton's shipmates and the memorial cross they erected to him in May 1922 at Hope Point. (Courtesy Scott Polar Research Institute.)

northern journey. Hussey departed on *Neko* for Rio de Janeiro to collect some of Shackleton's effects left at Montevideo and to report to John Rowett in the United Kingdom.

While at South Georgia, the expedition was able to continue with some scientific work. *Quest* sailed to Stromness Harbour where the expedition again met the Manager, Thoralf Sørlle, who had provided so much assistance to Shackleton in 1916. Then, after coaling at Prince Olav Harbour, *Quest* proceeded to the Bay of Isles, where ornithological and other investigations were made, some being recorded on cinematographic film. She then returned to King Edward Cove by way of Prince Olav Harbour and Leith Harbour.

There, a stone cairn surmounted by a wooden cross was erected at Hope Point by Shackleton's comrades, to form a memorial to him and to replace the one which he had erected in 1914 while surveying King Edward Cove. A compartment beneath it contained a photograph signed by all of them which was found when the memorial was moved a short distance to provide space for a gun emplacement during the Second World War. Shackleton had commented to Wild on 4 January 1922 about its absence as *Quest* sailed into the cove. The memorial still stands, having been occassionally repaired, and is a prominent landmark. The expedition departed, following a final visit to Shackleton's grave. They made a short halt at Moltke Harbour on 8 May; it was noted that one of the German International Polar Year Expedition's huts was still standing, and the Ross Glacier was resurveyed. *Quest* then left South Georgia.

Quest sailed to Tristan da Cunha and Gough Island where extensive observations and survey were conducted. Thence she continued to Cape Town where, after an enthusiastic welcome, repairs and supplies were arranged and contact established directly with John Rowett. *Quest* returned to the United Kingdom calling at St Helena, Ascension, the Cape Verde Islands and the Azores to arrive at Plymouth on 16 September 1922. Some investigations were made at these islands during the voyage. Many reports, several of which concerned South Georgia, were published using the data from the voyage and a variety of charts were prepared.

Near the time of Shackleton's death, two other notable Antarctic explorers also died. The Scotsman, Dr. W.S. Bruce, who met C.A. Larsen in the Weddell Sea in 1892 during their first Antarctic voyages, died on 28 October 1921. At his request, his ashes were scattered off South Georgia on 2 April 1923. On 7 December 1924 C.A. Larsen died while leading an expedition in the Ross Sea. His body was returned to Sandefjord, the principal whaling town in Norway, for burial.

Alberto Carcelles (1923, 1927 & 1929)

Early in 1923 the Argentine naval vessel, *Guárdia Nacionál*, made another call at South Georgia, having first visited the island in 1905. This was during the voyage to supply the meteorological station at South Orkney Islands for which the whale catcher *Rosita* was chartered. She was commanded by Captain Ricardo Vago who conducted a hydrographic survey in Cumberland Bay and geomagnetic observations at Royal Bay during the 16 days he was at the island awaiting the return of *Rosita* (5 to 22 February 1923). Alberto Carcelles, a biologist who was aboard her, made the first of a series of collections and observations on the birds of South Georgia on behalf of the National Museum in Buenos Aires. His later visits were in 1927 from 3 to 13 February aboard *Lancing* and from 8 October to 31 December 1929, when he arrived on *Ernesto Tornquist* and departed on *Astra*. During his last visit he made a tour of most of the island's bays aboard the sealer *Dias*. He prepared a report on birds, whales, whaling and various other aspects of the island. *Guárdia Nacionál* made a third voyage to South Georgia in 1924, arriving on 4 March. *Rosita* was again chartered to relieve the meteorological station while *Guárdia Nacionál* awaited her return.

The *Discovery* Investigations (1925–51)

Following the enormous increase in whaling activities at South Georgia, other parts of the Falkland Islands Dependencies and the Falkland Islands themselves, concern was expressed about the future of the industry, and the economic development of the Dependencies generally. An 'Interdepartmental Committee for the Dependencies of the Falkland Islands' was established in 1918 with members from the Board of Agriculture and Fisheries, the Colonial Office, the British Museum (Natural History), the Department of Scientific and Industrial Research and the Admiralty. The committee sat in London and its terms of reference were:

To consider what can now be done to facilitate prompt action at the conclusion of the War in regard to the preservation of the whaling industry and to the development of other industries in the Dependencies of the Falkland Islands; and to consider not only the economic questions above referred to and the scheme for employment of a research vessel, but also what purely scientific investigations are most required in connexion with these regions, and whether any preliminary inquiries by experts in this Country should be instituted.

The report of the Committee, Command Paper 657, a very comprehensive and detailed document, was presented to Parliament in April 1920. It included 36 recommendations, among which were:

1. A system should be devised for marking whales...
2. The food for whales, whether consisting of plankton or fish, should be carefully investigated...
6. An experienced zoologist should be deputed to work for some time at one or more of the whaling stations...
11. The utmost economy must be observed in the utilization of all whale products, and every effort should be made to prevent the capture of more whales than can be commercially utilized...
22. Study of the life history and habits of the seals of the Dependencies should form part of the work of the biologists attached to the proposed expedition...
27. A complete hydrographical survey of the Dependencies is necessary both in the general interests of navigation and in local interests of the whaling industry, and also to deal with the scientific questions referred to...
32. Two special vessels should be employed to carry out the researches proposed, to be provided with motorboats, and the crew and equipment recommended...
35. Expenses incurred in connexion with the economic development of the Dependencies, and in particular, with the preservation of the whaling industry, may properly form a charge against revenue raised in the Dependencies, and additional taxation may rightly be imposed upon the whaling and sealing industries for the purposes of meeting such expenditure...   (Colonial Office, U.K., 1920)

The Committee, who realised the importance of oceanographic investigation of the area, were able to draw upon the results and experience gained by the *Challenger* Expedition and the International Council for the Exploration of the Sea as well as some other sources. Much of their work was, however, of a pioneering nature. It was decided that the first operations should be an intensive survey of the waters in the vicinity of South Georgia. Captain

R.F. Scott's ship, *Discovery*, a three masted steam-assisted sailing vessel, built in 1900 at Dundee, was obtained for the work. The scientific programme was named '*Discovery* Investigations' and the Committee became the '*Discovery* Committee' after her. Dr S. Kemp was appointed Director of Research and Leader of the Expedition.

The extensive repairs, replacements and re-equipping which *Discovery* required delayed the commencement of the ship-borne part of the investigation to late 1925. A year earlier, however, a specially designed pre-fabricated building, later named 'Discovery House' was despatched to South Georgia to be deployed at King Edward Point. It was erected in January 1925 and opened on 20 February. It was not till April, however, that a staff of three zoologists, a hydrologist and a technician, led by Dr N.A. Mackintosh, commenced work in it, making studies of the whales caught at Grytviken. A wireless station and a new residence for the magistrate were built at King Edward Point at the same time.

Discovery, extensively refitted for her new work, sailed from Dartmouth on 24 September 1925 under the command of Captain J. Stenhouse who had previously been Master of the *Aurora* of Shackleton's 1914–16 expedition. She arrived at South Georgia on 20 February 1926 after having called at the Canary Islands, Ascension, Tristan da Cunha and Cape Town while conducting scientific work during the voyage. The first sight of the island was described as 'like the Himalayas seen from Simla'. The ship and land parties of the investigation then reunited and exchanged much information. The latter already had the results of measurements and examinations of 519 blue and fin whales. The enormous importance of Antarctic krill (*Euphausia superba*) in the diet of whales and in the food chains of the entire region was appreciated. Detailed studies had also been made of the birds, elephant seals and other seals of the island.

Discovery commenced operations around South Georgia by taking plankton samples and specimens of sea water for analysis. On the evening of the first day's trawling, enormous numbers of glowing krill were taken which gave rise to great enthusiasm as none was taken during the daylight hours despite much sampling. From many voyages which took regular samples and from data provided

The Royal Research Ship *Discovery*. (Courtesy Scott
Polar Research Institute.)

by the whaling companies, the Investigation was able to prepare maps of the distribution of krill, fin whales and blue whales around the island. Large numbers of the bottom-dwelling fauna were also obtained and preserved for later identification. A general account of whaling methods and the whaling stations as well as of the island was made before *Discovery* sailed for Port Stanley on 19 April 1926.

She sailed from the Falkland Islands on 20 May 1926 for South Africa, making oceanographical and other scientific observations on the way. At South Georgia, meanwhile, observations and measurements of whales continued at the Discovery House laboratories. At Simonstown Royal Naval dockyards, *Discovery* was fitted with false keels and some other improvements were made. She was met there by *William Scoreby*, a steam ship commissioned to join the investigations and commanded by Captain C.M. Mercer. *Discovery* sailed on 27 October 1926 for South Georgia by way of Bouvetøya and the ice front. She experienced

difficulties in navigating through some of the worst ice conditions since 1916, before arriving at King Edward Cove on 5 December 1926 where she rejoined *William Scoresby*. The latter ship had been on a whale-marking cruise around South Georgia, using brass marks bearing reference numbers and offering a reward for their return. This programme has continued to the present under the auspices of the International Whaling Commission.

The ships spent much of the summer of 1926/27 taking samples around the island. They made a 'network' of identical observations (in as short a time as possible) over the entire area of the South Georgia whaling grounds. These were arranged in eight lines taken from the coasts until a depth of 1800 metres was reached, which was well beyond the edge of the continental shelf surrounding the island. At each station on these lines, samples of water from different depths were taken for analysis and their temperatures logged. Planktonic collections and trawl samples were taken and

Discovery House in January 1980. (Author.)

preserved while various physical and meteorological data were being recorded. The behaviour of krill during light and dark periods was also studied. The locations of areas rich in microplankton and krill were determined and a theoretical basis for their presence, related to oceanic circulation, nutrients and the physiography of South Georgia, was postulated. These hypotheses have been largely confirmed by later investigation. *Discovery* left South Georgia on 4 February 1927 to continue her surveys in the vicinity of the South Orkney Islands, the South Shetland Islands and Graham Land. From there she sailed to Cape Town by way of the Falkland Islands and Gough Island.

During the 1926/27 season, Lieutenant Commander J.M. Chaplin conducted a major survey of South Georgia, especially of the harbours and anchorages. This was in connection with the *Discovery* Investigations but is discussed separately in this chapter.

Discovery had proven herself far from suitable for the work in which she was involved, although her voyages had yielded an enormous amount of scientific data. She was replaced by an oil-fired steam ship named after her – *Discovery II* – specially designed for the work. The first voyage of *Discovery II* was from 1929 to 1931, principally in the South Atlantic regions, in less contact with South Georgia than her prodecessor. *Discovery* transported the British, Australian and New Zealand Antarctic Research Expedition (1929–31), led by Sir Douglas Mawson, before retiring to London where she remains as a museum.

Discovery House was occupied by scientific staff until May 1927 when they returned to the United Kingdom. It was reopened early in 1928 for work associated with *William Scoresby* voyages and remained in use to 1931. Subsequently it has served many purposes and still stands, having been used by the British Antarctic Survey and, most recently, by the garrison at South Georgia.

As well as the work associated with the whaling industry, the scientific personnel accomplished many other investigations. Descriptions of the birds, elephant and other seals, several terrestrial invertebrate groups, and other animals were prepared. One of the scientists based at Discovery House, Dr L. Harrison Matthews, also wrote a comprehensive book about the island (Matthews,

1931) and prepared much cinematographic film of whaling and many other subjects.

Nothing comparable to the sustained efforts of the *Discovery* Investigations had previously been made in the field of oceanic exploration. The basic data obtained permitted a great increase in knowledge of the southern regions of the world, as well as much advancement in associated theoretical matters. The discovery of the Antarctic Convergence and the determination of its mechanism was a particularly notable achievement. The results of the investigations are published in a series of *Discovery Reports*, presently 38 volumes, with much scientific material relevant to South Georgia, and several general accounts were written (Hardy 1967, Ommanney 1971). The investigations continued making voyages up to the 1950/51 season with the sixth commission of RRS *Discovery II*. The later voyages, however, had far less association with South Georgia.

The *Meteor* expedition (1926)

A German expedition led by F.A. Spiess aboard *Meteor* made a very comprehensive oceanographical and hydrographic survey of the southern Atlantic Ocean. This included an enormous number of soundings, contributing greatly to the understanding of the bathymetry of the region. She called at King Edward Cove from 4 to 8 February 1926 on a voyage from the South Shetland Islands to Cape Town on the fifth profile of her 14 Atlantic crossings.

Lt Cdr Chaplin and *Alert* (1926–30)

As part of the *Discovery* Investigations, a hydrographic and general survey was made around South Georgia and elsewhere by Lieutenant Commander J.M. Chaplin from 1926 to 1930. The ships *Discovery* and *William Scoresby* were used for part of the operations. Much of the coast was surveyed from the motor launch *Alert*, which was able to enter many smaller bays and anchorages. At the time Chaplin began his survey, only King Edward Cove and Royal Bay had been properly surveyed (although a variety of plans of bays and of much of the rest of the island were available from 1775 when Captain Cook prepared the first chart of it).

The Surveys commenced at Undine Harbour

Shipping movements of the *Discovery* Investigations
from 1926 to 1938, drawn by Dr N.A. Mackintosh.
(Courtesy of the Institute of Oceanographic Sciences.)

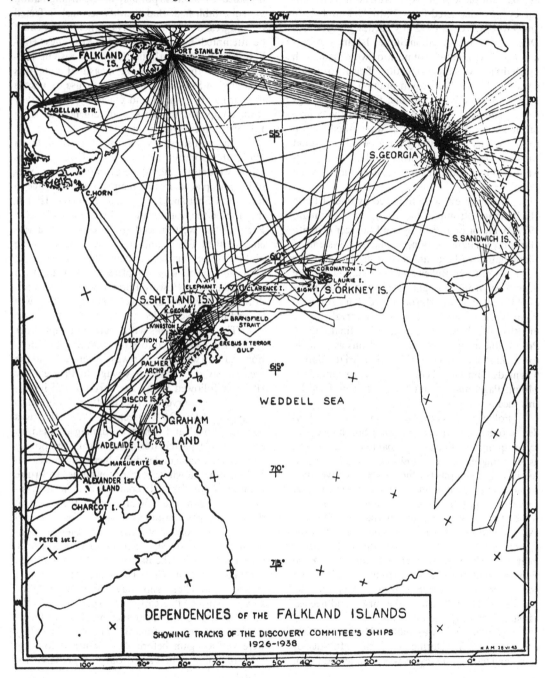

in the north-west and Larsen Harbour in the south-east in order to fix the two extremities of the island, as a base for subsequent work. At Undine Harbour, survey was conducted from 22 to 29 March 1926 and 9 to 17 December 1927 by parties, landed from *Discovery*, who camped on Survey Isthmus. Astronomical determinations of position were made with much difficulty as the proportion of clear to cloudy days was 1:6. A Norwegian pram, a rowing boat used by sealers, was available; from it soundings were made across Undine Harbour in the course of the survey. Larsen Harbour was visited from 6 to 13 January 1927 during a period of very good weather. Despite the distances involved, soundings were also made throughout the harbour from the pram.

These surveys were followed by one of Leith Harbour and Stromness Bay from 17 to 31 January 1927. A motor boat, lent by the Manager of Leith Harbour whaling station, enabled a system of triangulation to be established over the Bay. The positions of Cape Buller and Cape Saunders were determined from sun sights at this time. Tide poles established at Leith Harbour and King Edward Point were observed at hourly intervals for 2 months at each station to establish a tidal datum. Prior to the team's departure, additions were made to the chart of Cumberland Bay where triangulation was made from a number of points. The positions of Mount Paget and Mount Sugartop were fixed by this means.

Following this the *Discovery* Committee decided to organise a further survey of South Georgia, to be equipped with a motor boat and pram as well as camping equipment and provisions. A radio receiver was supplied which could receive time signals, thus greatly improving accuracy of the survey. Chaplin and the new launch *Alert*, with some of her crew arrived at Husvik on 13 October 1928 aboard *Orwell* where they met the rest of the survey team who had arrived on *Busen*. The headquarters for the survey were established at King Edward Point, where *Alert* arrived on the 15th and commenced sea trials. She proved very reliable, sailing over 5000 km during the survey and needed assistance on only one occasion.

The first survey was made of Husvik Harbour and its vicinity. During this, the attitude of skuas and terns to surveyors was commented upon;

individual and mass attacks were described. Soundings, positions of shores and some peaks were obtained to be incorporated in the triangulation system established. Bad weather seriously affected the progress and on many days no field work could be attempted. Survey continued in Stromness Harbour and Leith Harbour despite persistent problems with the weather. A measured mile was delimited in Stromness Harbour for use by vessels measuring their speed and efficiency. The survey team completed its work and returned to King Edward Point for Christmas.

The next sites surveyed were Maiviken, Jason Harbour, Godthul and Pleasant Cove. Camps ashore were established at these locations and a series of beacons placed around their shores. A voyage to the South Shetland Islands and Falkland Islands interrupted the work from 5 February to 18 March. Parts of the western end of South Georgia (including Bird Island and the Willis Islands) were plotted as the ship passed them. On the return, the Godthul survey was completed. The team then returned to King Edward Point to spend the winter, drawing charts in Discovery House. *Alert* was slipped to be overhauled.

The next season's work started in September 1929 when *Alert* sailed to Stromness Bay and on to Fortuna Bay where a camp was established and the area surveyed. A difficult voyage to Prince Olav Harbour followed where the harbour and its approaches were surveyed. Thence *Alert* continued to the Bay of Isles. On 2 December 1929, *Alert* was beached, holed and otherwise damaged when an anchor cable parted during a storm. Temporary repairs were made under difficult circumstances; on 12 December, a whale catcher towed her to Prince Olav Harbour for full repairs. These were completed by 21 January 1930 after which the survey of the Bay of Isles continued. This was finished by 9 February. An ascent of a local peak enabled connection to be made between the surveys of Prince Olav Harbour and the Bay of Isles. Work was then continued to Blue Whale Harbour, Elsehul, Right Whale Bay and Possession Bay. The efficiency of the survey improved greatly as experience in the methods and conditions prevailing at South Georgia was obtained. Fortuna Bay and Stromness Bay were visited again before *Alert* returned to King Edward Point to be hauled out and prepared for winter.

Short voyages of *William Scoresby* and the whale catcher *Southern Pride* enabled parts of the north and south coasts of the western end of the island to be charted. The coast from Bird Island to Cumberland Bay was passed in both directions many times, and fairly accurate plotting of it was achieved. For the last part of the operations on South Georgia, a circumnavigation of the island was made by *William Scoresby*. This allowed much improvement of the charts and, during a brief visit, the survey of Ocean Harbour. Chaplin also obtained information from the Masters of the sealing and whaling vessels which operated off South Georgia. They had an unsurpassed local knowledge which added greatly to the accuracy and utility of the charts. The survey team left South Georgia on 10 May 1930 aboard *Harpon*.

The Survey had, at its conclusion, charted approximately 580 km of coastline in 370 days' work on all vessels, 100 of which were unsuitable owing to the weather. Principal trigonometrical points were estalished at 112 stations for the basis of the charts, while 13 600 soundings were made during the survey of the various bays. These were incorporated in a series of 22 plans, published on 3 charts. The majority of these were completely new. A narrative account and description of the survey was published (Chaplin, 1932) which included many plates, notes on personnel and equipment, magnetic variation, weather and climate and kelp.

Governor Hodson (1927 & 1928)

A mail service between South Georgia and the Falkland Islands began in 1924 as a result of a contract between the Falkland Islands Government and the Tønsberg Hvalfangeri, which managed the whaling station at Husvik. The ship *Fleurus* operated it regularly up to 1933. Her voyage from Port Stanley to South Georgia, commanded by Captain Adamsen (from 2 to 6 August 1927, which departed on the 13th to return to Port Stanley on the 18th), carried Sir Arnold Hodson, the Governor of the Falkland Islands and their Dependencies. This was the first visit by a Governor to the island and to the Dependencies in general.

In 1928 the Governor made another visit also aboard *Fleurus*, commanded by Captain Carlsen. This formed part of a tour of the other parts of the Dependencies which called at Deception Island,

sailed through the Neumayer Channel, visited the Melchior Islands, passed Elephant Island, and called at Borge Bay in the South Orkney Islands prior to arriving in South Georgia. The Governor was on the island from 22 to 26 February 1928. On 24 February he unveiled the carved granite memorial, erected over Sir Ernest Shackleton's grave. The memorial, which was paid for from money raised by public subscription, stands prominently in the Grytviken cemetery with graves of sealers, whalers and others who have died there since 1846. The front bears a nine-pointed star, a symbol associated with the Shackleton family; the reverse a quotation from Robert Browning 'I hold that a man should strive to the uttermost for his life's set prize'.

While on the island the Governor also inaugurated the first South Georgian sports meeting at Grytviken and opened a rifle range at Hope Point. On 24 February, a reception was held at King Edward Point attended by the administrative personnel, all the whaling station managers and a detachment from *William Scoresby*. On the 25th he made a voyage aboard a whale catcher to Hound Bay where he landed and walked through Sörling Valley to Cumberland East Bay to be collected there. During the walk a giant petrel welcomed him to the island in the unpleasant manner characteristic of the species. During his visit the Governor met a party of men from the meteorological station on Laurie Island, South Orkney Islands, maintained by the Argentine government, and was interested to hear about their activities.

Consul Lars Christiensen's expeditions (1927–29)

A series of scientific voyages, organised and financed by Consul Lars Christiansen of Sandefjord, Norway, was made during the same period as the *Discovery* Investigations. C.L. Christiansen was the son of Christian Christiansen who had sponsored the voyage of *Jason* in 1892/93. C.L. Christiansen had also founded a museum and archive of whaling at Sandefjord in memory of his father. The voyages had the object of combining scientific research with development of the whaling industry; their results were to some extent complementary to those of the *Discovery* Investigations.

The first voyage was made by *Odd I*, a whale catcher, which sailed from Godthul South Georgia on 4 January 1927 to the South Shetland Islands, Antarctic Peninsula and Peter I Island. After accomplishing her observations she returned about a month later. The second and subsequent voyages were made by *Norvegia*. She left Sandefjord on 14 September 1927 for Cape Town under the command of Captain H. Hornveldt and carried two scientific personnel, H. Mosby and D. Rustad. The expedition explored Bouvetøya, which was claimed for Norway. The voyage continued south then went along the pack-ice margin. There, damage sustained at Bouvetøya was exacerbated which required her to divert to South Georgia. *Norvegia* arrived there on 22 January 1928 at King Edward Cove. She moved to Husvik where it was found that her repairs would require items unavailable on the island, so she remained until the following season while these were brought from Norway. Meanwhile, her scientific officers, with assistance provided by the whaling companies, conducted oceanographical and other research in several bays of the island. One of them was able to visit places on the west coast and collected fossiliferous rocks from Annenkov Island.

On 5 February 1928 O. Holtedahl and O. Olstad arrived at Prince Olav Harbour from the Antarctic Peninsula where they had been making biological and geological observations. Their work was continued at South Georgia. During February an extensive series of plankton sampling and hydrographic observations were made in Cumberland Bay and Larsen Harbour from *Norvegia*. All four scientists left the island in March aboard a whaling transport ship. During the 1928/29 season, after being successfully repaired, *Norvegia* sailed under the command of Captain Nils Larsen. She left South Georgia on 8 November 1928 to make an extensive voyage around Bouvetøya and thence westward almost to the Ross Sea. She returned to South Georgia on 30 March 1929 where she was laid up for her second winter whilst preparations for her third voyage were made. She departed from South Georgia for the last time, to join an expedition to Queen Maud Land after which she went to Cape Town. Her fourth and last Antarctic voyage took her around the continent during the 1930/31 summer season.

Kohl-Larsen expedition (1928–29)

A three-person expedition, whose principal object was to study the recent glaciation of South Georgia, worked on South Georgia during the 1928/29 summer. The leader, Ludwig Kohl-Larsen was accompanied by his wife, Margit Kohl-Larsen (one of C.A. Larsen's daughters, who was first on South Georgia in 1905), and Albert Benitz who prepared the first commercial cinematographic film of the island. The expedition was partly supported by the Notgemeinschaft der Deutschen Wissenschaft. In addition to the glaciological studies, maps were drawn for the first time of inland areas of the island and some biological collections were made. Ludwig Kohl-Larsen had previously been on two Antarctic expeditions led by C.A. Larsen in 1923/24 and 1924/25 and had first arrived on South Georgia in 1911 with the Filchner expedition.

The expedition left Sandefjord on 18 August 1928 aboard *Harpon*, a transport ship of the Compañia Argentina de Pesca. This was her first voyage of the summer and she was taking whalers and supplies down to start the whaling season. She called at the Canary Islands and arrived at Grytviken on 18 September. L. Kohl-Larsen was impressed by the numbers of whales, sea birds and quantity of kelp seen as they approached South Georgia. The expedition spent a short time at Grytviken and then went to the north-west end of the island.

The whale catcher *Tiburon* was made available to the expedition and they steamed along the north-west coast, describing the features and weather together with the inadequacies of the charts then available. After having sailed through Bird Sound, the night was spent in Coal Harbour where a camp was established. This was occupied until 20 November. Meterological data were recorded thrice daily; most of the weather observed was decidedly bad and caused many difficulties with the tents. Treks were made to Elsehul, Undine Harbour and elsewhere in the vicinity. Biological collections and observations were made, particularly concerning the life of the elephant seal. The expedition was equipped with a Norwegian pram which was used to make soundings in Coal Harbour.

From 5 to 9 November a sledging expedition was made towards Schlieper Bay. Benitz recorded parts of this on film. During the journey more

glaciological, geological and biological observations and collections were made, these included
more fossils from South Georgia. On 19 November
they were collected by *Tiburon* and, after a brief
landing near Cape Paryadin, continued to the Bay
of Isles and set up a second camp near the Grace
Glacier. This was occupied from 20 November to 7
January. The Grace and Lucas Glaciers were visited
and surveyed. Meteorological observations were
also maintained, with no improvement in the
weather recorded. The pram was again deployed to
make various excursions to bays and islands within
the Bay of Isles.

After this, the expedition returned to Grytviken and remained there to 16 January. Films
were made of activities at the station and specimens
of embryos, parasites, etc. from the whales being
processed there were preserved. On the 17th *Fleurus*
conveyed them to Husvik whaling station from
where they began charting and a glaciological
survey of the hinterland. Part of this involved the
exploration of a large ice plateau, later named the
Kohl-Larsen Plateau (but in 1979, after many years
bearing that name, it was renamed the Kohl
Plateau). Connections between the various glaciers,
the ice plateau, and the southern side of the island
were charted and photographed. On 29 January
they returned to Husvik, thence walked to Stromness and took a motorboat to Leith Harbour.
Investigations continued with an excursion on 5
February to Hercules Bay where a macaroni penguin colony was described and its population
estimated to be between 50 000 and 60 000. Two
days later they returned to Stromness and from
there to Grytviken.

A limnological investigation was made of the
lakes and ponds around King Edward Cove and
Maiviken, and a geographical description produced
of the area. Entomological and planktonic collections were obtained at various places around the
shores of Cumberland Bay. The botany of the area
was described and a comparison with that of
Lapland suggested.

The expedition departed for a voyage along
the south-east side of the island on 27 February
aboard the sealing vessel *Dias*. She steamed around
the southern end of the island visiting Drygalski
Fjord and Larsen Harbour. Their first attempt to
land on Annenkov Island was unsuccessful owing
to bad weather, so an exploration of several bays on
that side of South Georgia was made. Parts of these
bays had previously been seen from inland areas by
the expedition. The second attempt to land on
Annenkov Island was successful and they remained
there from 3 to 14 March. The birds were described,
film was made of the hatching of the wandering
albatross; photographs were taken of the mainland
and many more fossils were collected.

After returning to Grytviken, a trek was made
to the Lyell Glacier followed by excursions to St
Andrews Bay and Ocean Harbour. On one of the
landings the pram was destroyed. From 1 April the
expedition sailed counterclockwise around the
island aboard *Dias*, describing the various bays
encountered on the journey. At Diaz Cove the
remains of a sailing ship, ruins of a dwelling and
various artifacts were found and the Magistrate
informed. This is mentioned in the history of the
first epoch of the sealers (Chapter 3); some of the
ruins are still visible. Further investigations were
made at the southern end of the island, including a
description of the geology of the region. This is
of particular interest as it is different from that of
the rest of the island. They returned via Cooper Bay,
Wirik Bay, Iris Bay, and Gold Harbour to arrive
at Grytviken on 7 April.

St Andrews Bay was revisited on the 15 April,
to study the penguins there, unfortunately during a
period of very poor weather. Their last journey on
South Georgia was made from St Andrews Bay to
Royal Bay where the Ross Glacier was resurveyed.
They then returned to Grytviken.

The expedition departed from South Georgia
aboard *Busen* on 13 May 1929 to arrive in Sandefjord on 18 June. The specimens and data collected
were then distributed and a series of scientific
reports prepared. Most of the published work was
contained in the journal of the Senckenberg
Museum in Frankfurt. *An den Toren der Antarktis*
was written by L. Kohl-Larsen (Kohl-Larsen 1930)
and a cinematographic film prepared by Benitz. The
latter yielded, when titled and annotated, about
40 minutes of most interesting pictures of the expedition's activities, as well as of whaling and
wildlife on the island. The original nitrate film has
recently been copied to acetate base and is preserved. Margit Kohl-Larsen upon viewing it 50

years later, was delighted to see film of herself on South Georgia.

Throughout the expedition's duration, comprehensive meteorological records were kept. These showed that they suffered unusually poor weather. The numerous photographs taken have proved valuable for later comparative determination of glacier positions and other purposes. The surveying performed, especially of inland features (which they were the first to explore), resulted in great improvements to the charts of the island. Much of it was not superseded until the Carse expeditions from 1951 to 1957.

Erich Dautert (1931)

A two-man biological expedition, from the University and Museum of La Plata, Argentina, arrived on South Georgia early in 1931. A book about the island and their visit was published by one of them, Erich Dautert, a German, but this gives no data for the voyage and is decidely vague about several other matters. From some internal evidence, combined with the little direct information available, it may be deduced that, after spending a lot of time unsuccussfully trying to obtain promised transportation from the Argentine Government, they secured a passage in a ship belonging to the Compañia Argentina de Pesca. The ship *Tijuca*, an elderly three masted barque, sailed from Buenos Aires on 6 February 1931 and had a rough voyage to Grytviken. The book contains an excellent contemporary description of the whaling station and its activities, indicating that they came to know it well. The area within walking distance of Grytviken was investigated; many still and cinematographic photographs were taken. The expedition accompanied sealing, whaling and reindeer-hunting voyages. For the last part of the season, the Manager of the whaling station placed a sealing vessel at their disposal. This sailed around the island during a voyage which lasted 24 days. Many collections were made on the trip; these included the skin of an elephant seal (now on display at La Plata Museum). A landing on Annenkov Island was made during the voyage. The expedition left South Georgia at the end of the 1930/31 season to return to Buenos Aires, again aboard *Tijuca*.

A popular account of the expedition and South Georgia was published in 1935. This was translated and appeared in English 2 years later (Dautert 1937). Both editions are well illustrated but contain remarkably few chronological details. Apart from the Museum's specimens, there is little further information available.

Visits of *Penola* (1936 & 1937)

Two visits of *Penola* (of the British Graham Land Expedition) were made to South Georgia, in 1936 and 1937. The activities of the expedition were principally from several locations on the western side of the Antarctic Peninsula. *Penola* sailed from London on 10 September 1934 and established a station on Winter Island (one of the Argentine Islands), where she wintered during 1935. She made a return voyage to Deception Island in January 1936. Then, after establishing another station (at the Debenham Islands), sailed for the Falkland Islands to arrive on 24 March 1936. After spending part of the winter there, she continued to South Georgia, arriving at Stromness Harbour on 12 August after a very cold passage. The inhabitants of Stromness whaling station in August 1936 consisted of only five men and the manager's wife. Assistance from the other whaling stations was obtained; together with *Penola's* crew, they set about overhauling her. This involved lightening the vessel by removing almost everything aboard; after this she spent 18 days in the floating dock. The work was completed shortly before 2 October 1936, when she sailed for Port Stanley.

The expedition's ornithologist, Brian Roberts, was aboard *Penola* and he remained on South Georgia after she left, departing a month later on RRS *Discovery II*. In this month he travelled extensively around the island on sealing vessels investigating populations and other aspects of birds and seals, taking many photographs in the process.

Penola made a second visit to South Georgia where she arrived on 3 April 1937. This was at the conclusion of the expedition when the shore party returned to the United Kingdom aboard *Coronda II* (one of Christian Salvesen's ships), which departed on 17 April 1937. *Penola* sailed north early in May to arrive in the United Kingdom on 4 August 1937. During her time in Port Stanley at the beginning of the expedition on the way south, in December 1934 Duncan Carse was recruited to her crew from the *Discovery II* and remained with her subsequently.

He later organised a series of expeditions to survey South Georgia between 1951 and 1957; this is described later in this chapter.

HMS *Ajax* and *Exeter* (1937 & 1938)

British Naval ships visited South Georgia in the summers of 1937 and 1938. HMS *Ajax*, commanded by Captain C.S. Thomson, called at King Edward Point with the Governor aboard. She stayed a few days, during which the Governor went on a reindeer hunt. On her return voyage she was diverted to the South Shetland Islands to rescue the crew of the launch *Rapid* lost from the RRS *Discovery II*.

In 1938 HMS *Exeter*, commanded by Captain G. Fowler, called at King Edward Cove from 25 to 29 November 1938. During the visit she deployed two Walrus aircraft to take aerial photographs of the cove and various other parts of the island. This was the first time that aircraft had flown at South Georgia. The Walrus was a biplane with a 'pusher' engine, adapted to take-off and land on water, which has proved a very versatile machine in many parts of the world. The aerial photographs were particularly useful for local mapping and have also been valuable for comparative glaciological studies.

HMS *Ajax* and *Exeter* together with HMNZS *Achilles* were subsequently involved in the Battle of the River Plate, against ships of the German Navy, on

Walrus aircraft, deployed from HMS *Exeter* in November, 1938, with Mount Paget in the background. (Courtesy Royal Navy.)

13 December 1939. HMS *Exeter* sustained serious damage and was repaired at South Georgia where her wounded were hospitalised.

The Second World War (1939–45)

The Second World War interrupted civil expeditionary activities on South Georgia as well as causing great disruption to the whaling industry. All the British and many of the Norwegian floating factories were lost and most of their whale catchers were called up for a naval service. Whale oil was very important for the preparation of edible fats but it had neither the great military nor political significance that it held during the First World War. Other substances had replaced it as a raw material in the manufacture of explosives. Only Grytviken operated throughout the war. Leith operated in 1940/41 and briefly in 1942/43; all the other stations remained closed, in marked contrast to the First World War.

During the war, German ships operated in Antarctic waters against the allied whaling fleets (that of Norway being subject to the Norwegian Government in exile in London during the German occupation of Norway). In consequence, *Queen of Bermuda* was fitted out as an armed merchant cruiser in 1939 to patrol the coasts of South America and parts of Antarctica, during which she made two visits to South Georgia. The first reached the island on 22 January 1941 and she made an inspection of the northern coast before entering King Edward Cove on 25 January. She departed on the same day for Deception Island by way of the South Orkney Islands and areas further south in the Weddell Sea, where several floating factories of the whaling fleet were operating. At Deception Island action was taken to destroy all amenities of the abandoned whaling station there which could be of use to the enemy. After this she continued to the Falkland Islands.

Her second Antarctic voyage left Port Stanley for the whaling grounds passing South Orkney Islands to South Georgia, where she arrived on 29 March 1941. On 5 April she departed for Liberia on a zigzag course. The guns previously installed at Cumberland Bay and Stromness Bay were re-deployed to more effective positions during this visit.

During the Second World War another de-velopment took place which was to become a very important factor in the later history of South Georgia. In 1943 a secret Royal Naval programme, 'Operation Tabarin' (the code name was that of a Paris night club) was established to serve in the Falkland Islands Dependencies. After the war, control of the programme was transferred to the Colonial Office and it became the 'Falkland Islands Dependencies Survey'. This organisation, with several administrative changes, eventually became the British Antarctic Survey which is described in greater detail in Chapter 5.

Dr. Gibson-Hill (1946)

Dr. C.A. Gibson-Hill of the Raffles Museum, Singapore, made a private ornithological expedition to South Georgia from 1 January to 7 March 1946. He travelled aboard one of Christian Salvesen's whale catchers from Durban, South Africa. His headquarters were at Leith Harbour while he was on the island from where he examined the principal bird colonies between Fortuna Bay and Husvik Harbour. He also made short camping trips to Bird Island, Elsehul and the Bay of Isles; visited the whaling grounds twice; and travelled to 62° S with an oil tanker which was refuelling a floating factory. He made studies of the habits and behaviour of several species of albatross, penguin and of the giant petrel as well as bird observations at sea. Specimens of 14 bird species were collected for examination and over 200 photographs were taken. Dr Gibson-Hill prepared several scientific and popular articles concerning these observations.

Niall Rankin and *Albatross* (1946/47)

Another ornithological expedition, led by Niall Rankin, arrived at South Georgia shortly after the visit by Gibson-Hill. Rankin started planning this expedition before the Second World War, and completed preparations at its conclusion. A former Royal National Lifeboat, surplus from the war, was obtained and Christian Salvesen's Company agreed to tranport it on one of their floating factories to South Georgia. The lifeboat was named *Albatross* and preparations for her South Georgia voyage commenced. She was thoroughly overhauled, equipped with new engines and provisioned at rather short notice. The equipment for the expedition included some excellent photographic apparatus. Two Shetland Islanders, who proved very competent, were recruited as crew. On 7 October 1946 *Albatross* was taken aboard *Southern Venturer* to arrive at Leith Harbour, South Georgia on 26 November. There, a fourth crew member (a Norwegian) was recruited.

Albatross departed on 6 December for the Bay of Isles with provisions for 6 weeks. At first she anchored near Salisbury Plain, but later moved to Rosita Harbour where a severe storm and engine troubles almost terminated the expedition. Studies and photographs were made of king and other species of penguins. Their behaviour, diet, growth, incabation period etc. were described. Some of the observations complemented those made by R.C. Murphy in 1912 and by Brian Roberts in 1936. Subsequently Rankin was able to visit four of the other major king penguin colonies. Similar observations were made of wandering albatosses, mainly of

those breeding on Albatross Island in the Bay of Isles.

After a fortnight in the Bay of Isles, *Albatross* continued to Right Whale Bay and Elsehul, where a landing was made. The vicinity of Survey Isthmus, Undine Harbour and Coal Harbour were explored, and counts were made of various bird species there. An attempt to land on Bird Island was unsuccessful and after this, as supplies were running low, *Albatross* returned towards Leith Harbour. On the way, halts were made at Rosita Harbour, along the coasts of the Bay of Isles where more bird counts were made and at Prince Olav Harbour, where the deserted whaling station was visited. After visits to Possession Bay and Antarctic Bay, making more counts, *Albatross* arrived at Leith Harbour.

Albatross was restocked and one of her crew replaced before she returned to the Bay of Isles. There two king penguins were collected which

The 4 inch gun deployed at Hansen Point near Leith Harbour during the Second World War. Photographed in February 1980. (Author.)

formed part of her complement for several weeks. Another journey to Elsehul yielded an indication of fur seals in the vicinity (from the contents of the stomach of a leopard seal). Several live fur seals were later photographed elsewhere on the island. A sudden lull in the weather allowed an opportunity for a brief landing on Bird Island. An unsuccessful attempt to land on the Main Island of the Willis Islands was also made. *Albatross* returned to Leith Harbour after another call at Prince Olav Harbour. There she reprovisioned and sailed to Ocean Harbour where, with permission from the Magistrate, a successful reindeer hunt was made. She then sailed into King Edward Cove to visit the Magistrate and Grytviken whaling station. While in the Cumberland Bay area several further counts of bird species were made.

From King Edward Cove, *Albatross* returned to Ocean Harbour where she spent the night and continued to Larsen Harbour the next day. A landing was made, and (among other observations) studies of the long established Weddell seal colony were made. Treks to Esbensen Bay on the southwestern coast of the island and some other places were undertaken. Photographs of birds were taken from a hide established on a hillside above Larsen Harbour. *Albatross* next moved to Cooper Bay where a comfortable anchorage was found in which she remained from 29 January to 12 February. During this time a williwaw (a violent squall), estimated to be of at least $50\,\mathrm{m\,s^{-1}}$ (100 knots), almost wrecked her. Gentoo and chinstrap penguins as well other bird species were counted in the vicinity. Rankin found one colony of gentoos over 200 m above sea level. The expedition then returned to Leith Harbour.

After making a voyage on a whale catcher, Rankin and the other members of the expedition started collecting more penguins for the zoological gardens of London, Edinburgh and Glasgow. This was the last task of the expedition and was accomplished partly with assistance from whale catchers. The penguins were transported to the United Kingdom aboard the factory ship *Southern Harvester*. Appropriate accommodation arrangements were made for the birds and the operation was successful.

Albatross was taken aboard another returning factory ship on 20 April 1947, which sailed north with the expedition personnel. The results were published in a copiously illustrated book (Rankin, 1951). Rankin's photography was highly successful. Details of bird colonies around the island were compared with previous records, and have since been of importance in later comparative investigations. The book also included details of several species of seals and short accounts of the bird species of the island together with a description of whaling activities.

Steinar Olsen and fisheries investigations (1951/52)

Interest in commercial fisheries of South Georgia started in 1905 when many barrels of salted fish were sent to Buenos Aires as a test consignment by C.A. Larsen and sold well (the quality of the local fish was probably first noted by Fanning in 1800). Some other investigations were made but it was not until 1951/52 that the Tønsbergs Hvalfangeri of Husvik arranged for a scientific examination of the island's fisheries potential. This was at a time when the stocks of whales appeared to be diminishing and the company became interested in using fish for oil and meal production. Steinar Olsen, a Norwegian ichthyologist, assisted by a fisherman, used the whale catcher *Busen 6* to sample fish from many areas around the island and a motorboat or dinghy for similar work in some bays. *Notothenia rossii*, the Antarctic cod, was the principal species concerned but 15 species were collected for determination of size, sex, age, and (in some cases) weight. These data were related to the distribution of the fish and a report was prepared.

The conditions were sufficiently favourable for the company to despatch two 21.3 m fishing boats with experienced Norwegian crews in the next season. The quantity of fish taken was much less than expected which, combined with a depression in the price of fish and whale oil, led to a suspension of the project.

Royal Naval voyages (1948–55)

British Naval voyages to South Georgia became more frequent following the Second World War and the assertions of sovereignty over parts of the Falkland Islands Dependencies by Argentina and Chile which were made during the War. Many of these voyages were able to continue hydrographic survey and several were specifically for this purpose. Surveys of South Georgia were also conducted by Royal Naval officers from the Falkland Islands Dependencies Survey ships and other vessels during several summer seasons. The Governor of the Falkland Islands and Dependencies accompanied many of the voyages on tours of inspection. Often Naval voyages to South Georgia visited other parts of the Dependencies where they were engaged in similar activities. Up to the time of the arrival of the Royal Yacht *Britannia* these were:

1948 HMS *Snipe* (Commander J.G. Forbes) with the Governor aboard called at South Georgia while on a tour of the Dependencies.

1949 HMS *Glasgow* (Commander C.L. Firth) visited South Georgia in February and March to conduct a survey. The Governor was aboard as was a Colonial Service Auditor to inspect Post Office and other accounts.

1950 HMS *Bigbury Bay* (Commander G.R.P. Goodden.) called briefly in February and HMS *Veryan Bay*, with the Governor aboard, in May.

1951 Lt Cdrs D.N. Penfold and F.W. Hunt conducted surveys at King Edward Point and elsewhere on South Georgia from land stations and supported by the Falkland Islands Dependencies Survey ships, with assistance from the whaling companies.

1952 HMS *Burghead Bay* (Commander J.A. Ievers) called briefly in March.

1953 HMS *Snipe* (Commander D.G.D. Hall-Wright) called on the way to and from Deception Island from where two Argentine deportees had been taken aboard to be conveyed to their country of origin.

1953/54 HMS *Nereide* (Commander P.R.H. Harrison) with the Governor aboard, and HMS *St Austell Bay* (Commander B.C. Ward) continued surveys.

1954/55 HMS *Veryan Bay* (Commander L.R.P. Lawford) with the Governor aboard visited the South Georgia whaling stations and ran numerous lines of sounding in Cumberland Bay.

Duncan Carse and the South Georgia Surveys (1951–57)

The survey of the inland parts of South Georgia had generally been well behind the level of that of coastal and hydrographic survey since the discovery of the island. Up to the publication of the first Directorate of Colonial Surveys map (in 1950), only the Kohl-Larsen expedition had conducted any detailed inland mapping; the map was very poor for the areas that expedition had not visited. A series of four expeditions from 1951/52 to 1956/57 resulted in an enormous improvement in this and gave the basis for all subsequent charts of the island. The surveys were led by Duncan Carse who had previously visited South Georgia with *Discovery II* in 1933 and as a member of the British Graham Land Expedition in 1936 and 1937. After the end of the Second World War he was investigating the possibility of an Antarctic operation that he described as 'something small, inexpensive and useful'. Recommendations led him to consider a survey of South Georgia; the results of this certainly met the latter two requirements.

The first expedition, in the summer of 1951/52 was a simple programme of topographic survey and a familiarisation with the island, together with a chance to become acquainted with the problems likely to be encountered later. Six men arrived aboard *Southern Opal* of Christian Salvesen Limited. They were accommodated in the gaol at King Edward Point, which Carse later recommended highly. Survey was accomplished of the region from the Spenceley, Brøgger, Ross, Cook and Nordenskjöld Glaciers, to the area around Cumberland Bay and the Kohl-Larsen Plateau. Weather frequently presented problems for the expedition. A crevasse accident involving the geologist of the party, terminated the geological part of the programme. The first results were described as a moderate but encouraging success; a wide arc of the interior comprising 35–40% of the island was surveyed and considerable stretches of the coastline were redrawn. The expedition left South Georgia on 18 April 1952.

The second expedition was made up of four men, two of whom had served on the first. They arrived on 10 October 1953 to reoccupy the gaol. Surveying by three of them started at the Bay of Isles on 29 October. Weather conditions for travelling and a medical problem interrupted planned operations somewhat, as did the discovery of greater than expected inaccuracy in the existing maps. Errors of as much as 10 km were detected in some places. On 18 December, after spending 2 days at the field station at Ample Bay (occupied by N. Bonner and B. Stonehouse), they returned to King Edward Point. From there, the fourth man was repatriated owing to his medical condition. The rest of the expedition went to the southern end of the island where they landed at Wirik Bay on 11 January 1954. They returned to King Edward Point on 17 February, having surveyed much of the south-eastern part of the island and collected geological specimens from that most interesting region. For the remainder of the time they charted part of the coast and landed on Annenkov Island from the sealer *Albatros*. The expedition departed on 17 April 1954. During this season a film was made about the island.

The third survey, in 1955/56, was the most ambitious and included eight men. They arrived on

24 September 1955. The weather was comparatively favourable for a change, which permitted a large amount of work to be accomplished. This, at first, covered the area westward from the Kohl-Larsen Plateau and investigated the route Shackleton used in 1916. In a severe storm, two of the four tents were destroyed at a camp near the Grace Glacier. With some assistance from a sealing vessel, this section was extended to the western extremity of the island and closed with earlier work. The expedition then returned to King Edward Point and prepared for a southern survey.

This began at Royal Bay on 28 January 1956, from where the remainder of the south-west area was surveyed and closed with the earlier work. On 1 March, the sealer *Dias* was boarded at Larsen Harbour for return to King Edward Point. With only a short time available before leaving the island, the expedition endeavoured to survey the area south

Members of the South Georgia Survey sledging across the Kohl-Larsen Plateau in 1952. (Courtesy Duncan Carse.)

of the Kohl-Larsen Plateau. This almost ended in disaster following a period of very severe weather with poor visibility. The expedition members became separated, five eventually reached Husvik whaling station, while the other three remained in a precarious position in the mountains. Fortunately, *Southern Venturer*, a floating factory ship, had arrived and rescue was arranged using her helicopter. On 3 April they embarked on *Southern Garden* and sailed north.

The survey was completed by the leader, Duncan Carse, alone. During the summer of 1956/57 he was able to cover much of the coastline aboard the sealers *Albatros* and *Dias*. The results of this work, with addition of some features from aerial photography from HMS *Protector*'s helicopter, appeared in a completely new Directorate of Overseas Survey map in 1958 of the island, which was a remarkable improvement on its predeces-

sor and forms the basis of many later ones. Reports on sailing directions, place-names, equipment, trekking – including a travel overlay for the new chart, and a much greater appreciation of the island resulted.

The expedition was run privately with financial support from many sources (following much work raising funds by the leader). Accounts of expenditure were carefully maintained and show the remarkable economic efficiency with which the expedition was conducted. Another notable aspect was the application of place-names to the island. This was well organised and, owing to the area covered, the expedition was able to name more features than any other. Origins and derivations were recorded for all place-names submitted for approval. Of the 240 fixed points used to control the survey, 170 were estimated as having a probable error of 5 m relative to adjacent points. The remain-

The scientific station at Ample Bay, Bay of Isles, in 1954. (W.N. Bonner)

ing 70 were unlikely to be in error by more than 20 m. The accuracy of triangulated heights was within 3 m and spot heights within 6 m. The Patrick Ness award 'for setting a new standard of survey in Antarctica' was awarded to one of the participants, A.G. Bomford, for his part in the survey and preparation of the final chart (Carse, 1959).

Bernard Stonehouse and Nigel Bonner (1953–55)

A biological expedition worked at Ample Bay in the Bay of Isles, South Georgia, from a field station established there. The expedition consisted of two men; Bernard Stonehouse and Nigel Bonner. They reached South Georgia on 10 October 1953 aboard a whaling transport ship and arrived at Ample Bay on 29 October. The first work undertaken was the erection of the hut. During the early part of the summer preparations of embryonic king and gentoo penguins were made. Behavioural studies of these two species and some others were also conducted. Bonner began an investigation of elephant and fur seals. These investigations were continued through the 1954 winter, eventually making a full year's observations. Many birds were ringed in the course of that year.

The station at Ample Bay was closed in December 1954 and its occupants moved to King Edward Point. From there Stonehouse pursued some marine biological work while Bonner began a comprehensive study of the reindeer which had been introduced to the island in 1911. The expedition left South Georgia for the United Kingdom in February 1955. Several scientific publications resulted from their work including *Falkland Islands Dependencies Survey Scientific Reports* on reindeer, king penguins, the brown skua and other animals. The expedition recorded meteorological data over more than a year from the Bay of Isles. This was the beginning of Nigel Bonner's long association with South Georgia; he was appointed Government Naturalist and Sealing Inspector shortly after the expedition.

British South Georgia Expedition (1954/55)

For about six months during the summer of 1954/55, another private expedition was active on South Georgia. This was the five-man British South Georgia Expedition, led by George Sutton. Their principal object was mountaineering but they also accomplished some survey and glaciological studies. The expedition obtained transport (by somewhat audacious perseverance) on one of Christian Salvesen's ships, *Southern Opal*, which arrived at South Georgia on 30 September 1954. Camp was established in the gaol at King Edward Point after which some time was spent trekking in the local area.

The surveyor of the expedition, Harry Pretty, sustained an injury to his back and remained at King Edward Point while the others moved to the old Post Hut near Sörling Valley. Pretty later sailed around the island aboard the sealer *Albatros*, during which he made some geographical observations. The Sörling Valley part of the expedition attempted to climb Mount Paget by the Nordenskjöld Glacier. The highest point reached was 1490 m (at what was later named Sutton Crag), where a cairn was erected. A second attempt on Mount Paget was made from a bay near Larvik Cone on the south-western side of the island. The surrounding area was explored with much difficulty owing to the highly crevassed nature of the terrain, combined with a period of poor weather. The attempt to climb Mount Paget from that side was also unsuccessful.

The expedition then returned to King Edward Point aboard the sealer *Dias*. Shortly after their arrival they attended a reception for the Governor, Mr O.R. Arthur, who was making a tour of inspection aboard HMS *Veryan Bay*. Further trekking was undertaken in the vicinity and an unsuccessful attempt to climb Mount Sugartop was made. For the latter part of their time on South Georgia, the expedition was active at the south-east end of the island. Landings were made at Royal Bay and Cooper Bay from *Lille Carl* and other sealing vessels. A survey of the Ross Glacier and various other glaciological observations were made, together with counts of some bird colonies. The southernmost peak of the Allardyce Range (Mount Brooker, 1881 m) was climbed on 30 January for the first time. Trekking in the south-eastern area was adversely affected by poor weather for much of the time the expedition was in the vicinity, although some discoveries were made – notably of a large ice-dammed lake at the Twitcher Glacier. The last weeks were spent around King Edward Point, where further glaciological studies were conducted before the expedition departed aboard *Southern Opal* at the end of the whaling season. The leader later wrote an illustrated account of the expedition (Sutton, 1957).

The Duke of Edinburgh (1957)

HRH Prince Philip, the Duke of Edinburgh visited South Georgia on 12 January 1957. This formed part of a Royal tour of the minor colonies during which the Duke circumnavigated the globe. He arrived aboard the Royal Yacht *Britannia*, accompanied by the Governor and Sir Raymond Priestley, after visiting some of the British stations in the South Shetland Islands and Graham Land. RRS *John Biscoe* had carried him for part of that portion of the tour. Sir Raymond Priestley, who had served in the Antarctic on expeditions commanded by both Scott and Shackleton was then acting Director of the Falkland Islands Dependencies Survey. HMS *Protector* provided an escort for RRS *John Biscoe* and HMY *Britannia* during the Duke's time in the colony. The visit started at Leith Harbour where the Duke was received by the Magistrate, after which he inspected the station and continued to Husvik. There he boarded a whale catcher, *Southern Jester*, to travel to King Edward Cove. A demonstration of a harpoon cannon was provided during the short voyage. At King Edward Cove he was received through a great cloud of smoke from a 21 gun salute (fired by several whale catchers, from their harpoon cannons – using black powder). The Duke visited Shackleton's grave, Grytviken and King Edward Point, before departing aboard *Britannia*. The Governor remained on the island to depart later aboard *Protector*. During his voyage the Duke took many photographs of birds from *Britannia*, several of which were in the vicinity of South Georgia, and were later published.

The International Geophysical Year activities (1955–59)

The International Geophysical year of 1957–58 developed out of a suggestion made in informal scientific discussions in 1950. In 1882–83 and 1932–33 there had been International Polar Years and, initially, the proposal was to hold a third such cooperative programme from July 1957 to 31 December 1958. In subsequent discussions, the scope of the programme was enlarged to include the whole world but it was nevertheless intended that expeditions to the Arctic and Antarctic should form an important part of the undertaking. A Special Committee for the International Geophysical Year was established in October 1952; in December the Royal Society set up the British National Committee for the International Geophysical Year. Three parts of the programme concerned South Georgia: the visit of the Royal Society's expedition to Coats Land, the visit of the Commonwealth Trans-Antarctic Expedition to establish Shackleton Station, and some observations conducted on the island.

The Royal Society's Antarctic Expedition was to proceed to Coats Land to establish a scientific station, later named Halley Bay. This station has continued to operate; since the International Geophysical year it has been manned by the British Antarctic Survey. An advance party of the expedition departed from Southampton aboard *Tottan*, commanded by Captain Leif Jacobsen on 22 November 1955. She called at King Edward Cove on Christmas Day on her voyage to the Antarctic continent; this was reached on 11 January 1956. By 22 January the unloading of 220 tons of stores had been completed and *Tottan* returned to South Georgia on her way north.

The Commonwealth Trans-Antarctic Expedition, led by Dr (later Sir) Vivian Fuchs, the Director of the Falkland Islands Dependencies Survey, established an advance party on the Filchner Ice Shelf with the MV *Theron*, commanded by Captain Harald Marø. She departed from London on 14 November 1955 to arrive at King Edward Cove on 16 December. There Dr Fuchs met two colleagues, formerly from Stonington Base on the Antarctic Peninsula: the Magistrate R.E. Spivey and Manager of Grytviken, K.S. Pierce-Butler. Also on board was Sir Edmund Hillary, leader of the expedition's New Zealand team which established a base on Ross Island during the following summer. While at South Georgia, one of the expedition's two Auster aircraft was assembled, mounted on floats and test flown, so that it would be ready for ice reconnaissance, when the ship entered the Weddell Sea. *Theron* called at Leith Harbour for fuel and at Husvik for water, before continuing south. The passage was difficult because of ice conditions. She met HMS *Protector* at sea on 23 January 1956 and continued to Halley Bay where both expeditions were able to coordinate their activities. From there, *Theron* passed very close to the position where Shackleton's *Endurance* had been beset by ice in 1915. She continued south-west to the Filchner Ice Shelf where a station was established on 30 January 1956. This was near the place where, 40 years earlier, Shackleton had intended to begin the Imperial Trans-Antarctic Expedition. The new station was named after him and staffed by eight men. *Theron* departed on 7 February for Halley Bay and then returned to South Georgia in mid February and to London on 23 March 1956.

Both expeditions were supplied during the next season by the *Magga Dan*, commanded by Captain H.C. Petersen. She carried the equipment needed by the Trans-Antarctic Expedition in order to set out to cross the continent. On 15 November, *Magga Dan* sailed from London to reach King Edward Cove on 17 December 1956. Acquaintanceships were renewed and another test flight of an Auster aircraft was made, before she continued to Leith Harbour for fuel. They left South Georgia on 20 December 1956 and reached Halley Bay on 4 January 1957. *Tottan*, which left London on 17 November, had arrived 2 hours previously. An Otter aircraft, fitted with skis, was assembled; after some exploratory flights, it flew to Shackleton in advance of the ship. After unloading the expedition's supplies at Shackleton and sealing to obtain food for the sledge dogs, *Magga Dan* departed for Halley Bay on 29 January and then returned to South Georgia and further northwards. The Commonwealth Trans-Antarctic Expedition spent a second winter at Shackleton during which they also manned a small inland station, South Ice. Then, Dr Fuchs and 11 others set out to make the first overland crossing of the Antarctic continent. They reached the South Pole on 19 January and departed on 24 January to reach Scott Base at the Ross Sea on 2 March 1958. The expedition had travelled 3500 km across the continent in 99 days, making many scientific observations during the journey.

Studies associated with the International Geophysical Year were also made on South Georgia. A small laboratory was established at King Edward Point where seismic and gravitational observations were made. The meteorological observatory contributed further data. A tide meter was operated from July 1957 to December 1959, to obtain mean sea level data and to measure long waves of oceanic oscillation. A programme of glaciological studies was initiated in the 1957/58 summer by J. Smith. This principally concerned the small Hodges Glacier, although several others were also investigated. A geomorphological study of periglacial features was undertaken. The work was continued in 1958/59 by M. Stansbury. Present studies have made much use of the data accumulated and of markers placed during this period.

Another very important consequence of the International Geophysical Year was that the nations involved in the Antarctic programmes, in consideration of the benefits which resulted from their cooperation in the region, remained in close scientific contact after the conclusion of these programmes. Subsequent discussion formalised this relationship into the Antarctic Treaty which covers all lands below the 60° parallel of south latitude. It came into force in 1961, after ratification by all 12 nations concerned. A further 19 nations have since acceded to the treaty. Four of them are active in the Antarctic and have consultative status under it (May 1984).

The Bird Island Expeditions (1958–64)

During the 1958/59 season, a two-man South Georgia Biological Expedition was active on the island. It was led by W.L.N. Tickell who later returned on two of three similar expeditions. He and many of the other members of the expeditions had previously worked with the Falkland Islands Dependencies Survey on Signy Island.

The first of these expeditions arrived at Leith Harbour on 24 September 1958 aboard *Southern Opal* and, after a short stay at King Edward Point, went to Elsehul aboard the sealer *Dias*. At Elsehul a camp was established where they began studies of two species of albatross and a survey of other bird species breeding in the vicinity. Bird ringing was used in the course of this work. On 17 November the expedition, accompanied by Nigel Bonner, tried to reach Bird Island but weather conditions prevented their landing and they returned to King Edward Point. On 24 November, again with Bonner, they landed at Jordan Cove on Bird Island, set up camp and helped Bonner to deploy a field hut. At first they assisted with an investigation of the fur seals. This involved taking a census, and then weighing and tagging 1300 seal pups. After Bonner left the island on 20 December, the expedition endeavoured to ring all the yearling nestlings of two albatross species and the giant petrel; in all 6899 birds were ringed during that season. Another 400 seal pups were also tagged. The expedition also recorded meteorological observations and made a plane-table survey of the island. Two of the endemic South Georgia teal were collected live for transfer to the Wildfowl Trust at Slimbridge in the United Kingdom. On 6 March the expedition left Bird Island for King Edward Point from where, after a visit to Cooper Bay, they departed aboard RRS *Shackleton* to arrive in the United Kingdom on 16 May 1959.

The second expedition arrived early in 1961 and was sponsored by the United States Antarctic Research Program. Tickell and a companion arrived with Bonner at Bird Island on 22 January

aboard *Petrel*. For the rest of the month they assisted with fur seal tagging and then continued the 1958/59 ornithological programme. Altogether 10 196 birds were ringed by mid March, food and blood serum samples were taken, film and sound recordings were made, and censuses were conducted of many bird species. In order to obtain information about their movements at sea, 75 wandering albatrosses were dyed pink. Reports of pink albatrosses were subsequently received from ships near the South Shetland Islands, north of the Falkland Islands, near Bouvetøya as well as many around South Georgia.

The expedition was able to stay later than the first one and thus observed egg hatching of the wandering albatross. They were collected by HMS *Owen* in early April and taken to King Edward Point where accommodation was arranged in the gaol prior to departure aboard RRS *John Biscoe* on 10 April.

The third expedition in this series was led by

The Bird Island station in 1961. (W.N. Bonner.)

H. Dollman (Tickell's companion on the previous one), with an assistant. It too was funded by the United States Antarctic Research Program. They landed on Bird Island on 19 January 1962 and for the first week assisted Bonner and the new sealing inspector, Bill Vaughan, to tag 13 000 fur seal pups. During the next 8 weeks the previous operations were continued with routine observations of three albatross species, giant petrels, and the brown skua. In total 12 248 birds were ringed. Two skuas ringed in the South Orkney Islands and 24 wandering albatrosses ringed in Australia were found at Bird Island. The total breeding population of wandering albatrosses was counted (3237 pairs). On 24 March 1962, the expedition departed aboard RRS *Shackleton*.

The fourth and last expedition of the series was the most ambitious. It aimed at obtaining overwintering data about the biennial breeding cycle of the the wandering albatross. Other studies undertaken included meteorology and entomology. Tickell led a three man party which accompanied Vaughan, the Sealing Inspector, and landed on Bird Island on 5 December 1962. One of the first tasks was the erection of three buildings to accommodate the expedition over winter. A main building, Lönnberg House, 9 m by 4 m with four bunks, kitchen and laboratory facilities was the principal construction. A diesel dynamo set was installed to provide the station with electric power. On 12 February, HMS *Protector* arrived with the Governor, Sir Edwin Arrowsmith, who made a tour of inspection. When the ship sailed with Vaughan aboard, the expedition was left until August. The USNS *Eltanin* arrived on 24 August; it was the first visit by a United States ship to South Georgia since *Carnegie* in 1916. Several other ships visited during the summer season (including RRS *John Biscoe* and RRS *Shackleton*). The expedition's entomologist left in November and the station remained open to 2 April 1964, when the remaining two members left aboard RRS *Shackleton*.

Ornithological work accomplished included observation, filming and sound recording in designated albatross study sites. Shortly after the expedition's arrival, 1000 wandering albatross fledglings were ringed and their development was followed. Stomach contents were collected from 200 albatrosses in order to determine their diets. Studies

of the breeding cycles of different albatrosses allowed annual and biennial types to be determined. Field experiments performed during the second summer showed some species variation in chick and nest recognition by adult albatrosses. Large numbers of birds were ringed. From the South Georgia Biological Expedition in 1958–59 approximately 53 000 rings had been applied to seven species of birds.

Entomological collections were made from many habitats on Bird Island, including those of parasitic species. The entomologist, H.B. Clagg, left Bird Island aboard *Petrel* on 5 November 1963 to make collections in several other parts of South Georgia. Sixty-five terrestrial arthropod species were obtained from Bird Island and these, together with his other collections, were described in a comprehensive monograph largely concerned with South Georgia.

Climatological observations were recorded four times a day continuously from 1 January 1963 to 29 February 1964. In addition to the usual instrument and autographic records, grass minima and earth temperatures at several different sites and depths were recorded. An analysis of these data has shown that Bird Island has weather more representative of the region of South Georgia than those from the observatory at King Edward Point, although some local anomalies occur.

Some of the results of bird ringing and recovery experiments demonstrated the frequency of circumnavigation of the southern hemisphere by wandering albatrosses. The first long-distance recovery of a bird ringed in Australia was from South Georgia, at a distance of more than 11 000 km across the stormy oceans of the 'albatross latitudes'. A bird was ringed in 1958/59 which was recovered off the coast of New South Wales in July 1959. In 1961/62 it was recorded again at Bird Island and again off Australia in July 1962. There was no record of it for 1962/63 but it was in Australia in July 1963 and recorded at Bird Island in 1963/64. Since the expeditions to Bird Island many more wandering albatrosses have been reciprocally recorded between Australia and South Georgia, some on more than one occasion, and recoveries of other species are recorded elsewhere.

Several expeditions had previously landed on Bird Island, mostly for ornithological investigations. In 1956/57 Nigel Bonner landed there while searching for fur seals around South Georgia; he found 15 000 breeding animals. This discovery led to an investigation and regular visits to the island in subsequent summers. From this it has developed into a major centre for Antarctic seal and bird studies. From 1972, when the station was reopened, these have been made by the British Antarctic Survey and, for two isolated seasons and then continuously from 1978/79, in cooperation with the United States Antarctic Research Program (as in 1960–64). It is interesting to recall that the name 'Bird Isle' was applied by Captain Cook 'on account of the vast numbers that were upon it'.

HMS *Owen* and *Protector* (1960/61)

An officer of the hydrographic department of the Royal Navy was at South Georgia aboard RRS *Shackleton* in the 1959/60 summer to improve the survey of the northern approaches to the island. This was followed by an extensive programme of survey in the next summer. HMS *Owen*, commanded by Captain G.P.D. Hall with assistance from HMS *Protector*, commanded by Captain D.N. Forbes were involved. Priority was given to a thorough survey of the waters surrounding the western extremities of the island in order to facilitate the shortest passage between the whaling grounds and the whaling stations on the north-east coast. Other commitments were the landing of scientists, in cooperation with the administration, and, as opportunity offered, general improvement to the existing charts – with particular reference to the coastal shipping routes. HMS *Owen* had only 2 months available for this work. An advance party, fortuitously able to arrive almost 2 months earlier aboard HMS *Protector*, established sufficient ground control which greatly expedited the rest of the survey.

The advance party had landed and started work at Elsehul in mid December 60. With assistance from RRS *Shackleton* and the sealer *Petrel* they surveyed much of the north western extremity of South Georgia and offshore islands. HMS *Owen* arrived at Elsehul on 12 February 1960 where she continued the survey of that area. Duncan Carse, who had recently completed his South Georgia Survey series of expeditions was aboard for much of

the time to provide identification of land features and other advice. Landings were made at various places for botanical work by Dr S.W. Greene and others in the course of the survey.

Time available permitted operations to be extended around the island, although weather and ice made the work increasingly difficult. Much survey was accomplished from the ships' boats working in a pair in several areas. One of these, *Fantome*, was sunk in heavy seas in Bird Sound; fortunately her crew were rescued. After a final circumnavigation, during which Carse was visited at his hut near Ducloz Head, HMS *Owen* left South Georgia on 7 April 1961.

Many techniques had been used during the operations which included radar ranging and use of floating beacons which could be detected visually or by radar. Classical techniques and tellurometers were also employed. All the hydrographic charts of the island were extensively revised following the survey and an entirely new one was published.

Duncan Carse and his experiment on living alone (1961)

Duncan Carse made another visit to South Georgia in 1961. After many years interest in the problems of living alone in complete isolation he decided to perform an experiment at South Georgia – a place with which he was very familiar. He applied for a lease of land, approximately 4 ha, near Undine South Harbour on the south-western coast of South Georgia. This was granted by the Governor for a rent of one shilling a year and an advance payment for 10 years was accepted. On 23 February 1961 Carse was landed, with 12 tons of equipment from HMS *Owen*, at his selected site. He had a pre-fabricated hut which was established in a tussac area at the back of a sheltered cove. Early in April he received a visit from HMS *Owen* before she left the region. Before dawn on 20 May the hut was swept away by a surge wave while Carse was sleeping within it. He survived and was able to salvage the basic essentials to remain there for another 116 days before he was able to make contact with a ship. This was the sealer *Petrel* which brought him to King Edward Cove on 13 September. The site was later inspected, and some of Carse's possessions were salvaged by a party from HMS *Protector*.

British Services expeditions and the conquest of Mount Paget (1960 & 1964/65)

Three British Services expeditions have been made to South Georgia whose principal objects have been mountaineering. The first of these was a naval expedition in 1960. The others, in 1964/65 and 1981/82, were Combined Services Expeditions which included members of the three armed services.

The first, a party of 15 Royal Marines and a naval officer under the leadership of Captain V.N. Stevenson climbed the west peak of Mount Paget in December 1960. They were landed by helicopter from HMS *Protector* near Cape Darnley some 18 km south-west of Mount Paget on 10 December. After 3 days of difficult travel over glacier covered and highly crevassed terrain, two of them reached the summit of the lower peak of Mount Paget at 2915 m above sea level (19 m below the eastern summit). They descended in increasingly dangerous snow and ice conditions to be collected by helicopter on 15 December and re-embarked on HMS *Protector*. One of the men who reached the summit, Lt Cdr M.K. Burley, was to lead the next attempt.

The first Combined Services Expedition to visit South Georgia consisted of 10 men, led by Lt Cdr M.K. Burley. Their aims included climbing Mount Paget and Mount Sugartop, retracing Shackleton's trek and a survey of the Royal Bay region. Various scientific work was also to be undertaken. The expedition embarked on HMS *Protector* in Montevideo on 1 November 1964. She proceeded to the Falkland Islands, where the Governor boarded, and continued to King Edward Cove. Before she departed, HMS *Protector*'s helicopters were able to place several food and equipment depots for the expedition.

The expedition landed at King Haakon Bay from where it readily identified Shackleton's 'Peggotty Camp' site. While retracing Shackleton's route, a storm delayed them for 36 hours in the Murray Snowfield; three men were later caught in an avalanche but, with appropriate technique and equipment, they avoided disaster. The expedition reached Stromness whaling station, which was abandoned at that time. They continued to Leith Harbour, which was then leased to a Japanese company, to meet the Magistrate and station Man-

ager. The difficulties they experienced crossing from King Haakon Bay with excellent equipment, abundant food and a fresh start greatly increased their admiration of Shackleton, Crean and Worsley who made the same trek almost 50 years earlier under vastly different conditions.

The next part of the expedition's activities started with hauling sledges up the König Glacier to the Kohl-Larsen Plateau. Weather, which included 10 days of whiteout, severely hampered them for much of the time. From the plateau, a difficult trek took them to the depot above Larvik laid by HMS *Protector*. A base camp was established near it and two three-man parties prepared to climb Mount Paget and Mount Sugartop. The other four men remained at the base camp to be available in case of emergency. The Mount Paget party established a camp at 1060 m. After a couple of days of poor weather they moved up to a prominent plateau between the 'Far West' and 'West' summits on the peak, where another camp was made at 2651 m. At 02:55 on 30 December 1964, the final assault was started. By 03:40 the west summit was reached for the second time. From it, the lights of King Edward Point were seen far below. The mountain, 2934 m high, was conquered at 05:22 after the party crossed a 120 m deep saddle and trekked about 2 km. The Union Jack was unfurled and a 360° panorama photographed in almost perfect visibility.

The Mount Sugartop party had greater problems with the weather. Several reconnaissance treks were made while they awaited its improvement. One of which, on 4 January, examined a subsidiary ridge leading westwards from the peak. Conditions and inclinations led the party to continue the reconnaissance to the summit which was reached at 16:15. By coincidence, one of them happened to have a Union Jack with him. Climbs of two other peaks, Paulsen Peak and Mount Fagerli were attempted shortly afterwards. Despite getting close to the summits, these attempts had to be abandoned because of snow and ice conditions. A reconnaissance of a possible pass across the Allardyce Range between Mount Paget and Mount Sugartop was, however, made. This was then extended and a descent was found. Further very severe weather delayed the expedition near the pass for several days. Supplies were sledged over before another storm, which lasted 2 days, again halted operations. At the conclusion of it, the supplies were found to be buried under an avalanche; much digging was required to retrieve them. The expedition continued to King Edward Point where it arrived, after completing the first crossing of the Allardyce Range, on 29 January 1965.

A three-man party separated to conduct glaciological and geological observations in the vicinity of the Hamberg and Harker Glaciers in Moraine Fjord. Having taken less time in this than anticipated, they extended their work to investigate the Nordenskjöld and Lyell Glaciers. Igneous intrusions in these areas were also charted. Other geological observations involved measuring the slope of strata and collecting specimens.

Three stations were visited between King Edward Cove and Royal Bay, and one in the Royal Bay area. At all these lines of igneous intrusions were followed; angles and directions of slope were measured; collections, drawings and photographs were made; and mineral occurrences and geological irregularities were noted.

The expedition then moved, in two teams, to establish a camp in Moltke Harbour with the object of surveying Royal Bay. Fifteen trigonometric points were occupied during this work which, with six intersected positions, allowed interpretation of a series of photographs of the area. A chart at 1 : 50 000 was later produced with 30 m contours which covered about 50 km of coast. The German

The conquest of Mount Paget by the Combined Services Expedition on 30 December 1964. (Courtesy Commander M.K. Burley.)

International Polar Year Expedition of 1882–83 had produced the most detailed earlier map of this region; the accuracy of the old chart was well demonstrated by comparison with the new.

Botanical and ornithological collections and observations were also conducted during the expedition, which had the advantage of reaching some areas not previously visited and many where very few observations had been made. The expedition departed from South Georgia on 5 March 1965 aboard HMS *Protector* from Royal Bay. Several accounts of the expedition subsequently appeared and the Royal Engineers published the excellent map of Royal Bay from their survey.

Geodetic survey (1967–69)

A station for a British and United States geodetic survey was established at King Edward Point in 1967 when the United States army ship *FS 216* arrived on 31 December to land a party commanded by Major C.G. Nott-Bowers of the Royal Engineers. They were accommodated in the Customs House and continued satellite observations for more than a year before departing on the same ship. Their results, combined with those from many other stations operated during the survey, yielded an improved knowledge of the shape of the earth and its deviation from the theoretical geoid. Locally, they produced an exceedingly accurately

View from the summit of Mount Paget, looking north with King Edward Cove on the left. (Courtesy Commander M.K. Burley.)

fixed point where a concrete marker was established. The position is given in Chapter 1; several subsequent surveys have been coordinated with it. In 1978 the position was remeasured by visiting surveyors aboard RRS *Bransfield* using a satellite receiver.

Bill Tilman and the yacht *Mischief* (1967)

Bill Tilman reached South Georgia from the Antarctic Peninsula in January 1967, aboard the yacht *Mischief*, a very well built and maintained century-old Bristol pilot cutter. He remained to the end of the month before continuing to Buenos Aires. Several difficulties and disagreements aboard the yacht caused problems for the Magistrate. Tilman had previously navigated to several Antarctic islands, the Antarctic Continent, Patagonia, parts of the Arctic and much of the rest of the world while sailing approximately 240 000 km in small boats. He was also an accomplished mountaineer, having climbed in most of the Earth's great ranges.

This voyage was the first of a private yacht to visit South Georgia, and many more have arrived since. Unfortunately, in contrast to Tilman's, some of these have been far from properly equipped for such a voyage into Antarctic waters. Accounts of voyages to these areas have often encouraged others to venture to South Georgia (and elsewhere in Antarctic regions) inadequately prepared. In consequence, rescue has been required and supplies, together with other aid, are frequently solicited from many Antarctic research stations. These things can be provided only at inconvenience and expense to the stations with interruption to their scientific work. Unfortunately these problems have been progressively increasing over recent years.

The Royal Navy and HMS *Endurance* (1968–Current)

Many Royal Naval ships have been involved in hydrographic survey around South Georgia as well as in many other matters: including rendering assistance to the Falkland Islands Dependencies Survey and to its successor (the British Antarctic Survey). HMS *Protector*, which had hitherto the longest period of service in Antarctic waters, made her last call during the 1967/68 summer. Her first had been made 10 years previously, when she escorted the Royal Yacht *Britannia* with the Duke of Edinburgh aboard. From 1968/69 she was replaced by an ice patrol vessel named HMS *Endurance*

after Sir Ernest Shackleton's ship. She was extensively modified to be fit for her new work and accepted into the Royal Navy in June 1968. During most subsequent austral summers she has spent some time at South Georgia, often on Christmas Day when the church at Grytviken has been opened for the occasion. The British Antarctic Survey has received a large amount of assistance from her, especially from her helicopters (without which many field projects would have been exceedingly difficult if not impossible). A decision to withdraw her from early 1982, without replacement, was reversed following the events that year and her highly significant part in them; these are related in Chapter 9. Most of her work around South Georgia as well as in many other areas of the Antarctic has been hydrographic survey and she is responsible for a substantial proportion of modern charts of the region.

British Antarctic Survey (1969–Current)

From November 1969, a scientific station was operated at King Edward Point by the British Antarctic Survey which also assumed responsibility for the local administration. As well as this station, the Survey established several field stations including those at Bird Island, Sörling Valley, Elsehul, Dartmouth Point, and Schlieper Bay. During this period, somewhat fewer scientific expeditions arrived at the island, as research became effectively 'full time' and far more comprehensive. The British Antarctic Survey and its predecessor, the Falkland Islands Dependencies Survey, have been active on the island since December 1949; some details of their research and other activities are given in Chapter 5.

In April 1982 all of the Survey's personnel on South Georgia were removed from the island; either as prisoners aboard an Argentine naval ship or, later, aboard a Royal Naval ship taking them out of a war zone. The island's administration was subsequently attended to by the commanding officer of the British garrison established at King Edward Point. The scientific station there has been visited by the Survey and the Bird Island station was reopened in September 1982. At the time of writing, the future for the Survey on South Georgia is undecided.

Duncan Carse and Shackleton's trek (1973)

In 1973 Duncan Carse made his eighth expedition to South Georgia. He arrived on 17 December, landing from HMS *Endurance*, to try to identify and film a part of Shackleton's trek of 1916 which had not been definitely determined during his previous survey, or by the 1964–65 Combined Services Expedition. These were two passes across ridges near the mid section from the north-facing scarp of the Kohl-Larsen Plateau. Carse knew the region well and anticipated he could expect only three days in ten to be available for working because of the weather. Unfortunately, he experienced much more than the usual amount of terrible weather (including five days of severe gales, whiteout, and torrential rain which made a quagmire of the snow surface and which was followed by heavy snowfall and a blizzard). As he had to make a pre-arranged rendezvous with RRS *John Biscoe* and had experienced barely 3 hours of good conditions since his arrival on South Georgia, he regretfully abandoned attempts to identify the passes and proceeded to King Haakon Bay to meet the ship, leaving much valuable equipment at his last camp site.

Italian Antarctic Expedition (1974)

A lateen rigged private yacht, *San Giusseppe Dua*, owned and commanded by Captain Count Giovanni Ajmone Cat, visited South Georgia from 21 February to 19 March 1974. Her voyage was made in order to obtain information about conditions in Antarctica for the possible formation of an Italian Antarctic Institute. *San Giusseppe Dua* had sailed from Italy to the Antarctic Peninsula, the South Orkney Islands and South Georgia from where she returned to Rome. Journeys to various parts of South Georgia were made while she was there and some assistance was provided to the British Antarctic Survey. Her voyage was claimed to be the first Italian expedition to the Antarctic. It was followed by another private Italian expedition aboard *Rig Mate* in the summer 1975/76. Italy acceded to the Antarctic Treaty in 1981.

Economic survey of the Falkland Islands and Dependencies (1976)

Lord Shackleton, son of Sir Ernest Shackleton, arrived at South Georgia early in 1976 with another member of a committee conducting an economic survey of the Dependencies. He travelled aboard HMS *Endurance* (named after his father's ship), and landed first at Elsehul, where he was impressed by the abundant wildlife in the vicinity and was shown around by two British Antarctic Survey men stationed there. The voyage continued along the coast to King Edward Point, where he visited the scientific station, the whaling station, and his father's grave at Grytviken where a brief ceremony had been arranged by the Base Commander and Captain of *Endurance*. It was intended that he would fly over his father's trek between Stromness and King Haakon Bay, but unfortunately the weather deteriorated shortly after the helicopters took off and the journey was abandoned.

The report of the committee, of which Lord Shackleton was chairman, was principally concerned with the Falkland Islands, although some attention was given to South Georgia and to other parts of the Dependencies. Lord Shackleton was able to revise the report to produce an updated version in 1982 after the events of that year.

The yacht *Basile* and the ascent of Mount Paget (1980)

Of the many private yachts visiting South Georgia over the last few years the most significant was the French yacht *Basile*. Her complement consisted of four accomplished yachtsmen and four experienced mountaineers. They were properly equipped for the voyage to South Georgia as well as for climbing and trekking on the island. After obtaining approval and advice from the Magistrate, five of them made an ascent of Mount Paget from the Nordenskjöld Glacier reaching the summit on 1 February 1980. They also climbed some other peaks and trekked across several areas of the island before departing. Much cinematographic film and many still photographs were taken by the group who subsequently published an illustrated book.

Recent cinematographic visits (1979–82)

During the last three summer seasons, several television and other film parties have visited South Georgia. There has been a recent increase in interest in the island, its wildlife, the abandoned whaling stations, history and other aspects which has both followed and then given rise to this. Most of these visits have been substantially private or arranged by various television authorities. Many have been wholly or partly supplied with transport, accommodation, supplies and information by the

British Antarctic Survey. Film has also been taken recently for similar purposes by the Royal Navy and officers of the British Antarctic Survey. Two private film parties were on the island at the time of the Argentine invasion in April 1982; one obtained film of enemy activities at Leith Harbour and the other of events subsequent to the reconquest as well as much wildlife material. It is probable that the recent prominence of the island will lead to even greater interest and more such visits. It is interesting to recall that the first cinematographic pictures of the island were taken in 1914 by Sir Ernest Shackleton's expedition, the first commercial film was made in 1928 by Albert Benitz, and the first still photographs were taken in 1882 by the German expedition at Royal Bay.

The North Coast of South Georgia Expedition of Edwin Mickleburgh and David Matthews spent some time at Elsehul, Prince Olav Harbour, Stromness Bay, Grytviken, King Edward Point, and elsewhere during the 1979/80 season. They produced a comprehensive series of photographs of several of the whaling stations as well as much cinematographic film of the island's wildlife. Some of this was subsequently made into a film which also incorporated much historical film lent by Dr Harrison Matthews. A BBC television team led by David Attenborough filmed many areas of the island in 1980/81 to produce documentary films about Sir Ernest Shackleton and the wildlife. Miss L. Buxton and Miss A. Price of Anglia Television spent about 5 months at St Andrews Bay in 1981/82 where they filmed the king penguin colony, elephant seals and other animals. They remained with three of the Survey's men after the Argentine invasion of King Edward Point and were able to obtain some interesting film of the area after the reconquest.

Joint Services Expedition of 1981/82

The second combined services expedition made to South Georgia was in the summer of 1981/82. It consisted of 16 men, drawn from the 3 services and a civilian, and was led by Lieutenant R.E. Veal. They arrived aboard HMS *Endurance* on 12 December 1981. The Governor was aboard at the time and made a tour of inspection of several parts of the island. A base camp was established at Whale Valley near Royal Bay. Despite delays caused by some very severe weather, the expedition was able to pursue an extensive mountaineering and trekking programme principally in the south-west area of the island. Various scientific observations and collections were also made. As with the previous services expedition, they were also able to reach remote areas and had depots established by HMS *Endurance*'s helicopters to support them.

Most of the mountaineering was done in the rarely visited region to the west of the Salvesen Range which was reached by crossing the Ross Pass. Collections of high altitude plants were made in the region in the course of the exploration. Beaches near Trollhul were reached and several sealers' relics reported. Very severe weather at high altitude caused great difficulties, during which the expedition was confined to tents for 8 days on one instance. Nevertheless, on the rare really good days available, much ground was covered and many excellent photographs obtained. Another journey was made along the coast south-east from Royal Bay which reached the Twitcher Glacier; biological collections and observations were continued in the course of this. Mount Brooker (1881 m) was climbed. An attempt to conquer Nordenskjöld Peak, the second highest peak on the island, was unfortunately rendered impossible by the weather, as was an attempt on Mount Carse.

An investigation of birds, rats, reindeer and marine animals was made from the Base Camp in Royal Bay. Regular meteorological observations were also maintained there for the duration of the expedition. Treks were regularly made to near Calf Head, where regular weighings of albatross chicks were made to study their development. A glaciologist in the expedition was able to resurvey many sites between Cumberland Bay and the southern end of the island for which earlier data were available. He also established several new surveys and obtained numbers of ice cores. Some members of the expedition made visits to the abandoned whaling stations at Grytviken, Stromness Bay and Ocean Harbour to make photographic records.

The expedition left South Georgia aboard HMS *Endurance* on 16 March 1982. The experience they gained proved extremely valuable during the events which occurred later in that year.

5

Whaling, second sealing epoch and settlement

This chapter recounts the history of the activities of the people who have lived on South Georgia as distinct from visitors on expeditions. At times, however, the distinction becomes rather arbitrary. It covers the period from November 1904 when the first whaling station was established at King Edward Cove, to April 1982 when the settlement at King Edward Point was attacked, its population taken prisoner and removed by Argentine forces. This period includes the second exploitative epoch in the island's history. The possibility of a third such epoch, based on krill, is becoming increasingly likely and some aspects of this are discussed in Chapters 8 and 9.

Establishment of whaling on South Georgia (1904)

The history of South Georgia is in a large part the history of whaling in the Antarctic. The island was one of the most important places in the world for the whaling industry between 1904 and 1965. At its peak, in 1917/18 for instance, six shore stations were operating. In 1911/12, eight floating factories worked there; this included those using whole whales and others processing flensed carcasses from shore stations. This account of whaling on South Georgia is necessarily only a very restricted description of the industry which had such a great significance for the island.

The first remarks about the abundance of whales around South Georgia were made by Captain Cook and others aboard *Resolution* in 1775. The two visits to the island made by Captain Carl Anton Larsen in 1894 and 1902 were described in Chapter 4. Larsen had been involved in Norwegian whaling all his life and lived in Vestfold province, a part of Norway associated with it for much of the nineteenth century. He had been interested in extending the industry to Antarctica from late in the last century. On 5 December 1903, after the rescue of the Swedish South Polar Expedition of 1901–03 (led by O. Nordenskjöld) by the Argentine ship *Uruguay*, and their enthusiastic reception in Buenos Aires, Larsen made a speech at a banquet in the expedition's honour. A portion of this speech, rendered in his highly original English went: 'I tank youse vary mooch and dees is all vary nice and youse vary kind to mes, bot I ask youse ven I am here vy don't youse take dese vales at your doors, dems vary big vales and I seen dem in houndreds and tousends' (Bogen, 1954 & 1955). Three foreign residents of Buenos Aires (P. Christophersen the Norwegian Consul; H.H. Schlieper, a German-born United States national, and E. Tornquist, a Swede) expressed great interest in the matter. They formed a company which was registered in Buenos Aires on 29 February 1904; the Compañia Argentina de Pesca. Schlieper was the first president. Tornquist, a banker, became the principal share holder and Larsen was the Manager. Capital of 200 000 Argentine gold pesos was raised in Buenos

Aires. Larsen declined to buy the large share holding he was offered for himself as he could not afford it; or for persons in Norway as he declared that he had no wish to involve Norwegian capital 'because this attempt at whaling in the Antarctic regions should involve a risk for foreign capital' (Tønnessen & Johnsen, 1982).

Larsen returned promptly to Sandefjord via Stockholm, arriving in the spring of 1904. He ordered all the necessary equipment and procured three vessels: *Louise* and *Rolf* were sailing ships, the former a large transport vessel and the latter, a smaller one, to maintain a link with South America, and *Fortuna* a new steam whale catcher launched on 16 June 1904. Three pre-fabricated wooden houses

were also obtained for the management, employees, and factory. The expedition left Sandefjord on 20 September; *Louise* and *Fortuna* arrived in Buenos Aires on 28 October. From there they sailed directly to Grytviken which was reached on 16 November 1904 – a highly significant date in South Georgian history. *Rolf* was met there after sailing direct.

An enormous amount of work was required by the 60 Norwegians who arrived to prepare barges for landing the large amounts of heavy factory equipment, building sections, several hundred tonnes of coal, and many other items. They then constructed the three buildings, prepared a slipway, and built a factory with twelve blubber cookers. All this was accomplished in barely a month. One

Grytviken whaling station in 1914 from a photograph taken by Shackleton's expedition. (Courtesy Scott Polar Research Institute.)

building, the manager's residence and administrative office, had once been Christian Christensen's house near Sandefjord (it burnt down in 1916). The first whale, a humpback, was taken on 22 December 1904 and the factory became operational, producing its first oil, on 24 December. At first principally humpback whales were taken. These were mainly local inhabitants of the bays of South Georgia and kept close to land. Thus they were the first species to become very greatly reduced in population. The transport vessel *Rolf* left for Buenos Aires on 18 February 1905 with the first cargo (165 tons of whale oil). Whale stocks were so abundant, even within Cumberland Bay, that production was restricted only by the capacity of the factory and the number of barrels that could be made.

The problems concerning the authority for

Larsen's establishment of the whaling station in the Falkland Islands Dependencies are discussed in Chapters 4 and 9. This was regularised by the president of the company and an officer of the Argentine navy, who was associated with the company, making an application through the British Legation in Buenos Aires on 2 November 1905, for a whaling lease from the Governor of the Falkland Islands and Dependencies. The lease was granted, subject to various conditions, from 1 January 1906. When news of the successful establishment of the whaling station and the abundance of whales in the Dependencies was received in Norway and the United Kingdom, several other companies associated with whaling turned their attention to that part of the world. This was promoted by declining stocks and increasing

The shore depot at Godthul in 1981. Note the oil barrels and *Jolle*. (R. Edwins.)

restrictive legislation affecting the northern whaling grounds. Licences were granted for floating factories to operate in the South Shetland Islands and Falkland Islands in 1906, when many inquiries about South Georgian leases and whaling licences were received in Stanley. The second whaling expedition to South Georgia was aboard the *Fridtjof Nansen*. It came to a tragic end when she was wrecked at South Georgia on 10 November 1906, with the loss of nine lives.

The next arrivals at South Georgia were two companies from Norway; the Sandefjord and Tønsberg whaling companies, which established stations at Stromness and Husvik Harbours in Stromness Bay, on 20 and 24 December 1907. These were at first sites for floating factories only, when *Fridtjof Nansen II* accompanied by the catchers *Hercules* and *Samson*, and *Bucentaur* with *Carl* and *Mathilda* were deployed there. On the northern side of Stromness Harbour, the ruins of a dam and foundations of a small building (with the date 1908 carved on it) remain where the former floating factory anchored. Land stations were established at Husvik in 1910 and Stromness in 1912 although floating factories remained briefly afterwards.

The company Bryde and Dahl of Sandefjord, Norway, owned by Thor Dahl, purchased the South Georgia Exploration Company's lease for South Georgia. It applied for a whaling licence and was granted permission to deploy a floating factory. The *Aviemore* with the catchers *Edda* and *Snorre* was stationed at Godthul from 8 January 1908 and others later replaced her. No shore station was established at the site although a shed, some small tanks, a dam and ancillary works, and a depot for barrels were built there.

The first lease which required the utilisation of the whole carcass of whales was granted for Ocean Harbour (then named New Fortune Bay) to the Ocean Whaling Company of Larvik, Norway. This station was managed by Lauritz E. Larsen, a brother of C.A. Larsen, and began on 26 October 1909 when the ship *Ocean*, accompanied by two whale catchers *Penguin* and *Pelican*, arrived. The station was regarded as an ideal one for the most efficient utilisation of whales.

A British company, Christian Salvesen Limited, founded the third shore-based whaling station on South Georgia at Leith Harbour in November 1909 when *Starlight* accompanied by the catchers *Swona* and *Semla* arrived from the Falkland Islands with H. Henriksen as Manager and 57 men. The company had previously established a whaling station at New Island in the Falkland Islands and had others in the northern hemisphere. Hendriksen had visited South Georgia in the 1908/09 season to select a site for the whaling station. The equipment taken to Leith Harbour came mainly from an Icelandic whaling station the company owned and parts of the New Island station were incorporated after this closed in 1915. Christian Salvesen's companies had been established in Scotland since 1872 and had strong Norwegian connections. They had been involved with a whaling exploration expedition to the Antarctic in 1892/93 with *Balaena, Active, Diana*, and *Polar Star*. The lease granted to Christian Salvesen's company for Leith Harbour (and later Allardyce Harbour) required that the station utilise the whole carcass of the whales caught. Their first consignment of oil was despatched to Scotland aboard *Coronda*; where it arrived in March 1910.

From 1 January and 1 October 1909 leases were let for Jason Harbour and Rosita Harbour (then known as Allardyce Harbour) to the companies holding the Grytviken and Leith Harbour leases. They did not establish whaling stations at these sites although a strongly built hut was constructed at Jason Harbour. The two whale catchers each lease permitted were deployed at the companies' original stations.

The last whaling lease to be granted on South Georgia was for Prince Olav Harbour, from 1 July 1911, to the Southern Whaling and Sealing Company owned by Irwin and Johnson who had offices in Cape Town, South Africa, and South Shields, United Kingdom. The floating factory *Restitution* arrived with the catchers *TWI* and *COJ* (the initials of Messrs Irvin and Johnson) and was stationed there each summer until, in 1916, she foundered near the Scilly Islands on her voyage south. *Restitution* was the first floating factory to be equipped with wireless telegraphy apparatus on South Georgia. The whale catchers which first accompanied her were equipped with diesel engines and proved greatly inferior to steam powered ones. (Diesel engines were not successfully used in whale catchers until after the Second World War.)

This marked the end of what has been described as the 'gold rush' period of South Georgian whaling history. The files of the Governor's office in Stanley and of the Colonial Office in London concerning the Dependencies are remarkably thick for this period and an enormous amount of correspondence concerns applications for whaling licences and leases of land for whaling stations. The Governor adopted a policy which was supported by the British Government and largely based on the Magistrate's reports and recommendations. This was to avoid excessive exploitation of whales in the Dependencies by restricting the numbers of shore-based whaling stations and floating factories licenced, together with limiting the number of whale catchers these were permitted to use and protecting female whales accompanied by their calves. The

industry was further regulated by the introduction of requirements for full utilisation of whale carcasses to ensure the whales caught were treated in the most efficient, least wasteful manner. The Governor, Mr (later Sir) W.L. Allardyce, was a strong proponent of other conservation policies in the Dependencies. The whaling industry of the Dependencies remained substantially stable until floating factories became pelagic and worked on the high seas outside Government control. A list of sites leased to whaling companies on South Georgia is given in Appendix 1.

The whaling industry (1904–65)

The whaling industry was well under way at South Georgia less than ten years after it started, with shore stations and floating factories operating

Leith Harbour whaling station about 1912 from *The Whale Fisheries of the Falkland Islands and Dependencies* by T.E. Salvesen (1914). Note the large number of oil barrels.

in seven harbours of the island. The first type of floating factory used at South Georgia and other parts of the Dependencies (the only part of the Antarctic with a whaling industry at that time) was a converted ship moored in a suitable sheltered harbour where a supply of fresh water could be obtained. Ships suitable for conversion were cheap in the early years of this century and, at first, more whale processing was done from them than at the land stations of the island. Whales were towed to these factories and flensed while afloat alongside. The flensing was generally circular and the blubber was removed by a process not unlike peeling an apple, as opposed to longitudinal flensing used on shore stations (which might be compared to peeling a banana). A barge or 'jolle' was moored on the side of the whale opposite the factory and the flensers operated from it. The strip of flensed blubber was hoisted aboard by the factories' derricks, to be deposited in the cookers. Originally the remainder of the whale's carcass, the skrott, was cast adrift to rot in the harbours. This accounts for some of the very large quantity of whale bones deposited around several bays of South Georgia. Some floating factories were able to process skrotts, which were purchased from shore stations and which produced a profitable quantity of oil. Subsequently, as regulations requiring the utilisation of the whole carcass of whales were introduced, many of these old floating factories ceased operations. At South Georgia only two were equipped to process whales completely which, owing to the extra machinery required, was more efficiently done from shore stations.

At various times, 13 floating factories of the old type have worked at South Georgia and several have already been mentioned. Amongst the others, two worked from Grytviken and processed skrotts from the land station: *Nor* from 1909/10 to 1912/13, and *Ems* from 1913/14 to 1915/16. Stromness started operations with *Fridtjof Nansen II* in 1907/08 to 1912/13 and *Fulwood* assisted from 1909/10 to 1912/13. Husvik commenced with *Bucentaur*, 1907/08 to 1912/13 and was assisted by *Ems* 1909/10 to 1912/13 before she moved to Grytviken. Godthul was only a floating factory site with *Aviemore* 1908/09 to 1910/11, *Admiralen* 1909/10, *Thor I* 1910/11 to 1916/17 and 1922/23 to 1928/29, with *Vik* in 1911/12 and *Whale* in 1927/28.

Thor I spent the period 1917/18 to 1921/22 in the South Shetland Islands, and many other floating factories also spent seasons there. Leith Harbour used *Horatio* from 1912/13 to 1915/16 and *Neko* 1916/17 to 1917/18. *Horatio* caught fire at Leith Harbour on 11 March 1916 with a full cargo of 11 000 barrels of oil aboard and was towed out and sunk off the harbour. She became a total loss, taking a whole season's production with her. Prince Olav Harbour had *Restitution* from 1911/12 to 1915/16 but she foundered on her way south in the next season.

Both shore stations and floating factories originally transported their oil in oaken barrels and most early photographs show large numbers of these around them. They had many disadvantages including handling difficulties, leaks, and inefficient use of space, although the second problem was alleviated when metal drums replaced the barrels. Christian Salvesen's company registered a patent on 16 September 1910 for whale oil tanks but their introduction was comparatively slow and still incomplete in 1920. Not only did the whaling factories and transport vessels have to be appropriately modified, but the receiving ports and transport from them had to be adapted. The barrel remained, however, a unit of measurement and tank farms became a prominent feature of whaling stations. Exceptionally sperm whale oil was transported in barrels or drums for all the whaling period as it was produced in much smaller quantities than baleen whale oil and had to be kept separately.

One of the most important vessels in the whaling industry is the whale catcher. These were small fast boats used to pursue and harpoon whales, and then to tow them to the factory. Most were steam-driven with triple expansion engines until after the Second World War, when diesel engines were first successfully introduced. Originally they were coal-fired but oil firing was becoming common around 1930. Those used on South Georgia varied from about 30 m long and 160 tons registered to 50 m and over 500 tons and were exceedingly manoeuvrable. Their equipment included a strong steel mast with a crow's nest atop from where a look-out was maintained. An experienced observer could distinguish the species of a whale from its blow and announced its detection with a cry of: 'hvalblast' and an indication of the direction. The bridge was open and connected to the harpoon cannon's platform by a catwalk. Usually the Master was also the gunner, he steered the catcher close to its quarry before passing control to the mate and running to the cannon. The harpoon was fired into the whale at very short range, and usually killed the animal quickly and secured it. The whale was brought alongside to be inflated with compressed air to keep it afloat and either taken in tow or marked to be recovered later. At first, markers consisted of flags, then lights were added, then radar reflectors, and finally small radio transmitters. Whale buoys were positioned in several harbours around the island and were used to secure whales which were left for subsequent towing to the stations, by either whale catchers or tow boats. They also served as moorings for catchers (at night or in poor weather).

The harpoon and cannon were the essential apparatus for whaling. An explosive harpoon and suitable cannon to propel it were successfully developed and patented by Svend Foyn of Vestfold province, Norway, in 1872 and this is largely responsible for Norwegian ascendency in whaling. The Kongsberg Company of Norway and Bofors of Sweden made the majority of those used in South Georgia (and indeed elsewhere). Although many improvements were made, their basic design remained substantially unchanged. The harpoon consisted of a grenade, about 11 kg of cast iron, in which about 500 g of black powder was placed. This was screwed to a base, equipped with four pivoted flukes and a time fuse; which was in turn connected to a shaft of very high quality steel, with a longitudinal slit. The whole was about 2 m long with a line attached to the shaft by a ring splice which was running free in the slit. The harpoon was

Whale catchers moored at Stromness in April 1952 at the end of the whaling season. (Courtesy British Antarctic Survey.)

mounted in the cannon with a wad to separate it from the charge. The flukes were secured by light twine which was also connected and pulled the time fuse striker. The older cannons were muzzle-loaded but breech-loading was introduced quite early, as was a glycerin recoil system. On firing, the ring splice moved to the end of the shaft and the time fuse delayed about 4 seconds before detonating the grenade which should, by that time, have been in a vital part of the whale (where the opened flukes secured it). Normally this produced rapid death; if not, it was often possible to launch further explosive harpoons, without lines, into the animal. In any event, the whale was secured by the harpoon flukes to line which passed from the bows of the catcher,

around snatch blocks connected via the mast to a series of powerful springs (the accumulator), to a steam winch. The accumulator protected the line from sudden shock and the whale was thus hauled securely to the catcher.

Although the majority of whales were killed quickly, a small proportion were not and much cruelty was believed to be involved in these instances. Various attempts to reduce this and improve the efficiency of securing the whales were made with gas, curare, and electric harpoons. The electrical ones were the most hopeful development and first used in South Georgia in August 1931. The experiments were, unfortunately, unsuccessful.

One particular disadvantage with the explos-

A harpoon cannon mounted ashore at Grytviken. The 'old bunkhouse', in the background, was the first residence of the Magistrate. (Author.)

ive harpoon was the mixture of exploded guts with bacteria etc. which was introduced into the vascular system of the whale and distributed throughout its · body. This rendered post-mortem decomposition far more rapid; whales, once harpooned, quickly began to putrefy. Their blubber provided excellent insulation from the frigid Antarctic Ocean in death as in life, and a whale's insides could virtually cook themselves. The opening of a severely 'burnt whale' was an operation to be done very carefully by a flenser. Regulations were introduced to prevent the accumulation of whales awaiting processing; a matter which was, in any event, in the interest of the companies. Later attempts to introduce antibiotics into whales shortly after their death to delay putrefaction, were moderately successful. These at-

tempts were made when it was hoped to begin to use whale meat for human consumption.

Shore-based whaling stations, although initially far more costly, had the enormous advantage over floating factories in having much more space available in which to deploy their apparatus, accommodation, stores, etc. The comfort and safety of the employees were therefore superior and whales were processed with greater efficiency. All the shore stations were based on a very similar arrangement. The main feature of each station was the 'plan', a large boarded slipway, onto which the whales were hauled by a powerful winch, to be dissected. Around the plan were factories for processing blubber, meat, and bone, each of which required slightly different techniques. The blubber was re-

A Sperm whale on the flensing plan at Grytviken about 1916. Photographed by the Magistrate Mr E.B. Binnie. The child is Solveig Gunbjörg Jacobsen. (Courtesy Scott Polar Research Institute.)

moved first by flensers who made longitudinal incisions and separated it from the skrott. It was conveyed to a mincing machine (a hogger) using cables pulled by winches, from where it was passed into blubber cookers. The meat and guts were separated from the bones and conveyed to the lemmers who dealt with these. Cookers for meat and guts were somewhat specialised to prevent clots forming, to allow thorough cooking, and to ensure full separation of oil. Bone cookers were simple but, owing to the time taken and amount of residue left, larger numbers of them were required than for other cookeries. After cooking (which was at atmospheric pressure in the early days but later done in pressure vessels at about 4 atmospheres), the extracted oil was run off, residual liquor (glue-water) separated and solid matter removed. Although most oil was obtained from the blubber, meat and bone contributed substantial amounts

and that from the latter was generally of high quality.

Once hauled on to the plan, a fin or sperm whale was flensed in about 45 minutes and a sei whale in about 30 minutes. The numbers of whales processed in a day depended on many factors; an average of about 12 was a good rate but twice this was possible. Blubber, meat and bone cooking operations took different lengths of time and the apparatus was worked in relays to keep up with the flensing. The oil of each whale was individually accounted for and various statistics were recorded about it. A bonus, based on production, was paid to the employees in addition to their wages. A table of numbers and species of whales and seals taken at South Georgia during the period of the whaling industry is given in Appendix 2. A total of 175 250 whales were taken at South Georgia from 22 December 1904 to 15 December 1965, the duration

A male fin whale on the flensing plan at Grytviken about 1955. (Courtesy Trans-Antarctic Expedition.)

of the island's whaling industry. This may be compared with 1 432 862 recorded from Antarctica generally between 1904 and 1978 the duration of the region's major whaling epoch. The vast majority of the latter were taken by whale catchers working from factory ships. The fortunes of South Georgia whaling are well represented by the graph of total annual catches on page 123.

The whaling stations were powered mainly by

Pressure cookers in the blubber cookery at Grytviken. (I.B. Collinge.)

steam. This was produced in large boiler houses which were coal-fired at first but progressively converted to oil-firing from around 1930. Steam was used for the pressure cookers, steam saws, winches, pumps, donkey engines in machine shops, heating oil in pipes and tanks, as well as providing heat for accommodation and cooking. Electricity was a comparatively minor power source until much later, mainly after the Second World War. It was developed from up to five hydroelectric dynamos at Grytviken and one at Leith, reciprocating steam dynamos at Stromness (from the Bergen tramways system, made in 1888 and still in excellent condition) and diesel-driven dynamos generally. Electricity was at first required only for light; it was later used for wireless and for many other purposes.

When they were functioning, the whaling stations were remarkably independent establishments equipped to cater for all their varied requirements for many months without supplies from elsewhere. They had workshops for blacksmiths, pattern makers, carpenters, electricians, and radio operators as well as foundries and ship repair yards together with comprehensive stores of necessary materials. The vast majority of items required which were not held in stock, and much of the factory machinery, were made locally. Similarly messing, accommodation, laundry and other requirements for several hundred men were arranged and large stocks of food were held. A 'slop chest' acted as a general shop stocking clothing, tobacco, photographic materials, toiletries, stationery, and a large number of other items. Animal houses were run to provide fresh meat and eggs. Medical facilities included well-equipped hospitals – with large orthopaedic sections as skeletal injuries were the most common complaint. As well as all this, there were the facilities directly connected with the whaling factory: the boiler house; power house; administrative offices; laboratory; cookeries for blubber, bone and meat; separators; tanks; pump houses; and guano factory. A great range of employment was provided at the stations. As well as work directly connected with whaling operations, the associated tradesmen and others included: animal attendants, bakers, blacksmiths, boiler makers, butchers, carpenters, chemists, clerks, cooks, electricians, foundry men, laundry men, medical officers, stewards, stokers, storekeepers, and wireless operators.

The employees of the whaling stations were until the last three seasons mainly Norwegian; almost 80% of them came from Vestfold province. In later years an increasing proportion was recruited from the United Kingdom, South America, the Falkland Islands, and many other places but Norwegians retained their dominance until the final years, when the Japanese arrived. One Manager wrote of some employees arriving in the early days 'there are supposed to be seventeen nations, and that's counting all the negros as one'. After the World Wars, many displaced men joined the whaling and sealing industry and the technical vocabulary developed (especially by the sealers) is reported to have had very diverse origins. The population often exceeded 1000 during a whaling season, but dwindled to fewer than 200 in the winter (in the years when most stations were operating). Man-

agers and some other officers of the whaling stations were sometimes accompanied by their wives and even children. C.A. Larsen had most of his family there (including his wife, five daughters, and two sons) when HMS *Sappho* called in 1906.

For the seasons up to 1918/19, whaling at some stations continued throughout the year; but, owing to their habits, the ice and the weather, few whales were taken during winter. After 1918/19, whaling took place only during summer and a legally enforced season was adopted. The whalers worked long hours at South Georgia but they earned enough money during the Antarctic summer to last through the Norwegian one, before they returned. Many became wealthy from whaling and Vestfold was a prosperous province. Some men over-wintered and spent long periods continuously on South Georgia. Many others returned regularly, in some cases for almost 50 years. Facilities were improved as time passed and the church, libraries,

Whale blubber ready for the factory at Leith Harbour. (Courtesy Trans-Antarctic Expedition.)

clubs for films, recreation and welfare, as well as better conditions were introduced. All the whaling stations had football pitches and Grytviken was equipped with a ski jump. Nevertheless, for most whalers, South Georgia was principally a place for hard work, the profits of which were to be enjoyed elsewhere.

The processing of the whales yielded oil, gluewater, grax, meat residue, and bone residue. The oil was usually filtered although later purification was done by centrifugal separators. Gluewater also yielded some oil when centrifuged. The meat and bone residues were dried in rotary kilns, sometimes with the addition of gluewater and grax to increase their protein content. These formed meat and bone meal which was bagged and known collectively as guano. Although oil was the primary product of the industry a substantial profit was also made from guano.

In the early days, the baleen plates ('whalebone') of many species of whales were also a substantial article of commerce, but their value later declined because plastics replaced them in many applications. Teeth of sperm whales were taken, although many were distributed locally. Ambergris, also from sperm whales, was occasionally encountered; it was a rare but valuable product. Skin and endocrine glands, used for medical purposes, were also separated but only towards the end of the whaling epoch. Sperm whale oil has properties distinct from those of baleen whale oil and was kept and sold separately. During the last years of Husvik, Leith Harbour, and Grytviken, production of whale meat extracts and frozen meat was tried and found to be economically worthwhile.

The main products of the whaling industry were oil and guano. Baleen whale oil was principally used in the manufacture of margarine and cooking

The bone loft at Leith Harbour showing the cutting of a spinal column with a steam saw. (Courtesy Trans-Antarctic Expedition.)

fats. Glycerin was also a major product from it for the First World War. Sperm whale oil was used for cosmetics, candles, and as a lubricant. For the last purpose, it is superior to mineral oils in some specialised applications because it remains as a surface film on metals whereas mineral oils tend to drain away. Both types of oil also formed the basis of soaps. pharmaceutical, chemical and other products. Meat and bone meal were sold as stock food and as fertilisers. The increase in protein content which resulted from incorporation of gluewater and grax during drying, enhanced their value for both purposes.

From the beginning of whaling on South Georgia, improvements in techniques for catching and processing whales and a strong demand for whale oil yielded a period of rapid expansion during which very substantial profits were made. The First World War increased demand even further as whale oil was used to manufacture explosive as well as for edible oils and other products. Oil prices increased by as much as four-fold during this period. Many regulations, including those for full-carcass utilisation, were relaxed for its duration and much of the whale bones around the sites of factories, previously

working in accordance with this regulation, date from then. After the War demand persisted and was promoted by improvements in margarine manufacturing processes. The price of whale oil was about half that prevailing during the war.

In the 1925/26 season *Lancing* was successfully operated in Antarctic waters. She was a factory ship rather than a floating factory and equipped with a slipway which enabled her to take whales aboard for processing. She could operate almost anywhere in the high seas rather than being confined to sheltered harbours and was generally in areas outside the regulation of governments. By 1930/31 there were 41 factory ships operating in Antarctic waters with 232 catchers attending them. Their efficiency improved rapidly and many new techniques were introduced, including the deployment of aircraft for locating whales. The ability of a factory ship to work in areas where whales abounded gave them an insurmountable advantage over shore-based stations and their development was one of the major factors in the enormous reduction of the stocks of whales and consequent decline of South Georgia's whaling stations.

The immediate consequences for the whaling

Annual catch of all whales species at South Georgia during the whaling era (1904–65.)

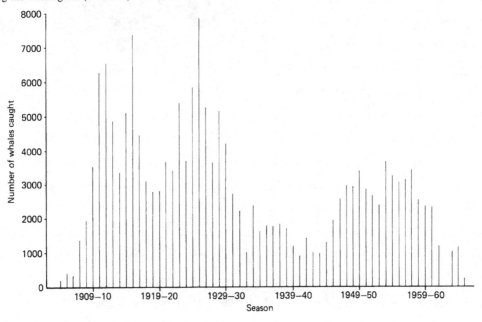

industry of the great increase in factory ship numbers was overproduction; the price of whale oil fell to less than one-third of what it had been when *Lancing* first operated. This resulted in the closure of all whaling stations on South Georgia, but for Grytviken, from the 1932/33 season. Leith Harbour alone reopened the next season and the two continued operating until the Second World War. During this war, 30 factory ships were lost, including the entire British and Japanese fleets, while the demand for whale oil increased greatly (mainly for edible fat production). Leith Harbour closed for 1940/41 as the whale catchers were requisitioned for war service and sent to South Africa, then operated for about 2 months in late 1941. Grytviken alone continued throughout the war (see Chapter 4). Oil prices remained high when peace resumed and therefore both Leith Harbour and Husvik reopened. The scarcity of whales was becoming severe; catchers which originally barely needed to leave the bays and later were able to secure most of their whales from near the island, had to steam to areas up to 300 km away for whales. The economics of this became rapidly less rewarding. Husvik last operated in 1959/60; Leith Harbour last operated under Christian Salvesen's management in the 1960/61 season. Grytviken was last operated by Albion Star in 1961/62 and closed for

Grytviken whaling station from Mount Duse, December 1980. (Courtesy British Antarctic Survey, C.J. Gilbert.)

the next season after 58 continuous ones. Important reasons for the greater success of Grytviken and Leith Harbour were the elephant seal industry (a more constant resource than whales) at Grytviken, and the use of Leith Harbour as a forward base for factory ships.

Japanese companies had first become active in pelagic whaling in 1934, when one purchased a factory ship from a Norwegian company. Their proportion of the industry then increased very rapidly. Japanese Antarctic whaling operations were entirely pelagic until the last seasons of whaling on South Georgia when they took sub-leases of Grytviken and Leith Harbour whaling

stations. They had a much greater interest in obtaining whale meat than did the Norwegian or British companies. Grytviken was operated by a consortium of three Japanese companies for half a season in 1963/64, from 7 October to 2 December 1963, and again in 1964/65 (from 1 October to 4 December 1964). That was the end of Grytviken's whaling which had started on 16 November 1904. Leith Harbour was sub-leased to one Japanese company and worked for full seasons in 1963/64 and 1964/65, with a half season in 1965/66 (from 1 October to 15 December 1965, the beginning of the pelagic whaling season), when it too finished whaling. The company paid for a sub-lease for the

Plan of Grytviken whaling station at the time of its closure. (Author.)

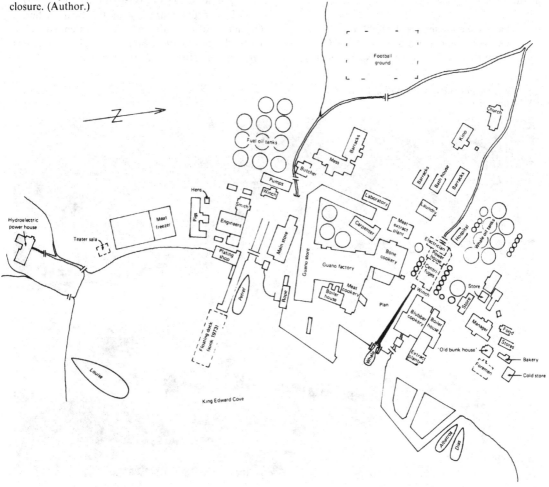

1966/67 season but did not use the station. Caretakers remained at Grytviken to 15 April 1971 and at Leith Harbour to 15 January 1966, after which the stations were abandoned.

At Grytviken the British Antarctic Survey has maintained a dam and watering facilities, one of the piers, the football pitch, and the church. The rest of the station, as well as all of the others, are rapidly succumbing to the depredations of age, climate, and an enormous amount of vandalism from the crews of some visiting ships. Grytviken has also sustained a serious fire. A large quantity of archival material has recently been removed from the stations with the permission of Albion Star and Christian Salvesen, and donated to the Scott Polar Research Institute, Cambridge.

The second sealing epoch (1909–64)

The second epoch of sealing on South Georgia lasted from 1909 to 1964. Elephant seals were taken throughout this period for extraction of oil together with 755 leopard seals (taken prior to 1927) and 97 Weddell seals (taken prior to 1916) and one fur seal (taken accidentally in 1915). The industry was controlled to ensure it operated at or below the maximum sustainable yield and that no danger was presented to the survival of the species. It came to an end when whaling ceased on the island as the amount of seal blubber obtained was insufficient to justify the operation of the whaling station to process it alone. The Government later advertised a sealing lease with the additional requirement that all the seal's carcass must be used. Although the *Run* Sealing Company took five seals for experimental purposes in 1969, nothing resulted. The sealing lease was last advertised on 25 April 1972.

Under the authority of the Seal Fisheries Ordinance of the Falkland Islands extended to the Dependencies in 1909, South Georgia sealing was organised into three and later four geographical divi-

Grytviken whaling station operating in 1961.
(W.N. Bonner.)

sions and reserves were also established. The divisions were generally worked in rotation; each year the seals in one division were left undisturbed. It was prohibited to kill fur seals and, from 1916, this protection was extended to Weddell seals. The first licence was applied for by and granted to the Compañia Argentine de Pesca, the pioneer whaling company on South Georgia. Others were issued: to Captain Cleveland, who did not comply with most of its requirements; to the South Georgia Company in about 1958, for a small number taken at Hound Bay; and to a Japanese company at Leith Harbour which took 59 seals for experimental processing in 1965. Some provisions of the sealing regulations included a closed season as well as limits on numbers, sex, and age of seals which might be killed. The closed season was for the seals' breeding

period and its times were occasionally modified as local conditions required. Although there was only one closed season, sealing operations were undertaken during two periods of the year (during the rest of the year the seals were pelagic). The numbers of seals that were permitted to be taken was limited to 6000 for many seasons, but this was raised for several years from 1948. The provisions also required the sealers to provide the Magistrate with details of the numbers of seals taken.

Generally, the vessels used for sealing were whale catchers which had been superseded by faster, more modern ones. The last three (*Albatros*, built in 1921, *Petrel* in 1928 and *Dias* in 1906) have sunk at their moorings at Grytviken. Many of the more famous small vessels used at South Georgia (*Undine, Lille Carl, Granat, Don Samuel* and various

An elephant seal breeding beach at Doris Bay, South Georgia, 18 October 1959. (W.N. Bonner.)

others) had been sealers for parts of their careers. The crews of these vessels were about 16, half of whom were sealers. Their national mixture was probably greater than anywhere else on the island and thus very diverse. The sealing voyages lasted for perhaps a week, after which they returned with seal blubber for processing. Three sealing vessels were usually operating during the season and their Masters were well accustomed to entering and landing in nearly all bays of the island. They had by far the best local knowledge and managed without much reliance on the charts, about which they made many critical comments. Only two vessels involved in this epoch of sealing were lost; *Granat* in 1925 and *Don Samuel* in 1951. Radio was introduced to sealing vessels after the second loss.

Techniques of sealing varied to some extent but followed the same general lines. Work started early in the morning when the vessel sailed to and anchored in a suitable bay. The sealers went ashore in a specially built rowing boat referred to, by its

Norwegian name, as a pram. The design of a pram permitted it to land on hard stony beaches through surf when necessary. A motor boat assisted by towing the pram where practical and carried tow ropes connected with the sealing vessel. The pram, which was either hauled up on the beach or waited beyond the breakers, carried to the beach hooks, rope strops, a long line and other requirements, as well as the sealers. Each sealer was equipped with a very sharp knife with a sharpening steel and various hooks which were used for handling the blubber. The gunner carried a rifle and soft-nosed bullets. A beater, bearing a metal condenser tube or similar rod, drove selected bull seals to the water's edge. There they were killed by a shot through the head. A knife or special lance was thrust into the heart to ensure death and to relieve the pressure of the blood by causing internal bleeding. Flensers started separating off the blubber with a series of incisions around the flippers and head, followed by a longitudinal one. After this the blubber, grasped by

Sealing at South Georgia, 1957. Driving a bull elephant seal to the water's edge. (W.N. Bonner.)

hooks, was removed from one side. The corpse was then turned and the blubber removed from the other side. The whole process took, on average, three to four minutes. Some variation was necessary with particularly large seals.

The blubber and skin thus removed was in one piece, between 3 and 20 cm thick, roughly circular with two holes where the fore-flippers passed. These pieces were attached by strops to a line and hauled by the motor boat to the sealing vessel where they were stowed in the hold and, when that was full, on parts of the deck. The flensed carcasses were left at the water's edge to be consumed by birds or rot. This was the most wasteful aspect of the industry. Attempts to use the whole carcass had proven unremunerative although the skrotts contained about a quarter of the available oil as well as being a

potential source of meat and bone meal. After obtaining a full cargo the vessel returned to Grytviken where the blubber was processed at the whaling station. As the seals were flensed within such a short period after death and very little meat or blood was left with the blubber, the oil obtained from them was of the highest quality. It has similar properties to baleen whale oil and was used for identical purposes, although sold as a separate item. The sealing was very profitable while the whaling stations operated as, for comparatively little investment, up to 2000 tons of high quality oil could be obtained in each season.

In 1951, an investigation of the elephant seals and sealing industry on South Georgia was initiated by Mr R.M. Laws of the Falkland Islands Dependencies Survey. This indicated that further mea-

Sealing at South Georgia, 1957. Flensing an elephant seal. (W.N. Bonner.)

sures were required to conserve the species and that additional data were desirable. By means of age-structure studies linked to pup counts, he was able to model the population and estimate stock levels at that time. From several other sources, including calculations of catch per unit effort, he showed that the male population had been declining since 1930. He recommended new regulations, which were implemented in the 1952 season. These included catch quotas calculated for the four divisions of the island, the imposition of a minimum length for seals taken and the monitoring of the age composition of the catch.

The latter was possible as a result of his discovery, made during earlier work in the South Orkney Islands, that the teeth have annual growth layers. When properly prepared, the teeth reveal the age of the animal and allow determination of when it reached maturity. A new regulation required that a lower canine tooth should be collected from every twentieth seal killed. A Sealing Inspector was later appointed to continue the management research and control the industry and was present until it finished. The detailed results of Laws' work on age, growth, reproductive physiology and behaviour were published in four monographs in the *Falkland Islands Dependencies Survey Scientific Reports* series.

Three main improvements followed the introduction of the new management measures. There was a sharp rise in the catch per unit effort, marked improvements in the yield of oil per seal, indicating larger average size of the animals taken, and an increase of the average age of the catch which levelled off, as predicted, at $7\frac{1}{2}$ to 8 years. These results indicated that a sustained yield was being achieved with an annual take of 6000 adult males. This all meant that the rational management of the industry continued with a much stronger scientific basis.

During the second epoch of South Georgia sealing, approximately 260 000 seals were taken which yielded 84 000 tons of oil. The annual number of seals taken and oil production is included in Appendix 2.

Prominent persons in whaling and sealing at South Georgia

Very large numbers of men served on South Georgia in the period of the whaling and modern sealing industries. Many of them became well known and achieved considerable historical importance, both with respect to the island and to their industries. Brief biographical details, which concern their association with South Georgia, follow for some of these men. Difficulties in making such a selection are great; records are sparse and discontinuous, and there are many men who could be mentioned. The notes are thus only a selection of some persons greatly involved with the history of South Georgia and is by no means comprehensive.

Captain Carl Anton Larsen (1860–1924) of Sandefjord, Norway, is the most significant person in the history of whaling and modern sealing on South Georgia. His establishment of the first whaling station in 1904 and some expeditions in which he took part are described in Chapter 4. He was the first manager at Grytviken and served to 1914 after which he left the company, in part owing to disagreements about his plans to improve conditions for men working at South Georgia. Lauritz Larsen (1855–1924), his elder brother, also arrived at Grytviken at its foundation and acted as manager during his brother's absences. In 1909 he became manager of the new whaling station at Ocean Harbour. Fridthjof Jacobsen (1874–1953) also arrived in 1904; from 1914 to 1921 he was the second manager of the station. Two of his children were born on South Georgia. He later became a vice-president of the company. Another 1904 arrival was Captain Wictor Esbensen (1881–1941), who became the third manager (from 1921 to 1930). Ludwig Allum arrived at South Georgia in 1925 as plan foreman at Grytviken. He became secretary there and served until the end of Norwegian whaling from the station. Einar Strand arrived on the island in 1917 as blacksmith at Ocean Harbour. When this station closed he transferred to Grytviken where he also served to the end of Norwegian whaling there. Ole Hauge was Master of the sealer *Albatros* for many seasons and became an expert on navigating in coastal areas of the island. Because of this he was able to be of great assistance to Duncan Carse during the South Georgia Surveys.

The first manager at Husvik was Søren Berntsen (1880–1940), who arrived with the floating factory *Bucentaur*. He was responsible for the establishment in 1910 of the shore station which first operated from 1910 to 1931. Nils Olsen was one of the managers after Husvik reopened in 1945. He was a noted artist and many examples of his cartoons decorated the Husvik manager's villa.

Thoralf Sørlle (1875–1939) was the first manager of the shore station at Stromness, a post he held until 1931 when whaling finished there. His association with Shackleton and the assistance he provided in 1916 are recounted in Chapter 4. Harald Studsrød was a blacksmith at Stromness in 1959; he began his career at the New Island whaling station in the Falkland Islands in 1913 and transferred to Stromness when this closed.

Henrik Henriksen (1866–1925) established the whaling station at Leith Harbour in 1909 and managed it to 1916. In the previous season he had selected the site while visiting the island aboard the catcher *Semla* from New Island. From 1916 Leganger Hansen (1883–1948), who had arrived in 1911, was manager. He held the post for the next 20 years to retire in 1937; during this period he became one of the best known persons on the island. Finn Dahlberg also visited the island aboard *Semla* and was present when Leith Harbour was opened. He continued working there until his retirement in 1954 as a whale-catcher chief engineer. Gunnar Gulbrandsen (b. 1905) started as a pattern-maker at Grytviken in 1927 then worked as a carpenter at Stromness before joining the management at Leith Harbour in 1946. Nochart Nielsen (b. 1894) was first a gunner at Grytviken and transferred to Leith Harbour in 1949. He was the company's senior gunner and master of *Southern Jester* in 1957 when he took the Duke of Edinburgh on a brief cruise and demonstration of a harpoon cannon.

Although some of the whaling companies were registered in countries other than Norway the industry was predominantly Norwegian, as the names above testify, until the Japanese arrived in 1963. The first British Manager of a South Georgia whaling station was Captain Sinclair Begg who managed Leith Harbour from 1947 to 1950. Major K. Pierce-Butler managed Grytviken from 1954 to 1958 during which he introduced many improvements and modernisations. From 1946 to 1951 he had been a member of the Falkland Islands Dependencies Survey and from 1951 to 1954 he was the South Georgia Magistrate.

Much more biographical detail about these men, and about many others associated with the island, is available from the sources included in the bibliography at the end of this book.

The Church

Captain C.A. Larsen was largely responsible for the establishment of the Church at Grytviken, the first in the Antarctic. Support came from the companies operating the floating factory *Nor* and Ocean Harbour whaling station as well as from many whaling employees. Larsen was assisted by the newly appointed pastor and the Norwegian Seaman's Mission which maintained an active interest in it. The Church originally stood by Strømmen, in Norway; it was dismantled and taken to Grytviken to be re-erected in late 1913. It was completed in time to be opened and consecrated on Christmas Day, 1913. Two bells, cast in Tønsberg, are hung in its steeple which were first rung at midnight on that Christmas Eve.

In February 1910 Ivar Welle from the Norwegian Seaman's Mission in Buenos Aires, made a brief visit to Grytviken and discussed the appointment of a pastor with C.A. Larsen. Following this Kristen Loken, a newly ordained Norwegian Lutheran pastor from Lillehammer, was recruited for the post. He arrived on 1 April 1912 and remained to 4 June 1914; during this time he had an active interest in whaling and the island. His photographs have been preserved; these show many aspects of the island, the whaling stations and life there. Despite the new Church and his activities he had to conclude that 'Christian life unfortunately does not wax strong among the whalers'. His place was taken by a theological student, Frithjof Zwilgmeyer, who arrived on 14 March 1914 and remained to mid 1916. There was then almost a ten years gap until the arrival of Fredrik Knudsen, an ordained pastor, who was present from October 1925 to May 1926. The last Pastor at the Church was Sverre Eika, of the Norwegian Seaman's Mission, and reportedly a keen footballer, who served from September 1929 to April 1931. Thereafter, as in previous years, various chaplains from ships calling at King Edward Cove have opened the Church, many from the Royal Navy. The Dean of Christchurch Cathedral, Stanley, who arrived on 17 December 1932 to consecrate Sir Ernest Shackleton's grave, was among them.

On 17 September 1913 the Bishop of Kristiania, now Oslo, dedicated and despatched a church record book to pastor Loken at Grytviken. Details of pastors, deaths, and births on the island were recorded in it for the years when a pastor was present. Some inclusions from other years were added, presumably by the Managers of Grytviken. The book remained in service until 1931. Pastor Loken wrote reports to the Bishop about the island, its inhabitants, and his work; in these he expressed his opinions of the quality of some of the men recruited to work at the whaling stations which was, in many cases, far from favourable.

The church at Grytviken. (R.I. Lewis Smith.)

During the first year of Pastor Loken's residence on South Georgia a typhus epidemic occurred at Grytviken and nine men died during the winter of 1912. This arrived with a ship from Buenos Aires and infected 17% of the whaling station's personnel. Kind Edward Point and Grytviken remained in quarantine throughout the 1912 winter. These and many other deaths are listed in the church book. Although the Magistrate was the official registrar of births, marriages and deaths, the church records give some additional details including those of the first two births. Each whaling station had a cemetery; some were established where sealers had buried their shipmates in the previous century. In total there are almost two hundred graves on the island from 1820 to 1983 and many men were lost or buried at sea in the vicinity (from the Viceroy of Chile in 1756, onwards).

The Church is at present the soundest building in Grytviken and has been maintained by the British Antarctic Survey, Royal Navy, and the garrisons occupying King Edward Point who still occasionally use it.

Administration and King Edward Point

Although various administrative actions of the Governments of the Falkland Islands Dependencies in Stanley had concerned South Georgia prior to 1909, the island's administration is most conveniently described from that year when the first resident Magistrate arrived. Mr James Innes Wilson, the acting Treasury Clerk at Stanley (and former itinerant school master), was appointed Stipendiary Magistrate of South Georgia on 20 November 1909. He proceeded to the island aboard *Coronda* of Christian Salvesen Limited to arrive at Leith Harbour on 30 November and continued to Grytviken on 2 December. Accommodation was first provided for him by C.A. Larsen in a small cottage close to the Manager's villa at the whaling station. This building still stands although it is now rather dilapidated. Two rooms were available for him and accommodation existed on an upper floor for a constable designate to lodge, if necessary. Larsen refused payment at first, despite the Magistrate's protestations; however several months later he was prevailed upon to accept payment for providing board and lodging.

James Wilson had a great variety of duties to

perform. The establishment of the Post Office and conduct of the first census were some of his earliest; these are described in Chapters 1 and 6. He made an inspection of all whaling stations in December, travelling aboard *Undine*, then prepared reports on the whaling and sealing industries. He also investigated and prepared maps of the island and the distribution of whales around it, received 23 ships in his first 6 months, investigated irregular sealers, and various other matters. He made regular detailed reports to the Governor Mr (later Sir) W.L. Allardyce, with accounts of events and various recommendations for the island. Labour troubles between some employees who disputed their contracts, and the manager of Leith Harbour whaling station, required his intervention early in 1910. He swore in two special constables, who served from 31 December 1909 to 5 and 22 February 1910, to assist him in this. The duties of Coroner required him to investigate several cases of shipwreck, fire and accidental death. In one instance, a skeleton with a bullet hole through the skull was found during the building of Ocean Harbour whaling station, probably a relic from a disagreement between sealers. He also reported the first of a series of landslides affecting the Leith Harbour whaling station at its old site on 2 July 1911. These were sufficiently frequent and serious for the station to be moved and for the most severely affected area to become known as Jericho.

The Governor desired that an administrative post be established separate from any whaling station. Wilson recommended and it was resolved, that this should be on King Edward Point where a meteorological station was located. A prefabricated building was ordered from Norway and transported to South Georgia early in 1912. Erection took place during the subsequent winter but was protracted owing to the epidemic of typhus. The building was completed and occupied on 17 September 1912. It served for many years, first for the Magistrate, then as Government officers' quarters from 1925 until it burnt down accidentally on 16 March 1946. A Customs shed, also prefabricated, was built next to the residence. This concrete, wood, and iron structure was completed on 15 March 1913. It still stands and is now the oldest building on King Edward Point. Part of it was converted in 1914 to a gaol; the escape of the

first prisoner was described in Chapter 1.

Many minor problems for the Magistrate and major ones for the whaling companies were caused by occasional acute over-indulgence in alcohol by whaling employees. The companies requested the Magistrate to have the sale of alcohol legally controlled and he made a detailed submission for this to the Governor. As a result, on 27 November 1911 'An Ordinance to Regulate the Sale of Intoxicating Liquors on South Georgia' was enacted which provided for licensing of the sale of alcohol on the island. Although this was fairly successful, it was often noted how much alcohol used to 'evaporate' from certain compasses, that particular brands of

The first Magistrate's residence at King Edward Point, 1912–46. From a photograph taken by the Magistrate, Mr E.B. Binnie, about 1920. (Courtesy Scott Polar Research Institute.)

patent medicine and toiletries were most favoured, and that certain ships became particularly well known for smuggling. Despite vigilance, some liquor used to get through unlawfully, due to much ingenuity. Stills, often very carefully constructed and concealed, were also sometimes discovered; the remains of some are still occasionally revealed in the deteriorating building of the whaling stations.

A Customs Officer was appointed to assist the Magistrate and assumed duties in July 1912. He was often able to visit the Stromness Bay whaling stations to enter and clear ships, a matter of great assistance to the whaling companies. On October 1915 he reported the arrival of 12 negros who had stowed away at St Vincent and were suffering very severely from the cold of South Georgia. The whaling companies' vessels often had to apply elaborate procedures to detect stowaways, prior to departing from certain ports. These occasionally involved the use of the steam hoses.

During the early years of whaling, the Magistrate was involved in securing the use of the whole carcass of whales rather than the blubber only by the whaling factories. The Governor was particularly interested in this and appropriate regulations were progressively introduced. Whaling at the South Orkney Islands and South Sandwich Islands was also subject to the administration of the South Georgia Magistrate, who reported on it regularly. The Magistrate made several recommendations and took the advice of the whaling companies about the installation of various lights, beacons, a bell buoy, and other aids to navigation around South Georgia. All of these were installed and maintained by the whaling companies, sometimes as part of their lease requirements. Many remain today although only the light at King Edward Point now functions.

James Wilson remained as the South Georgia Magistrate, apart from periods of leave, until 19 October 1914 when he left the island and Edward Binnie was appointed to succeed him. Binnie served until April 1927 and his successors are listed on page 137.

Two shipping mysteries were investigated in 1916. A 537 ton steamship, *Argos*, left Buenos Aires on 7 May 1916 carrying coal and empty oil barrels for South Georgia and was never seen again. Some

timber, recognised as part of her cargo, was later found in King Haakon Bay during the 1916/17 whaling season. When the first ship arrived at Prince Olav Harbour to reopen it, the station was found to have been broken in to. Seven rough beds were discovered in the 'Guano Shed' with a note written in Spanish. Unfortunately, the finder of the note could not read it and lost it. Later that summer a partly decomposed body was found a couple of kilometres from the whaling station. It was presumed that the *Argos* sank not far from South Georgia, that some of her crew reached Prince Olav Harbour, then set out along the coast to seek an inhabited whaling station and perished on the way.

On 26 July 1916, during the First World War, a large four masted barquentine was seen in pack-ice near the island by, *Saima* a Leith Harbour whale catcher. She was thought to be possibly a supply ship for a German commerce raider and did not acknowledge signals. An attempt to contact and investigate her the next day was unsuccessful owing to the extensive pack-ice. On the day following a very severe storm blew up. She was not seen after the storm abated and may well have foundered during it.

Labour troubles began again during the winter of 1918 and were greatly exacerbated the following summer. Some Russian employees of Grytviken, recruited in Buenos Aires, reportedly agitated for South Georgia to become the second Bolshevik Republic. The situation deteriorated even more in 1919/20, when the Magistrate and his staff were seriously endangered and other stations became involved. HMS *Dartmouth* very conveniently arrived, as related in Chapter 4, and the provisions of the 'Peace Preservation South Georgia (Aliens) Ordinance 1919' were applied.

Electricity was brought to King Edward Point in 1923 from Grytviken, where it was generated at hydroelectric or diesel stations, and the buildings were lit electrically. A major building programme began in 1924/25 as a result of the needs of the *Discovery* investigations and the installation of a radio station. Three buildings (a new Magistrate's residence, the radio station and operator's accommodation, and Discovery House) were constructed that summer. Two substantial radio masts were also erected. Two of the buildings still stand, as does the small dam built to provide water for the station and laboratories. A petrol or paraffin driven dynamo was deployed with a large battery bank to power the radio and supplement lighting power. Even a water-borne sewage system and a light railway were installed. A new jetty and boatshed were built at the end of King Edward Point where the shallop found by Andersson in 1902 had previously served. The present jetty is based substantially on the 1926 one and the shallop is buried in its approaches.

The Magistrate had to attend to the defences of South Georgia when the Second World War broke out. Two four inch guns, made in 1918, were despatched to the island and were deployed at Hope Point and on a hill above Leith Harbour whaling station. A volunteer force from the whaling employees was to operate them. Many Norwegians were, to some extent, experienced in gunnery owing to use of the harpoon cannon (it is recorded that

these were also deployed), and all were loyal to their King who was in exile in London during the German occupation of Norway. When a gunnery expert arrived aboard *Queen of Bermuda* both guns were redeployed, to near Susa Point and on Hansen Point, so as to more effectively cover the heads of Cumberland Bay and Stromness Bay. Training was put on a more formal basis, accommodation was established at the Susa Point battery and a telephone line run from it to King Edward Point. The guns and their ancillary works are still present although somewhat dilapidated. They have never fired in anger. Owing to the shortage of shipping and the destruction of much of the Norwegian whaling fleet by German raiders, communications and supplies to South Georgia were greatly reduced. Only Grytviken whaling station continued to function throughout the war. Interestingly, in the North Atlantic, many whale catchers were deployed for

The settlement at King Edward Point in 1936 photographed by the British Graham Land Expedition. (Courtesy Scott Polar Research Institute.)

submarine hunting as it was claimed, by the Admiralty, that chasing submarines was not greatly different from chasing whales. After the War many patrol boats were converted to become whale catchers.

During the War, a particular problem was envisaged at South Georgia with Argentina, whose territorial pretensions had been extended to the island a short time previously. While British and Allied resources were already extended to their limits, it was considered that Argentina might invade South Georgia and even the Falkland Islands. Argentina's neutrality was uncertain until the outcome of the War became apparent (she declared war against Germany in March 1945) and her prompt encroachments into other parts of the Falkland Islands Dependencies were cause for concern. Thus, it was considered possible that the guns might open fire on Argentine ships and that a scorched earth policy might need to be adopted if defences proved inadequate. Some Argentine employees of Grytviken whaling station were considered as potential enemy agents by the War Office, and an appropriate course of action was decided in case they became active.

On 2 September 1941, the Magistrate William Barlas died following the effects of an avalanche which knocked him into the sea. It fell on the track between King Edward Point and Grytviken where he was proceeding on duty. He had been Deputy Magistrate at South Georgia since 1920 and Magistrate since 1928, as well as serving in other posts in the Falkland Islands and Dependencies. He is buried in the Grytviken cemetery.

Things rapidly returned to normal after the War which, despite much concern, had not seriously affected South Georgia. Whaling resumed at Leith Harbour and Husvik and continued at Grytviken. Lost ships were replaced, many by new vessels, and other vessels returned to their former service. The world's whaling companies were coming to depend increasingly on pelagic floating factories by this time, and some of these visited South Georgia. Problems of diminishing whale numbers were also becoming increasingly serious.

During the 1950s the settlement at King Edward Point included the Magistrate, a Customs Officer and his assistant, a couple of policemen, one or two diesel mechanics, two or three wireless operators, a meteorological forecaster and three observers, a cook, and a steward. They were principally recruited from the United Kingdom, the Falkland Islands and Ireland. Several men were accompanied by their wives and a number of children also lived there. A Government Naturalist and Sealing Inspector was appointed in 1956; this post existed until 1966. During the decade, the most significant event was the official visit of the Duke of Edinburgh on his tour of the minor colonies which is related in Chapter 4 together with several other events affecting King Edward Point. One inhabitant, Nan Brown, wife of a wireless operator, wrote an account of life at the settlement at this time which was published as *Antarctic Housewife* (Brown, 1971).

From 1959/60 the whaling stations of South Georgia began closing because of the decreasing abundance of whales. Japanese companies took over for the last few seasons but whaling was finished before 1966. Great changes came about at King Edward Point and facilities were improved to enable it to function quite independently. These included expansion of the power station and workshops, provision of greater food storage facilities (including large cold storage rooms), construction of a new Magistrate's residence, and a new building to provide hospital and accommodation facilities for the other Government staff. Most of these projects were complete and in operation by early 1964. The new building is a large three-storey structure which was constructed near Hope Point and named 'Shackleton House'. Its construction was arranged by the Crown Agents and prepared in a pre-fabricated form for rapid erection on the island. It has stood almost 20 years, is in very sound condition, and presently provides the main accommodation on King Edward Point.

The last stipendiary Magistrate remained until November 1969, almost 4 years after the closure of the last whaling station. In that month the settlement at King Edward Point was taken over by the British Antarctic Survey. The Base Commander was appointed Magistrate and, as the total population of the island thenceforth rarely exceeded 24, he has continued to perform both functions, being responsible to both the Survey's Director and the Governor of the Dependencies respectively. The stipendiary Magistrates appointed to South

Georgia were:

J.I. Wilson	20 November 1909 to 19 October 1914
E.B. Binnie	20 October 1914 to 1 April 1927
F.B. Allison	22 August 1927 to 24 November 1927
W. Barlas	27 September 1928 to 2 September 1941
A.I. Fleuret	17 April 1942 to 15 March 1951
K.S. Pierce-Butler	15 March 1951 to 19 April 1954
R.E. Spivey	20 April 1954 to 3 April 1957
J.W. Matthew	4 April 1957 to 25 June 1959
D.J. Coleman	26 June 1959 to 13 November 1969

Generally the Customs Officers deputised for the Magistrate while he took leave and between appointments.

Comprehensive administrative files were maintained by the South Georgia Magistrates. These included reports to the Governors, correspondence with the whaling companies, meteorological statistics from the whaling stations, post office matters, shipping records and many other items. These were donated to the Scott Polar Research Institute by the Governor and collected in 1975. They provide a virtually complete record of South Georgia administration from 1909 to 1969.

The Falkland Islands Dependencies Survey

The Falkland Islands Dependencies Survey was developed from Operation Tabarin after the Second World War, as was related in Chapter 4. Its activities were in the South Shetland Islands, Antarctic Peninsula, and South Orkney Islands until 24 December 1949, when they were extended to South Georgia. The Survey then took over the meteorological observatory of the Compañia Argentina de Pesca at King Edward Point from 1 January 1950, and operated it to 1952. The Survey's first research programme on the island was an investigation of the elephant seals. This was conducted by Dr R.M. Laws, he became the first Base Leader at South Georgia and is now the Director of the British Antarctic Survey. This followed similar research on Signy Island in the South Orkney Islands and is described in this chapter with the second epoch of sealing. Since then the Falkland

Islands Dependencies Survey and its successor, the British Antarctic Survey, have been responsible for extensive scientific investigation of South Georgia and its surrounding oceans. Together they have conducted substantially more research than all other investigations combined.

As well as the meteorological observations and Dr Laws' work, the Falkland Islands Dependencies Survey was involved on South Georgia with the Stonehouse and Bonner expedition at the Bay of Isles, the International Geophysical Year, Lance Tickell's three and one other Bird Island expedition, and the botanical expeditions from 1960. The Trans-Antarctic Expedition and Royal Society Expedition of the International Geophysical Year also called at South Georgia on their ways to the south of the Weddell Sea; their work was closely associated with that of the Survey. Most of these expeditions have already been described in Chapter 4. Some of the Bird Island expeditions were notable for being cooperative projects between the Falkland Islands Dependencies Survey and the United States Antarctic Research Program, as well as being associated with the work of the Government Naturalist.

A glaciologist attached to the Survey, Jeremy Smith, who contributed to the International Geophysical Year research, also made a detailed series of botanical observations on the island. This laid the foundation for a comprehensive botanical investigation by the Survey. From 9 December 1960 to 7 April 1961 the first purely botanical expedition was active on the island. Although this had as its main objective a floristic study of the island's bryophytes, collections and observations of all groups of land plants were made. The expedition, led by Dr S.W. Greene, had its headquarters at King Edward Point and was able, with the assistance of HMS *Protector* and her helicopter, to visit places all around the island (many of which had not been previously investigated botanically). The results of these collections and observations, with a detailed account of earlier collections, appeared as *The Vascular Flora of South Georgia* and *A Synoptic Flora of South Georgian Mosses* published by the Survey (Greene, 1964).

On 1 January 1962 the Falkland Islands Dependencies Survey was renamed, to become the British Antarctic Survey. This was in accordance with the impending change in the division of the Dependencies into the British Antarctic Territory and newly defined Dependencies (consisting only of South Georgia and the South Sandwich Islands and some isolated rocks). This division was made on 3 March 1962 owing to administrative requirements for the Antarctic Treaty. At that time, all of the Survey's scientific stations were in the Territory (although it later extended operations into the Dependencies).

The British Antarctic Survey

Continuous British Antarctic Survey research conducted on South Georgia began in the 1967/68 season with botanical, geomorphological, and geological investigations. Research contributing to the first part of an International Bipolar Botanical Project on South Georgia included studies of plant growth rates and primary production with a series of microclimatological measurements in the plant communities under investigation. Similar experiments were performed at Disco Island, off north-west Greenland, and the results were compared. A comprehensive investigation of the genus *Acaena* was commenced concurrently. Following the diversion of a geomorphological investigation from Deception Island in the South Shetland Islands, where a severe volcanic erruption occurred on 4 December 1967, an investigation of the Cumberland Bay region of South Georgia was made. The results of this included a report and a detailed geomorphological map of the area. Geological observations and collections were also made by the Survey in the Prince Olav Harbour region.

During the latter years of the whaling industry on South Georgia the problem of maintaining an administration by the Falkland Islands Dependencies Government arose as the demands upon this would become greatly reduced. By the late 1960s the Survey's biological programmes were well established on the South Orkney Islands and provided an impetus to the extension of activities to South Georgia. It was therefore resolved that the Falkland Islands Dependencies Government should lease the settlement and equipment at King Edward Point to the Survey for a scientific station. In return the Survey was to perform any necessary administrative functions, particularly in regard to

visiting ships, of which there were a considerable number every summer. Accordingly, in 1968/69 a summer party of builders set to work on Shackleton House to make appropriate modifications and to install laboratories. On 11 November 1969, a party of 11 Survey officers arrived. On 13 November an official hand-over take-over from Captain D.J. Coleman of the Government of the Falkland Islands Dependencies to Mr E.J.Chinn of the British Antarctic Survey took place. The island was administered by the Survey on behalf of the Government until the 1982 Argentine invasion. On 19 November 1969, the former civil staff departed; a formal 20-year lease was drawn up between the Foreign and Commonwealth Office and the Natural Environment Research Council in 1972 for the Dependencies Government and Survey.

Since then, scientific research on South Georgia has had the enormous advantage of being conducted on a virtually continuous basis. The Survey took over meteorological observations, opened an ionospheric observatory, and continued the botanical investigations which were extended to include a lichenological survey. Phase two of the International Bi-Polar Botanical Project was instituted which involved crop plants grown in the open under specified conditions to act as phytometers for comparison with similar experiments on Disco Island, as previously. In 1970/71 a long-term investigation of the effects of reindeer on South Georgia commenced, as did stratigraphic studies which formed the first part of a comprehensive geological investigation of the island. The South Georgian part of the International Hydrological Programme began when helicopters from HMS *Endurance* lifted about 10 tonnes of equipment to the foot of the Hodges Glacier, to establish a glaciological observatory. The Survey conducted two glaciological investigations for this programme, one at South Georgia and another at Spartan Glacier on Alexander Island, 1800 km farther south. These sites are two in a programme extending from Alaska to Antarctica. The first of many visits by tourist ships was also made in that season when *Lindblad Explorer* arrived in December 1970.

The station at Bird Island was reopened from 11 November 1972. The first activities there, after it had been prepared for habitation, concerned the fur seal; about 100 originally tagged between 1957 and

1964 were recorded, the oldest of which was aged 14 years. About 400 pups were tagged and weekly counts of bulls, cows, and pups were made. The fur seal population of Bird Island was estimated as 22 000 compared to 11 500 in 1963. Early in 1973 ornithological investigations recommenced there also. Some albatrosses, ringed as chicks between 1958 and 1964, were returning to breed at this time. Censuses of the population of several species were made and studies were begun of their breeding biology. Bird Island has since been occupied during every summer season and its research programmes have been increased (in the cases of species originally studied) and extended to several others.

On 10 September 1973 Sir Vivian Fuchs, the first Director of the Survey, retired and was succeeded by Dr R.M. Laws. Three years later, the Survey moved to a new headquarters in Cambridge, UK. Previously, it had functioned from several sites around the country where various research divisions and other sections were located.

Research on South Georgia continued and increased during the 1970s although this changed subsequently owing to matters described later, in Chapter 9. Atmospheric research incorporated a solar radiation programme and magnetometers were deployed at the King Edward Point observatory in 1973. King Edward Point also became a National Meteorological Centre on 1 October 1974 following the substantial reduction of the Survey's meteorological office in Stanley, Falkland Islands. Operational instructions for preparing meteorological data at all the Survey's stations were revised to bring the procedures more clearly into agreement with the World Meteorological Organisation's recommendations. In late 1974, a comprehensive geological examination of the island began when teams of geologists started mapping the

Carlita Bay field hut deployed in 1974 for use by persons travelling between Stromness Bay and Cumberland Bay. (Author.)

southern igneous zone and later extended this to cover the whole island. During the same season, investigations of fish and marine invertebrates began; the latter involved sampling of krill and was the beginning of a very extensive investigation of this animal. A South Georgia Benthic Survey was also conducted which required the establishment of diving facilities. Seal research was greatly aided by comprehensive photographic surveys of fur seal breeding beaches conducted with the assistance of HMS *Endurance's* helicopters. Botanical and ornithological studies also continued although the botanical survey of the island was completed after almost all coastal areas and offshore islands had been visited.

From the late 1970s, shipping at King Edward Cove began to increase substantially and this trend has been maintained. The United States Antarctic research vessel *Hero* visited the island with a geologist of the Survey aboard; bird and seal studies were conducted from her and she assisted the Survey in several other projects. A Polish Antarctic research ship *Professor Siedlecki* called and has since made several more visits as have various ships of the Russian Polar Institute. HMS *Endurance* spent 4 weeks one summer providing helicopter assistance for geologists in making landings on the more inaccessible places. Geological mapping was thus completed from the coasts to the main ranges; this resulted in many

The British Antarctic Survey's boatman's workshop at King Edward Point, January 1980. (Courtesy British Antarctic Survey, C.J. Gilbert.)

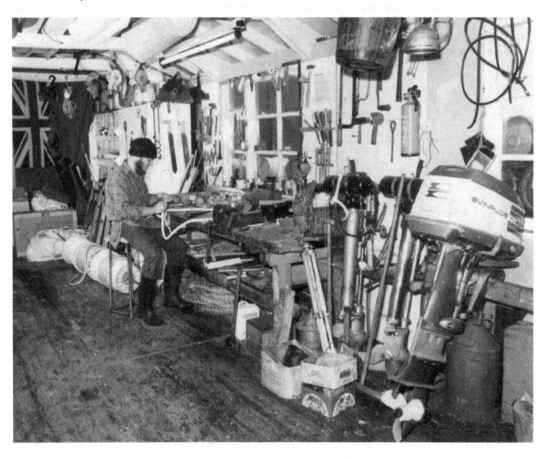

improvements to general as well as geological maps. A comprehensive examination of salt and fresh water, land and marine plants, sediments, and many animals was made around King Edward Cove to determine the effects of pollution from the activities of the Grytviken whaling station more than a decade after its closure.

Four South Georgia Reference Sites were designated for long-term terrestrial biological research in early 1977. These were established in different plant communities of the island for the intensive investigation of functional processes involving living and other components of the selected communities. A fifth site was added in 1979. Studies undertaken have included mineral cycling;

The British Antarctic Survey's carpenter's workshop at King Edward Point, January 1980. (Courtesy British Antarctic Survey, C.J. Gilbert.)

entomology; arthropod responses to seasonal changes; decomposition processes; activities of fungi, protozoa, and bacteria; and micrometeorological observations, which also provided some data for other investigations. Other terrestrial research included a detailed investigation of the introduced rat and of the mice discovered by a geological party in 1977. Phenological data on the vascular flora were also collected.

A two-year investigation of elephant seals was made from several field sites, especially Elsehul and Dartmouth Point. This studied population structure and social organisation. It showed that the South Georgian population was no longer affected by its earlier exploitation. The study of over 4000

confrontations between bulls demonstrated how stereotyped these were. Physical contact occurred in 4% of cases and in only 1.8% did biting take place; dominance was often established by roaring alone. Investigations of fur seals were also commenced at Schlieper Bay and Bird Island field stations. Owing to their activity and to the possibility of attack by these seals, elevated gangways were constructed so that observations could be made of seal beaches from safe positions above them.

In 1978, a substantial marine biological investigation (the Offshore Biological Programme), was initiated around South Georgia. This was principally ship-borne and RRS *John Biscoe* was specially

modified for it. Because of declining stocks in fishing grounds in other regions of the world, combined with increasing demand for food, it is probable that krill (one of the principal resources of the Antarctic Ocean) will become the object of large-scale exploitation. Already, fishing in the region is well developed and overexploitation is apparent. The effect of large krill harvests, especially on the other components of the marine ecosystem, was substantially unknown before the programme commenced. It is intended to discover more about krill biology and the possible effects of its exploitation prior to the establishment of major krill fisheries. The programme has worked

A biological laboratory in Shackleton House, King Edward Point, January 1980. (Courtesy British Antarctic Survey, C.J. Gilbert.)

partly in cooperation with similar international research in the region.

Although various scientists from institutions other than the Survey have worked with it from 1969, there has been a substantial increase in their numbers since about 1978. As well as scientists from other British organisations, several from Norway, West Germany, the United States of America, and South Africa have recently been active on the island. Some private expeditions, substantially supported by the Survey, have visited for the purpose of making cinematographic films of the island in general, and of its whaling stations and wildlife in particular. Survey officers have been able to produce similar films and much material has recently become available. An artist, David Smith, travelled on one of the Survey's ships in 1979/80 and was able to spend about 6 weeks on South Georgia during which he produced pictures of many aspects of the island and the Survey's activities.

Since 1980, during a period of general economic recession, the Survey had begun to be affected seriously by financial resources insufficient to allow the continuation of all its stations, ships, research programmes, and headquarters functions. Expenditure was reduced as far as practical, in an endeavour to maintain normal activities. This eventually proved inadequate and consideration of other measures to reduce expenditure began. The closure of a scientific station was considered to be the most expeditious way of relieving the problem. It was decided that King Edward Point should close as, however unfortunate, its loss would least seriously disrupt the Survey's research activities. Although the Offshore Biological Programme relied heavily on the station's facilities, it was believed that it could become entirely ship-borne and therefore continue. The Governor of the Dependencies and the Foreign and Commonwealth Office, who considered South Georgia from many

The scientific station at Bird Island in 1982.
(P. Goodall-Copestake.)

more aspects than the scientific, were advised of this intended action. They were able to arrange for a grant of money to the Survey from the Dependencies' revenue to maintain a reduced scientific station at King Edward Point. About half the usual complement was to remain and many other reductions of expenditure were to be made. These changes began early in 1982 and were well underway before the interruption described in Chapter 9 occurred. Changes being made were the preparation for the closure of Shackleton House (which the Survey had applied for permission to demolish), redeployment of accommodation facilities to the former Magistrate's house, establishment of restricted laboratory facilities in the former Customs House, and other removals. At the time of writing the future of the Survey at King Edward Point is yet to be decided.

The small scientific station on Bird Island had been manned for several months each summer since 1972/73 for ornithological and seal research. Accommodation was available for only four men and, although space was occasionally provided for more, this was a limiting factor on any investigations conducted. In 1981/82 a new building was deployed there, with improved facilities and accommodation for eight men. It was intended that three should spend the 1982 winter there but the events of that year precluded this. The station opened again for the 1982/83 summer and remained manned for the rest of 1983. It is proposed that it will also be

RRS *John Biscoe* in Cumberland Bay, February 1980.
(Courtesy British Antarctic Survey, C.J. Gilbert.)

manned continuously from 1984. Some of the recent research conducted by the British Antarctic Survey at Bird Island has been in association with the United States Antarctic Research Program, as in the early years of the station.

The men employed at South Georgia by the Survey spent either summer-only periods there or served a full contract. The former rarely lasted more than 3 months and made up about a third of the summer population. The latter lasted over 2 years, covering 3 summers and 2 winters, in common with the majority of the Survey's wintering contracts. Several men have returned to the island after an initial tour of duty and some have, for various reasons, extended their original one. Wintering officers have recently included biologists, boatmen, builders, diesel mechanics, divers, electricians, general assistants, ionosphericists, meteorologists,

physicists, radio operators, and technicians. As well as the duties connected with one's post, officers have a great variety of others which include; cargo handling, building maintenance, mooring ships, cleaning and domestic duties, steward's functions and relief cooking. These are performed by everyone at the station.

The ships of the British Antarctic Survey and HMS *Endurance*

Most of the transport of men and supplies as well as landings around South Georgia are made by the British Antarctic Survey's ships RRS *Bransfield* and RRS *John Biscoe*, both registered in the Falkland Islands, with much assistance from HMS *Endurance*. Since the Survey took over King Edward Point, these have been the most frequent and regular visiting ships and a brief description of them is given below.

RRS *Bransfield* in King Edward Cove, February 1979. (C. West.)

RRS *John Biscoe* is named after a British Antarctic explorer who circumnavigated the continent between 1831 and 1832. She is a 1584 ton ice-strengthened vessel, 67.1 m long which was launched in 1956. She was extensively modified for the Offshore Biological Programme in 1979 when new diesel electric engines, bow thrusters, and a variable pitch propeller were fitted. Her complement is 32 officers and crew with 28 supernumeraries.

RRS *Bransfield* is named after a British naval officer Edward Bransfield who claimed the South Shetland Islands for the Crown in 1820 and mapped part of Trinity Land, thus making the first chart of the Antarctic continent. She is a 4816 ton ice-strengthened vessel, 99.1 m long, launched in 1970, and equipped with diesel electric engines with a variable pitch propeller. Her complement is 37 officers and crew with 59 supernumeraries. She also carries a helicopter-landing platform.

HMS *Endurance* is named after Sir Ernest Shackleton's ship of the Imperial Trans-Antarctic Expedition of 1914–16. She was formerly *Anita Dan*, launched in 1956 and commissioned into the Royal Navy in the 1968 as an ice patrol vessel. She is a 3650 ton vessel, 83 m long, with a complement of approximately 120, and carries two helicopters. Her principal task is hydrographic survey and she is only lightly armed as a warship.

These ships are distinctively painted with red hulls and white superstructures to be visible in pack-ice. They are frequent visitors to the Falkland Islands, the Falkland Islands Dependencies, and British Antarctic Territory.

HMS *Endurance* and helicopter off King Edward Point, March 1982. (Author.)

6

Travel and communications

South Georgia is one of the remotest inhabited places in the world; its communications are greatly influenced by this as well as by its mountainous terrain and weather. 'Communications' is used in a broad sense in this chapter; land, sea and air travel together with radio and the postal services are discussed. To many readers, knowledge of South Georgia and the Falkland Islands Dependencies comes from philatelic studies, so particular attention is given to this – an important, and perhaps the best known, part of the island's communications.

Surface travel on the island

The mountains and glaciers of South Georgia make land communications very difficult. Only one motorable road is established, a rough track between King Edward Point and Grytviken, approximately 1 km long and it is open for about 7 months of the year. Tractors and trailers form most of its wheeled traffic. There are several rough pedestrian tracks including one which connects Husvik, Stromness and Leith. Within the whaling stations and the scientific station, narrow gauge railways once operated. These were mainly winch-powered as well as moved by gravity but some were man-powered or had locomotives. The introduction of tractors in the 1950s resulted in much reduction of railway use. A narrow gauge steam locomotive (0-4-0) remains at Ocean Harbour, on its side but substantially intact.

Around King Edward Cove, and in some other places, various tracked vehicles have been used for travel over snow. The frequent steep slopes and snow drifts greatly limit the range and utility of such vehicles.

Generally, overland travel is on foot, using snowshoes, skis or crampons, where appropriate. The glaciers present formidable barriers as many are highly crevassed; in winter, with most crevasses filled or bridged by snow, access to many areas becomes easier. From King Edward Point it is possible to trek with few difficulties to regions between Royal Bay and Fortuna Bay if a launch is used for access to the east and west shores of Cumberland Bay. This is the longest distance one can proceed without equipment for glacier travel. In the author's opinion, trekking on South Georgia (although only for short distances) is easily comparable to that in the Himalayas and in the best parts of the Andes, and is thus among the best in the world. A series of field huts has been established by the British Antarctic Survey for various purposes; these, together with a variety of other refuges, provide accommodation at convenient halting places. As most summer trekking is done in the course of duty, an officer's employment largely determines how much is possible for him. Fortunately a large area of good country for short excursions is easily accessible from King Edward Point.

Glacier travel requires a party with equipment for crevasse rescue as well as for crossing ice and snow fields together with tent accommodation, food, fuel, etc. Suitable training and experience are essential for this type of travel; as are arrangements for rescue, if required. Problems most likely to be encountered are crevasses or other ice features and prolonged periods of very severe weather. Nevertheless, if all goes well, most parts of the island may be reached across the glaciers, snow fields and ice-caps although changes in glacier forms owing to their recession are presently making this increasingly difficult. Several fully equipped expeditions have travelled extensively on the island (some of which are described in Chapter 4). The British Antarctic Survey work does not usually involve travel across such areas and, apart from some brief recreational excursions in winter by men able to spare the time, little is done.

The contrast between trekking over unglaciated areas of South Georgia in summer and winter is great. Much summer trekking is over rocks, scree, or vegetated areas. The greatest difficulties are presented by extensive boggy areas and many streams. Melt streams from glaciers and ice caps are often deep and fast-flowing. As their water is 'glacier milk' – an opaque suspension of very finely ground rock (rock flour) – it is very difficult to ascertain the form of the stream bed or the depth before putting one's foot in. Depths may vary from roughly 1 to 150 cm and men have been surprised after very over-cautiously stepping into the former to be subsequently, quite unexpectedly, submerged in the latter. In winter, snows cover everywhere and their surfaces are very variable. Deep soft snow may give way to a hard and slippery glazed surface, to good compact snow, to a thin glazed fragile surface, and to other forms in a matter of only hundreds of metres. This may make travelling very hard work. Cornices, windscoops, and similar formations are commonly developed and, at the high tide line small unstable snow and ice cliffs may make it difficult to travel along some coasts at other than low tide. In any season reindeer trails provide useful tracks in certain regions.

The Salvesen Range near the head of the Brøgger Glacier showing ice surface conditions. (B.C. Storey.)

Sea communications

Until very recently, the only way personnel and supplies could reach South Georgia was by ship. The historical chapters relate various voyages recorded since 1675 and the first landing in 1775. Several of the vessels taking part in the first epoch of sealing became regular visitors to the island. The beginning of reasonably frequent communications came with the whaling era and the subsequent establishment of a local administration. The times and destinations of most ships then operating were directly related to the needs of the whaling industry. Therefore arrivals tended to be from September and departures from April with several ships in summer and none or very few reaching the island during the winter. Most of the vessels were operated by individual whaling companies and travelled from or to Norway and the United Kingdom, often with calls at the Dutch East Indies, for oil, or the Cape Verde Islands (where stowaways presented great problems) and at various South American ports.

Most ships worked an annual schedule for many years, supplying particular whaling stations. Many became well known and associated with South Georgia.

Various other vessels also arrived, including Royal Naval ships, ships of expeditions, and ships operating for the Government. During the whaling era, and even sometimes later, King Edward Cove had much more shipping pass through it than had Port Stanley. King Edward Cove was declared to be a 'Port of Entry' for the Falkland Islands Dependencies in 1912 and is the only one on South Georgia. For the convenience of the whaling companies the Customs Officer or Magistrate occasionally went to other whaling station harbours to enter or clear ships.

The vast majority of the passengers on the whaling companies' ships were their employees; occasionally Government officers or members of private expeditions were also carried by arrangement. The first regular shipping service to carry

The 60 cm gauge steam locomotive at Ocean Harbour, March 1979. (Author.)

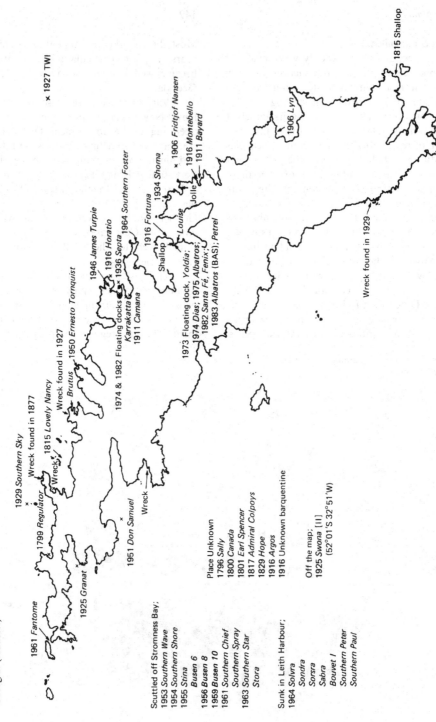

Wrecks, hulks and other vessel remains at South Georgia. (Author.)

passengers to South Georgia was that of *Fleurus* from 1924 (£10 return in the saloon, £5 in a cabin) which operated from Port Stanley. Since she was withdrawn from service in 1933, Royal Naval and Government ships have called at the island more frequently but, owing to their routes, it has often been quicker to travel between South Georgia and the Falkland Islands by way of Punta Arenas, Buenos Aires, or Montevideo (or conceivably Norway or the United Kingdom, at the end of a whaling season) than to await a direct service. For the period after the end of the whaling era, and especially after the withdrawal of the civil administration, the ships of the British Antarctic Survey have been the most regular visitors, calling several times a year during the austral summer.

The port at King Edward Cove is still remarkably busy as increasing numbers of ships from the fishing fleets of Poland and the Soviet Union have called over the last 10 years. Up to two dozen have in one season recently been received, when 71 were working in the island's vicinity. Their principal purpose in calling is to take water, of which abundant supplies of high quality are available at Grytviken. Others call to transfer cargo and use the protected harbour to make repairs when necessary. The opportunity for their complements to go ashore, after generally long periods at sea, is also appreciated. Visits of ships of various other nations, many involved in Antarctic or fisheries research are also common. Two tourist ships, *Lindblad Explorer* and *World Discoverer* are regular visitors; the occassional private yacht also arrives. Since the mid 1970s, vessels registered in the United Kingdom, Falkland Islands, West Germany, East Germany, Poland, the Soviet Union, France, Norway, Panama, Argentina, Peru, the United States of America, South Africa, Italy, and Australia have called at the island. There are approximately 50 recorded shipwrecks and otherwise abandoned vessels around the island from the sealer *Sally* in 1796 to the launch *Albatros* in 1983. These include shallops, sealing vessels, whale catchers, launches and a submarine. The positions of many of these are shown on the accompanying map.

Some famous ships of South Georgia

Many vessels have gained a strong historical association with South Georgia and some of these

remain at the island. Brief details of a number of these are given below.

Sailing Ships

Bayard: a riveted iron-hulled three masted sailing vessel is still at South Georgia, aground on the southern side of Ocean Harbour. She was built in April 1864 by T. Vernon and Son of Liverpool and is 67 m long, 1335 tons. On 6 June 1911 she was moored at the coaling pier on the north side of Ocean Harbour when a severe gale caused her to break loose and run aground across the harbour, where she was holed. Attempts by two whale catchers to tow her off were unsuccessful and she remains there. During the First World War, when ships were scarce, it was proposed that she should be refloated and put into commission again, but no action resulted.

Brutus: another three masted iron-hulled vessel is at Prince Olav Harbour. She was built in 1883 by J. Reid and Co. of Glasgow and first named *Sierra Pedrosa* (which may still be discerned on her hull) while serving the Sierra shipping line of Lima. At Prince Olav Harbour she was a coaling hulk and arrived under tow by four whale catchers from Cape Town. Her length is 75.9 metres and tonnage 1686. Brutus Island is named after her.

Louise: a copper-clad three masted wooden vessel, she was one of C.A. Larsen's first fleet which arrived at King Edward Cove on 16 November 1904, where she remains beached on the southern side. She was built in 1869 by G. and C. Bliss of Freeport, Maryland, United States, as the *Jennie S. Barker*. In 1880 she was sold to a Sandefjord company which used her in the Baltic timber trade (hence her baulk ports for loading long timbers). After transporting materials and men for the first whaling station (Grytviken) she provided accommodation until barracks were completed. Subsequently she became a coaling hulk until oil replaced coal for fuel. She is 52.8 m long and of 1065 tons. Luisa Bay [sic] is named after her.

Tijuca: was also a wooden vessel 52.3 m long of 846 tons, launched in Nantes in 1866 and one of A.D. Bordes fleet for many years. She carried supplies from Buenos Aires and Montevideo to Grytviken from 1907 to 1942. In 1924 she was fitted with an auxiliary engine. Subsequently she was laid up but resumed service later in the Second World War between Buenos Aires and Cape Town until wrecked on 26 July 1946 off southern Brazil. Tijuca Point is named after her.

Steam ships; whale catchers, sealers, and transport ships

Albatros, Petrel, and *Dias:* are still at the Grytviken whaling station jetties, sunken at their moorings. *Albatros* and *Petrel* were originally whale catchers and later were used for sealing when superseded by faster vessels. *Dias* was a sealer for all her service on South Georgia. Their machinery consisted of triple expansion coal-fired steam engines, later modified for oil firing. *Albatros*, 32.8 m long, 210 tons, was built in 1921 at Sevlik, Norway. *Dias*, 33.1 m long, 167 tons, was built in 1906 at Beverley, near Hull. *Petrel*, 35.1 m long, 245 tons, was built in 1928 at Oslo; a plan to salvage and restore her was proposed in 1983. The scientific

station's launch, *Albatros* bears one of these names, Diaz Cove [*sic*] and Petrel Peak the others.

Karrakatta: also remains at South Georgia, high and dry on the slipway at Husvik. She was built at the Akers Makanisk Vaerksted, Christiania (now Oslo) in 1912 for the West Australian Whaling Company (Chr. Nielsen and Co.). Her length is 32.4 m, tonnage 179 and her engines developed 69 horsepower. She was last used at the slipway as a boiler – a lagged steam pipe left her bows, near where her whale cable ran, to lead to an engineering workshop powered by steam. An adjacent slipway allowed other vessels to be hauled up for repairs. Her boiler remained coal-fired and a hole cut in her hull provided access to the boiler room. Despite the ravages of weather she is remarkably intact. Her bell is at King Edward Point – hanging on the flag pole.

Lille Carl: (originally the *Duncan Grey*) was probably the best known of South Georgia's whale catchers and arrived in 1907. She was built in 1884

The hulk *Bavard* at Ocean Harbour, March 1979. (D. Sanders.)

in Christiania (now Oslo) was only 29.6 m long and 77 tons. It was once reported that she secured a whale slightly larger than herself. Another of her distinctions was to have taken the last Kaiser out on a whaling voyage from Tromsfylke in the extreme north-east of Norway. The name *Lille Carl* was applied to her as a distinction from the later, larger catcher *Karl*. She was first used for towing flensed whale carcasses from Grytviken to the floating guano factory *Nor* in King Edward Cove. Then she became a sealer mainly working the beaches from the Bay of Isles to St Andrews Bay. After leaving the island in 1960 she was reported as serving in Paraguay on the Paraná River. Carlita Bay, a Spanish diminutive feminine form, is named after her.

Fortuna: the first whale catcher at South Georgia remains, in part, on the beach near low water just north of Hope Point. She was another of C.A. Larsen's first fleet, constructed in 1904 by Framness Mekanisk Værksted, in Sandefjord specially for the newly formed Compañia Argentina de Pesca. This Værksted subsequently built many other vessels for South Georgia. Her length was 30.3 metres, tonnage 164 and she arrived in November 1904. On 14 May 1916 at 06:00 she ran aground at full speed while her helmsman was reading a letter. The damage sustained was severe, only her boiler was salvaged to be used in the whaling station. Parts of her bulkheads and propeller shaft may still be seen on the beach and Fortuna Bay is named after her. *Rosita* was her sister ship, launched in 1905 to become South Georgia's second whale catcher.

Undine: was an iron-hulled steam yacht built in 1884 at Leith, Scotland. From 1888 to 1906 she served in the Royal Navy and once, as RY *Lady Aline*, carried Queen Victoria. In 1907 she was purchased by C.A. Larsen who made several voyages of exploration aboard her from South Georgia where she arrived on 22 September 1907. He also lent her to the Filchner expedition for this purpose. She continued service as a sealer until sold to an Argentine fishing company in 1919 and, after

The hulk *Brutus* at Prince Olav Harbour. (D. Sanders.)

various changes of ownership, capsized in 1960 and sunk. Undine Harbour and Undine South Harbour are named after her (Undine is a classical water spirit).

Coronda: a transport ship of Christian Salvesen's Company which originally served at New Island whaling station in the Falkland islands. On her first voyage to South Georgia, to the newly established Leith Harbour whaling station in November 1909, she carried James Innis Wilson, the first Magistrate. She was built in 1892 as *Manica* by W. Grey at Hartlepool and purchased by Salvesen's in 1908. Her length was 94.6 m, tonnage 2733, and her engines developed 279 horsepower. In 1917, during the First World War, she was sunk by enemy action. Coronda Peak is named after her and subsequently another of Salvesen's ships carried her name.

James Turpie: an iron steamship built in 1881 by A. Leslie and Company of Newcastle was another Leith Harbour ship purchased by Christian Salvesen's Company in 1909, which arrived at South Georgia on 17 May 1910. She was used as a coal hulk until oil replaced coal as a fuel. In 1946 she sank at her moorings in Leith Harbour, where she remains. Her length was 82.5 m, tonnage 1732 and her engines, hardly ever used at South Georgia, developed 172 horsepower.

Ernesto Tornquist: a 6620 ton ship, 137.5 m long, formerly *Craftsman, Hampstead Heath* and *Kommandøren I*; she was built in 1897 by C. Connell and Co. of Glasgow. She was originally a passenger liner but was converted to a floating factory in 1924 and operated out of South Georgia from 1932. On 16 October 1950 she ran aground in a gale and blinding snow storm at 'Windy Hole', South Georgia and was lost. Her complement and passengers, 260 people, were fortunately rescued, largely by *Petrel*. Subsequently 'Windy Hole' was named Tornquist Bay. Ernesto Tornquist, a Swedish banker, was one of the founders of the Compañia Argentina de Pesca.

The hulk *Louise* at Grytviken, June 1979. (Author.)

Fridtjof Nansen: of 2563 tons, was another floating factory which foundered on South Georgia. She was built in Newcastle in 1885 and converted for whaling by the Sandefjord Whaling Company in 1906. On 10 November 1906, on her first voyage to South Georgia, she struck an uncharted reef off the Barff Peninsula, broke into three sections and foundered in about 7 minutes. Nine of her complement of 58 were drowned and the others rescued, by her two accompanying whale catchers *Norrona* and *Sudero*. The reef she thus discovered was subsequently named after her, as was the second floating factory used by the same company.

Fleurus: was the first ship to operate regularly between the Falkland Islands, South Georgia and other parts of the Falkland Islands Dependencies. She was built in Savannah, Georgia, United States in 1919, was 42.7 m long and of 355 tons. The Tønsberg Whaling Company, which operated the Husvik whaling station, had contracted with the Falkland Islands Government to provide a mail service and purchased her at Boulogne in April 1924 for this. From that year until 1933 she made over 30 voyages from South Georgia which included two with the Governor aboard in 1927 and 1928. Subsequently she served in the North Sea, became HMS *Thorodd* during the Second World War and was scrapped in 1953 in Norway.

Albatros (port) and *Dias* (starboard), sunken at their moorings at Grytviken, January 1980. (Courtesy British Antarctic Survey, C.J. Gilbert.)

Aircraft

Air communications with South Georgia have all been, until very recently, from ship to shore. The first were carried by those pioneer communication aviators throughout the world, carrier pigeons. They transmitted advance news of arrivals and catches from whale catchers to Prince Olav Harbour, prior to the advent of radio. One imagines that a substantial proportion of them formed food for skuas. The first aircraft to arrive and fly at South Georgia were two Walrus float planes, aboard HMS *Exeter* in November 1938 (described in Chapter 4). Several floating factories of the whaling companies and the Trans-Antarctic Expedition subsequently deployed similar aircraft on the island. The former used them further south for spotting whales. Helicopters were first flown on South Georgia from the floating factory, *Southern Venturer*, in 1956. The Royal Naval ships; HMS *Protector*, *Endurance*, and several others (especially in 1982) have used them extensively for landings and survey, in the course of which they have provided much assistance to the British Antarctic Survey. During the summer 1981/82 the first aircraft crash occurred on the island – to be followed by four others before winter (two of which were shot down Argentine aircraft) – a total of five helicopters crashed in 6 months.

The first land-based aircraft to reach South Georgia were C130 Hercules of the Argentine Airforce which made several flights over the island prior to the 1982 invasion. One later made the first direct air delivery to the island: on 31 May 1982, a bomb, aimed at a Royal Naval Task Force ship, missed its target. The Royal Air Force has subsequently made several deliveries of mail for the British Forces Post Office and other items by parachute from Hercules flying from Stanley. Although the first airmail arrived at the island on 12 October 1982 by this means, none has been des-

The whale catcher *Karrakatta* on the slipway at Husvik, February 1982. (Author.)

patched owing to lack of appropriate facilities. Deliveries were made virtually fortnightly by air drop from Hercules aircraft operated by the Royal Air Force during much of the 1982/83 summer and at least monthly thenceforth.

One of the proposals of the 1982 Shackleton Report was for the construction of a landing ground on South Georgia. This has received serious consideration following the events of that year; it had already been investigated by the British Antarctic Survey in 1974. Problems are very great owing to scarcity of suitable land; the very strong and gusty winds as well as other meteorological conditions experienced; transport required to King Edward Point from the landing site; effects of winter; and possible bird strike. Several sites have been proposed and investigated. Certain types of helicopter and amphibious aircraft have the capacity to reach and land at South Georgia from the Falkland Islands but it is an expensive operation, with only limited loads possible. A serious problem for any flights is the lack of alternative landing grounds closer than the Falkland Islands and the very rapid changes of weather which South Georgia experiences.

Radio

The first wireless telegraphy facilities at South Georgia were those aboard the ship *Deutschland* which arrived on 31 October 1911 (described in Chapter 4). She made no radio contacts, however, probably as her apparatus was inadequate and few other stations were established in that region of the world at the time. Her Wireless Officer, Walter Slossarczyk, was lost at sea while she was at South Georgia. Whaling companies had pioneered the use of wireless in Antarctic regions and used it to keep catchers in contact with their base (a land station or floating factory). This allowed communication of progress reports, advice of movements, and locations of whales.

The use of radio by whaling ships was well developed before the Government considered establishing a station on South Georgia. This was first proposed for Cape Buller, in 1913. The 'Wireless Telegraphy Ordinance of 1903' for the Falkland Islands was extended to the Dependencies in that year. The first Government Wireless Station was not, however, established on the island until 1925. This was built at the same time as the marine

laboratory and accommodation for the *Discovery* Investigations. Official transmissions started on 1 April 1925 with Mr H.R. Prickett as operator. The station was equipped with two high aerials of steel lattice construction which remained as prominent landmarks until early 1954. Their foundations are still apparent and the old wireless station still stands, although closed on 13 January 1970, providing an emergency radio room. The original apparatus was provided by the Marconi Company. This was replaced with 1935 equipment during the Second World War. Some of this remained in place and could still be operated in 1982.

South Georgia received its international radio call sign, 'ZBH', when the station opened (it has no numerical suffix as only one government station now operates). On 1 January 1926, the Magistrate was able to report the reception of the first broadcast programmes of 'British Official Wireless News Messages' in Morse from the Rugby transmitter in the United Kingdom and (in 1932) the 'Voice of the Empire' from Daventry, the precursor of the BBC World Service. King Edward Point became a registered coastal station for ships' communications shortly afterwards and traffic, with consequent paperwork, increased enormously. All early transmissions were in keyed Morse before Morse tapes, teleprinters and, later, voice became practicable. The most recent activities of ZBH, until the 1982 invasion, principally involved British Antarctic Survey communications. Morse was generally used for meteorological signals and teleprinters for most of the other traffic. Radio telephone services have not been extended to the island and the next development is expected to be facsimile transceivers. The Post Office also accepts telegrams and the Falkland Islands Dependencies are represented at the International Telecommunications Union by the British Post Office.

Various officers of the British Antarctic Survey and other persons during the whaling era have operated in an amateur capacity from South Georgia with Falkland Islands licences. Contact and exchanges of 'QSL' cards have been made with virtually everywhere on Earth but, as South Georgia is a very rare contact, operators often report getting swamped with incoming transmissions. When atmospheric conditions are appropriate, a remarkable amount of communication becomes easy with South Georgia's effective antipodes, Japan.

Atmospherics and the remote location of South Georgia often cause communications difficulties, this was especially true in the early days. At times when the ionosphere is passing rather than reflecting radio emissions and during periods of magnetic storms, communication with the island may be possible in Morse only and even that may be difficult. The local topography of the King Edward Point region causes additional complications as Mount Duse rises immediately to the north of the radio station. Although the main aerials are disposed in three directions, communications difficulties at certain distances and directions are com-

mon. In the future it is expected satellite communications will be used, which will alleviate these problems (for long distance radio at least).

Postal communications

Although the basic and most important task of the Post Office is to provide communications for the inhabitants of South Georgia, it also has other significance. The existence of a postal service, especially one associated with and recognised by the Universal Postal Union, has always been an important demonstration of the effective government of a region and thus of the sovereignty of that government over the region. The Post Office at King Edward Point, South Georgia, Falkland Islands

The Radio Room at King Edward Point, January 1980. (Courtesy British Antarctic Survey, C.J. Gilbert.)

Dependencies, has functioned since 1909 with only one brief interruption when, in April 1982, King Edward Point was attacked. As well as this, the Post Office has raised considerable revenue for the Dependencies, especially over the last decade, owing to the desire of philatelists to collect its stamps and study their history.

The officer-in-charge is designated Deputy Post Master and is responsible to the Post Master of the Falkland Islands in Stanley. The British Post Office represents that of the Falkland Islands Dependencies at the Universal Postal Union. Generally, the magistrate is the Deputy Post Master although it has, at various times, been delegated. During the whaling era, mails were received far more frequently than in later years. They arrived from the United Kingdom, Norway, the Falkland Islands and Argentina, together with a few exceptional ones from elsewhere, mainly on ships of the

Government and the whaling companies. Several difficulties were experienced in establishing a regular mail connection between the Falkland Islands and South Georgia for some years until a regular service was inaugurated by *Fleurus* as related earlier in this chapter. She also carried various supplies, including live animals.

Since the conclusion of the whaling era early in 1966 and up to early 1982, most of the mails have been carried by the ships of the British Antarctic Survey. Thus they were received during the summer season only. The first usually arrived around the end of November and was followed by about four others before the end of season in April. There were then no postal communications for at least 7 months. Mails were received from the Falkland Islands and the United Kingdom. They were despatched to these places with the occasional one going to a South American or other port if the

The Post Office at King Edward Point in October 1982. The windows are taped as a precaution against blast damage. (Author.)

shipping opportunity became available. Airmails were accepted for conveyance by sea to Port Stanley or elsewhere for forwarding. There was no direct airmail service to the island and incoming airmails were held at Stanley to await the next available ship which could convey them. The Argentine postal link, arranged by the 'Joint Declaration' of 1971, used an Argentine military airline between Comodoro Rivadavia and Stanley. Unfortunately this service proved rather unreliable and it was not uncommon for items passing through Argentina, both inward and outward, to be 'lost in the post' and sealed items were sometimes received violated. This link was, however, more rapid than alternatives available during the time it operated. Since the removal of the invaders in 1982, postal arrangements have been different. This and some other recent changes are described later in this chapter and in Chapter 9.

Mail deliveries and collections on the island depended very much on circumstances prevailing at the time; it has always been, substantially, an opportunistic process. Shortly after the establishment of the whaling stations on the eastern side of Barff Peninsula (Godthul and Ocean Harbour) a refuge for persons carrying the post and other items was established near the beach in Cumberland East Bay at the foot of the valley leading to Ocean Harbour. Former users of this have carved their names and the date in the woodwork, the earliest being 1912. The post refuge was reached by sailing or rowing about 12 km from Grytviken. To operate, this service required a courier who was used to trekking as well as being an accomplished boatman.

This type of communication did not last for long, being replaced by quicker and safer transport by whale catchers or other larger vessels. In the late 1950s, a Customs House was established at Leith Harbour which functioned as a sub-Post Office there when occupied. The offices of the whaling stations had letter boxes for collecting mail to be despatched to the Post Office or to be dealt with by the Deputy Post Master when he visited the stations. Receipt, sorting and delivery of mails was sometimes done on the ship carrying them, where this was more efficient for delivery.

Since the closure of the whaling stations, most mail arriving (with the exception of philatelic items) has been for British Antarctic Survey personnel and is distributed on receipt at King Edward Point. Mail for Bird Island, and some other field stations, is carried onwards by the Survey's ships or HMS *Endurance*, the latter often makes deliveries with her helicopters. Occasionally, mail is carried by 'postal courier', persons trekking in the course of performing other duties, to field parties in appropriate places (for instance in 1981/82 to St Andrews Bay and Royal Bay). A *Poste Restante* service is also available which is occasionally used by persons aboard some of the visiting ships.

One of the early duties of James Innis Wilson, the first Magistrate appointed to South Georgia, was to establish the Post Office, and he maintained remarkably accurate records of its operation. This was done on 4 December 1909 when a Post Office Notice was promulgated. The first mail was despatched on 23 December 1909 aboard *Cachalot* to Port Stanley. This comprised 1000 letters and 389 postcards of which 845 letters and 339 postcards were for Norway, 51 and 10 for the United Kingdom, 28 and 14 for Sweden, 25 and 9 for the United States of America, 11 and 2 for Finland, 8 and 5 for Russia, 8 and 3 for Argentina and 11 letters were for Denmark. The remainder were addressed to: Chile, Belgium, France, Canada, Iceland, China, Holland, Brazil and Germany in descending frequency. They were marked with one of two small metal stamps reading 'SOUTH GEORGIA' to indicate their origin, for their stamps were cancelled by a Falkland Islands frank taken to South Georgia for that purpose. A second despatch of 1134 letters and 406 postcards was made on the same ship on 19 February 1910. The composition was: 872 letters and 361 postcards to Norway, 82 and 8 to the United Kingdom, 38 and 6 to Sweden, 28 and 10 to Argentina, 24 and 8 to Denmark, 21 and 1 to Germany, 20 and 4 to the United States of America, 15 letters to the Falkland Islands and the remainder to 13 other countries. The third despatch was made on 2 April 1910 to 20 countries and comprised 682 letters and 177 postcards. The first South Georgia franking stamp and the requirements for a registered letter service were received by the Magistrate on 17 May 1910. These items went into service directly. Although these mails were the earliest official ones, photographic postcards had been despatched from South Georgia as early as February 1905, aboard *Rolf*. One of these, sent to Norway by C.A. Larsen, had a plate of *Fortuna* and partly constructed Grytviken.

At first the Magistrates were required to pay for stamp supplies personally and in advance. J.I. Wilson brought supplies of ½d, 1d, 2d, 2½d, 3d, and 6d stamps with him and sold £12 worth for the first despatch. These were used for ½d book post, 1d Imperial letter, 2d registration (when introduced) and 2½d foreign letter. Norwegian currency was accepted for them at the rate of one Krona for one shilling. Higher denomination stamps of 3 and 5 shillings, together with stamped post cards, were ordered in March 1911.

Postage stamps of South Georgia and the Falkland Islands Dependencies

Falkland Islands stamps were used at South Georgia and other parts of the Dependencies until 1944. A provisional stamp for South Georgia, depicting the profile of King George V, was designed in 1911 at the suggestion of the Magistrate and with the Governor's recommendation, but it was not adopted. The only way to determine whether a particular Falkland Islands stamp was used on South Georgia prior to 1944 (and in a few cases afterwards) is from the postmark. Various Falkland Islands stamp issues were used on South Georgia during this period and four anomalies were produced locally to meet shortages of particular denominations. These are described later in this section.

The first stamps for the Dependencies of the Falkland Islands were issued at South Georgia on 24 February 1944. These were a series of eight stamps of the Falkland Islands definitive set first issued in 1938, overprinted for the four Dependencies with Post Offices: South Georgia, South Orkney Islands, South Shetland Islands, and Graham Land. They were of denominations of ½d, 1d, 2d, 3d, 4d, 6d, 9d, and 1/-. Neither stamps in the same series with denominations of greater than 1/- nor the short-lived 2½d stamp were overprinted; ordinary Falkland Islands ones remained in service for higher values until 1954. Three of the overprints had themes relevant to South Georgia: 6d with RRS *Discovery II*, 9d with RRS *William Scoresby*, and 1/- with Mount Sugartop. The others had Falkland Islands scenes and one of Deception Island in the South Shetland Islands.

These were replaced on 1 February 1946 with a series of eight stamps produced for the Dependencies (½d, 1d, 2d, 3d, 4d, 6d, 9d, 1/-). All were of the same design, a map depicting the Dependencies. There were two noticeably different printings, the first had rather thick lines in the maps and contained a large number of minor errors; the second, introduced on 16 February 1948, had finer lines and was better produced. In 1949 a 2½d stamp was added to the series.

In 1946, the first commemorative issue for the Dependencies was produced. This was the 'Victory 8 June 1946' issue of two stamps released in South Georgia on 4 October 1946. The 1d and 3d stamps had the same vignette showing the Houses of Parliament. The theme was also used in many other parts of the British Empire to commemorate the official celebration of the end of the Second World War and was the first Dependencies stamp in the Crown Agents' 'omnibus' series. These series consist of stamps of similar designs and denominations commemorating events deemed appropriate and issued, as far as possible, simultaneously in a variety of British dominions and colonies. They generally have little direct connection with their place of issue. In 1948 another omnibus pair was issued on 6 December for the Silver Wedding Anniversary of King George VI and Queen Elizabeth. The stamps were valued at 2½d and 1/- and show profiles of their Majesties. The 75th anniversary of the Universal Postal Union in 1949 was commemorated by four stamps with appropriate themes for denominations of 1d, 2d, 3d, and 6d; this was also an omnibus issue released on 10 October.

The Coronation of HM Queen Elizabeth II on 2 June 1953 was commemorated with a 1d stamp, issued on 4 June, showing a vignette of the Queen. Identical stamps were issued in all other parts of the Empire and Commonwealth. The new reign resulted in the issue of a very attractive set of definitive stamps for the Dependencies. These were issued on 7 February 1954, and consist of 15 denominations depicting the ships involved in the exploration of the Dependencies: ½d, *John Biscoe*, in service from 1947 to 1952–another ship of this name came into service in 1956; 1d, *Trepassey*, 1945–47; 1½d, *Wyatt Earp*, 1934–36; 2d, *Eagle*, 1944–45; 2½d, *Penola*, 1934–37; 3d, *Discovery II*, 1927–37; 4d, *William Scoresby*, 1926–46; 6d, *Discovery*,

1929–37 and in 1901–04 when she carried Captain Robert Falcon Scott's first expedition; 9d, *Endurance*, 1914–16 (HMS *Endurance*, presently in service, is named after her); 1/-, *Deutschland*, 1910–12; 2/-, *Pourquoi Pas?*, 1908–10; 2/6, *Française*, 1903–05; 5/-, *Scotia*, 1902–04; 10/-, *Antarctic*, 1901–03; and £1, *Belgica*, 1897–99. Many of these ships have visited South Georgia. Four of these stamps (the 1d, 2½d, 3d, and 6d denominations) were overprinted for the Trans-Antarctic Expedition 1955–58 and first issued on 31 January 1956 when the expedition arrived at the Filchner Ice Shelf.

On 3 March 1962, the limits of the Falkland Islands Dependencies were redefined, when British Antarctic Territory was separated from them to accord with administrative practices necessitated by the Antarctic Treaty. The 1954 definitive stamps were thenceforth only used at the Post Office on King Edward Point, South Georgia and were replaced on 17 July 1963 by an issue for South Georgia. This was also of 15 stamps but of different denominations to correspond more closely with postal rates. They were; ½d, Reindeer; 1d, map showing South Georgia and the South Sandwich Islands; 2d, sperm whale and giant squid; 2½d, chinstrap and king penguins; 3d, bull and cow fur seals; 4d, fin whale and whale catcher; 5½d, two bull elephant seals; 6d, sooty albatross; 9d, whale catcher; 1/-, leopard seal and its prey – a penguin; 2/-, Shackleton's memorial cross and Mount Paget; 2/6, wandering albatross; 5/-, elephant and fur seals; 10/-, plankton and krill, and £1, blue whale. The £1 stamp was replaced on 1 December 1969 by one depicting two king penguins, a matter that gave rise to some adverse criticism in the philatelic press.

On 15 February 1971, decimal currency was introduced in the Falkland Islands Dependencies. The fourteen 1963 definitive stamps of value less than £1 were overprinted with decimal values and further adjustments to accord with postal rates were made. The overprints were: ½p on ½d, 1p on 1d, 1½ on 5½d, 2p and 2d, 2½p on 2½d, 3p on 3d, 4p on 4d, 5p on 6d, 6p on 9d, 7½p on 1/-, 10p on 2/-, 15p on 2/6, 25p on 5/-, and 50p on 10/-. The £1 stamp remained unchanged and by 31 May 1972 pre-decimal stamps were invalidated. The series was the last British definitive issue printed by the intaglio method. With the exception of the bichrome 2/- or 10p, their designs are monochrome, simple

and unembellished. They demonstrate strongly a whaling theme which was of great importance to South Georgia at the time of their issue. In 1977, a reprinting was made on a glazed paper. This was issued until 5 May 1980, when a new set of definitives for the Falkland Islands Dependencies replaced it.

During the currency of these stamps several multicoloured commemorative and special issues were produced for South Georgia. On 5 January 1972, the 50th anniversary of Sir Ernest Shackleton's death at South Georgia was commemorated by a set of four stamps: 1½p showed *Endurance* beset by pack-ice in the Weddell Sea; 5p illustrated the launching of *James Caird* for the journey from Elephant Island to South Georgia; 10p depicted the route taken and the 20p had a portrait of Shackleton with *Quest* in King Edward Cove (where he died while leading an expedition to the Weddell Sea). A commemorative medallion was issued concurrently. This set was the first special series to be made for South Georgia or the Dependencies with a theme relating to the region and its history. Its predecessors, and about half of the subsequent special issues, formed parts of the Crown Agents' omnibus issues.

On 20 November 1972 two stamps of 5p and 10p were issued to commemorate the Silver Wedding anniversary of HM Queen Elizabeth II and HRH the Duke of Edinburgh as part of an omnibus edition. The designs were identical, with the Royal portraits surrounded by an elephant seal and king penguins against a background of mountains and a frozen ocean. The Royal Wedding of Princess Anne and Captain Mark Philips was celebrated by an omnibus issue on 1 December 1973, comprising a 5p and 15p stamp, both depicting their portraits. The centenary of the birth of Sir Winston Churchill was commemorated on 14 December 1974 by another omnibus issue of a 15p and 25p stamp with Churchill's portrait against a background of the Houses of Parliament and a battleship respectively. These were the first South Georgian stamps to be issued also in a miniature sheet – a commercial device to raise revenue from some collectors.

A most important South Georgia event was commemorated in 1975 with an issue on 26 April. This was the 200th anniversary of Captain Cook's exploration of South Georgia during which he took possession of the island for King George III (on 17 January 1775). Three stamps were produced: 2p, with a portrait of Captain Cook; 8p, with HMS *Resolution* amid icebergs off South Georgia; and 16p, with a reproduction of Hodges' plate of Possession Bay (which illustrated Cook's description of his voyage, published in 1777; it is included in Chapter 2). In the next year, another South Georgian event was commemorated: the 50th anniversary of the *Discovery* Investigations which commenced on the island in 1924 (although it was 1926 before *Discovery* arrived there). The four stamps issued on 21 December 1976 were: 2p, *Discovery* and her biological laboratory; 8p, *William Scoresby* and a water sampling bottle; 11p, *Discovery II* and a plankton net; and 25p, Discovery House (the laboratory, stores and accommodation building which still stands on King Edward Point) and an Antarctic krill.

The 25th anniversaries of the Accession and Coronation of HM Queen Elizabeth II were commemorated on 7 February 1977 and 2 June 1978, respectively, by omnibus stamp issues. The former comprised three stamps; 6p, showing the Duke of Edinburgh at Shackleton's Cross during his visit in 1957; 11p and 33p showed HM the Queen against Westminster Abbey and the Coronation Procession, respectively. The latter issue was another rather unfortunate artificial creation, the style of which has fortunately not, so far, been repeated. It consisted of miniature sheets of six stamps in *se-tenant* strips, of three different designs and separated by a horizontal gutter margin. All were of the same denomination (25p) and a note in the gutter described them. Very few were actually used for postal purposes. The designs included a heraldic animal, HM the Queen, and a fur seal in what must have been a most uncomfortable pose.

The last issue of stamps produced for South Georgia was to commemorate the 200th anniversary of Captain Cook's voyages (1768–1771, 1772–1775, 1776–1779) and this too was an impractical creation for a postage stamp. One was embossed and all were unnecessarily large. When used for post cards, film packages, and most small envelopes, they had to be affixed in irregular places or sideways. There were four denominations: 3p, *Resolution*; 6p, *Resolution* together with a very stylised chart of Cook's voyage to South Georgia

and the South Sandwich Islands; 11p, reproducing a plate of a king penguin drawn by George Forster; and, 25p an embossed cameo medallion of Captain Cook.

Subsequent to these, on and from 5 May 1980, stamps used on South Georgia reverted to the previous designation of Falkland Islands Dependencies. A new set of definitives was issued on that date consisting of 15 denominations with a pictorial theme of the Dependencies reproduced from colour photographs. The design of this issue was rather a prolonged affair which began in 1971; the results are virtually miniature postcards with a series of views of the Dependencies. They consist of: 1p, map of the Dependencies, a great improvement on most earlier ones; 2p, Shag Rocks; 3p, Bird and Willis Islands; 4p, Gulbrandsen Lake, an ice-dammed lake adjacent to the Neumayer Glacier; 5p, the settlement at King Edward Point with the Allardyce Range in the background; 7p, Sir Ernest Shackleton's grave; 8p, Church at Grytviken; 9p, *Louise*, one of the first vessels which founded Grytviken; 10p, Clerke Rocks; 20p, Candlemas Island, one of the South Sandwich Islands, with a penguin colony; 25p, Twitcher Rock and Cook Island, also in the South Sandwich Islands; 50p, £1, and £3 show the three ships which most regularly visit the Falkland Islands Dependencies (RRS *John Biscoe*, *Bransfield*, and HMS *Endurance*), all are shown in or near King Edward Cove.

From the introduction of these new definitives up to the Argentine invasion in 1982, four special issues have been made, three with biological themes and one commemorative. On 5 February 1981 a 'Plants issue' of six stamps was released. These comprise somewhat stylistic renderings of some of the island's vascular plants: 3p, *Lycopodium magellanicum* (Magellanic club-moss); 6p, *Phleum alpinum* (Alpine cat's-tail); 7p, *Acaena magellanica* (Greater burnet); 11p, *Galium antarcticum* (Antarctic bedstraw); 15p, *Rostkovia magellanica* (Brown rush); 25p, *Deschampsia antarctica* (Antarctic hair grass, which is also the only known vascular plant from the South Sandwich Islands).

The Wedding of HRH Prince Charles to Lady Diana Spencer on 29 July 1981 was commemorated on 22 July 1981 by three stamps of an omnibus issue, which consisted of the following denominations; 10p, bouquet representing some of the island's plants; 13p, Prince Charles; and 52p, Prince Charles and Lady Diana.

Two issues with a biological theme were made early in 1982. On 29 January a series on the reindeer was released which showed four scenes: 5p, reindeer with fawns during spring; 13p, a stag during the autumn rut; 25p, reindeer and mountains in winter; 26p, reindeer feeding on tussock grass in late winter. The 'Insects' issue, made on 16 March 1982, should properly be called 'Arthropods' as only half of it represents insects. These are a very well drawn series and show: 5p, *Gamasellus racovitzai* (a mite); 10p, *Alaskozetes antarcticus* (another mite); 13p, *Cryptopygus antarcticus* (a springtail); 15p, *Notiomaso australis* (a spider); 25p, *Hydromedion sparsutum* (a beetle); and 26p, *Parochlus steinenii* (a fly). The first three also occur on the South Sandwich Islands. The official first day covers of these two issues are accompanied by a concise descriptive note, an excellent innovation which provides information about the animals portrayed and the production of the stamps. The use of these notes is to be continued. The insects set was issued on 17 March 1982, only 22 days before the island was invaded.

After the retaking of South Georgia, two more stamp issues were made in 1982 which made a total of four for that year, twice the previous maximum. On 7 September 1982 an omnibus issue to celebrate the 21st birthday of the Princess of Wales was made. This consisted of four stamps: 5p, 17p, 37p, and 50p. The first was the escutcheon of the Falkland Islands Dependencies and the others multicoloured portraits. The date of issue for this set and that for the Falkland Islands was delayed owing to the Argentine occupation and thus are later than the others in this omnibus set. Most of the first day covers were cancelled by men of the garrison occupying King Edward Point on behalf of the Post Office, as was the next issue. In order to raise money for rebuilding, following the damage after the Argentine invasion and subsequent defeat, specially surcharged stamps were issued for the Falkland Islands and Falkland Islands Dependencies. That of the Dependencies was one large stamp with a map of South Georgia of £1 denomination with a £1 rebuilding surcharge. The first day covers, principally cancelled by the garrison, were dated 13 September 1982. The equivalent Falkland Islands stamp was very similar and issued on the

same day.

A Crown Agents' omnibus edition was issued on 23 December 1983; 200 years of manned flight. This consisted of four stamps; 5p, showing a Westland Whirlwind helicopter; 13p, a Westland Wasp helicopter; 17p, a Walrus aeroplane; and 50p an Auster aeroplane, together with a descriptive note. The first flight at South Georgia was made by a Walrus aircraft from HMS *Exeter* in 1938 and the Trans-Antarctic Expedition test flew an Auster from King Edward Cove in 1955; these operations are described in greater detail in Chapter 5. Wasp helicopters formed part of the equipment of HMS *Endurance* for several years and have assisted the British Antarctic Survey in many ways on South Georgia. Additionally, they were of great service during the Argentine invasion of and subsequent removal from South Georgia in 1982; as related in Chapter 9. Whirlwind helicopters served on HMS *Protector* during her surveys and other operations at South Georgia since 1955 and were used extensively during the comprehensive survey of the South Sandwich Islands from HMS *Endurance* in March 1964.

Another biological issue was made on 23 March 1984 consisting of four stamps illustrating Antarctic crustacea: 5p, *Euphausia superba* (Antarctic krill); 17p, *Glyptonotus antarcticus*; 25p, *Epimeria monodon*, and 34p *Serolis pagenstecheri*, also accompanied by a descriptive note. These are accurate representations of some of the more significant invertebrate inhabitants of the waters around South Georgia, the other parts of the Falkland Islands Dependencies, and Antarctica generally.

The future stamp-issuing programme for the Dependencies, managed by the Crown Agents Stamp Bureau, is expected to include two special issues annually for the rest of the 1980s. A list of the 31 series of postage stamps issued at King Edward Point is given in Appendix 3. Additionally, it is interesting to note that Argentina has issued several stamps with maps and other features of South Georgia to emphasise her pretension of sovereignty over the region. These have never been used on South Georgia, even during the brief Argentine occupation of King Edward Point.

Three instances of stamp shortages were solved by Deputy Post Master's initiative. In October 1911, supplies of 1d and 2½d postage stamps were

A selection of postage stamps issued on South Georgia
(with permission of the Postmaster, Falkland Islands
and Dependencies.)

exhausted and, owing to the slow speed of communications at the time, there was no possibility of replenishment until January 1912. The postal requirements of the German South Polar Expedition abroad *Deutschland* exacerbated this deficiency. The Acting Deputy Post Master, E.B. Binnie, solved the problem by producing a rubber handstamp reading 'Paid At , SOUTH GEORGIA', he endorsed the amount and initialled it. These were then cancelled in the usual way. About 2000 were produced between October 1911 and January 1912.

In March 1923 a shortage of 1d stamps, for Imperial letters, led the Deputy Post Master to create a bisect. The $2\frac{1}{2}$d stamp, cut diagonally, was substituted. Unfortunately this led to a shortage of $2\frac{1}{2}$d stamps for foreign letters. This problem was solved in the same way; a small number of 6d stamps were bisected to serve as $2\frac{1}{2}$d ones. Fortunately new supplies were received before this process went even further. These bisects were not authorised by higher authority but were accepted through the Stanley Post Office. They were used for a brief period only.

From 3 to 23 February 1928 an overprinted stamp was used to meet another shortage. 2d stamps were revalued at $2\frac{1}{2}$d with a locally produced metal die and 1179 were so treated. The use of this procedure was fully authorised and the metal die was defaced after it was no longer required.

Postal cancellors and cachets

Other aspects of South Georgian philately which received extensive study are the various cancellations and related marks. At first, shortly after the arrival of the first Magistrate, a Falkland Islands cancellor was used, supplemented by a metal stamp reading 'South Georgia'. Since July 1910, circular metal cancellors with 'South Georgia' at first and, from 1944, this with 'Falkland Islands Dependency' have been in use. The various types are catalogued in specialist publications. A total of five types of metal stamps have been used with one continuously for about 25 years. The latest metal stamp introduced was only in service from 17 March 1982 until it was disposed of immediately prior to the Argentine invasion. It was introduced to be in conformity with those of the British Antarctic Territory and the Falkland Islands, but was regarded unfavourably by the South Georgia Post Office. It took several times as much effort to make a clean cancellation with it than its predecessors did, so most impressions it produced were decidedly messy or weak. It was replaced by a similar cancellor on 14 December 1983.

Rubber cancellors have been in use since the introduction of the first overprinted South Georgia stamps in 1944. Recent types had a self-inking mechanism, which made the work of cancelling large numbers of stamps very much easier. This especially applies to first day covers where many thousands have to be franked. Owing to their comparatively short life, an equal number of rubber cancellors have been used in the last 5 years as metal ones ever since the Post Office opened. Variations on the style and wording are fairly common on them. Special rubber cancellors are used to indicate that an issue is supposed to have been made on the first day of issue. Some of these have been used for several issues, but in the case of the issues for the 25th anniversary of Her Majesty's coronation and the wedding of the Prince of Wales, special cancellors were provided.

As a general principle, the rubber cancellors do not appear to have the same appeal to philatelists as the metal ones; they do, however, result in clearer and more rapid cancelling and are strongly preferred by the Post Office. One recent instance of forgery of rubber cancellations is known. In order to try to reduce the workload of making special cancellations, an automatic machine was introduced in 1977. Unfortunately, despite many attempts at adjustment, it refused to mark both stamps and envelopes correctly – or went to the other extreme of producing very over-inked impressions. It was eventually despatched to the Stanley Post Office and never went into service.

First day covers have generally been produced as special items at the request of collectors of South Georgia and Falkland Islands Dependencies issues. Envelopes with appropriate designs and 'Official First Day of Issue' or similar marks have been used with the stamps of most issues since 1972. These have been produced in large numbers, especially over the last few years. These numbers are much greater than can be cancelled by the staff available even with co-opted volunteers on the first day of issue. Preparing them can often take many weeks or longer to complete in what spare time the members of the scientific station have. Thus the date of the post-mark may not be relied upon as the actual date of cancellation. Usually most of these are cancelled before the date of release; however, for some issues, cancelling may continue for months after the date – especially during the winter season. This back-dated cancelling has not continued beyond the next outward mail from the island after the 'first' day of issue until very recently. However, the date stamps of cancellors are reliable, saving accident, on ordinary mail and other philatelic items.

Various ships and other cachets have also proved of interest to collectors of postal history. The first of these was used by the ship *Deutschland* in 1911. She had a special adhesive stamp made, perforated around the edges and very similar to a postage stamp. It had a vignette of the ship and the text 'Deutsche Südpolar Expedition'. Those used on postcards of the expedition at South Georgia were 'cancelled' with a cachet reading 'Polarschiff Deutschland, Süd-Georgia, Grytviken, November 1911'. The expedition also had pictorial postcards with an etching of the ship which were printed in Hamburg and Buenos Aires. The popularity and great use of these cards was one factor resulting in the Post Office running out of some denominations of stamps and the Post Master making an appropriate remedy.

Subsequently various ships' cachets, but no ships' stamps, appeared on South Georgia mails. All of these are, however, the responsibility of the Master of the vessel concerned and do not form part of the Post Office functions. The most common recent ones are from RRS *John Biscoe* RRS *Bransfield* and HMS *Endurance* which are usually changed annually. No account or record of these is kept by the Post Office as they are private marks. Some visiting yachts have reported that philatelic dealers have supplied them with various cachets in return for exclusive use of or provision of large numbers of covers with them. There are similar rubber stamps used to identify the British Antarctic Survey station at King Edward Point and some of the expeditions to South Georgia. Several fabrications of these types of cachets are known.

The postal instructional cachets of South Georgia are numerous and, as most are rubber, have been replaced as necessary. Many of them exist for various postal instructions. The majority concern redirection for different reasons and several publications of a specialised nature have described them.

Falkland Islands Philatelic Bureau and Dependencies Philately

Although an officer is designated Deputy Post Master on South Georgia, this is only one of his duties. Since 1969, the principal employment of personnel on the island has been with the British Antarctic Survey and the most recent person delegated as Deputy Post Master was the author, a member of the Terrestrial Biological Section (which conducted botanical, invertebrate, and related studies on the island). In general, no great conflict with time arises as only one day, at the most, is required to effect each despatch and maintain the postal records. Appropriate guidance and advice is provided by the Stanley Post Office when matters substantially different from the normal routine arise. When a new issue is to be made, a postal clerk from the Stanley Post Office is sometimes seconded to South Georgia and the philatelic officer from Stanley has assisted during his occasional visits. Virtually all official first day covers are received at South Georgia with the stamps already affixed by the Crown Agents Stamps Bureau, so they need only cancelling on the island.

In order to meet requests and answer enquiries for Falkland Islands stamps, a Philatelic Burea is established in the Stanley Post Office. This provides similar services for the Post Offices of the Dependencies and British Antarctic Territory. Thus mint stamps and official first day covers are sold there, information sheets and similar items are distributed, accounts are operated for some clients, and correspondence is attended to. Some orders for Falkland Islands Dependencies stamps are also fulfilled by the Crown Agents in the United Kingdom. On South Georgia, only normal postal facilities are available. Most postal items which philatelists desire to be franked on the island are received in a form ready for this. The South Georgia Post Office does not normally undertake to affix stamps to envelopes and it is unable to fulfil many of the more outlandish requests made, such as that letters be flown, dived with or carried by courier to various field stations and appropriately certified. Indeed considerable amusement is supplied by some correspondents who have exceedingly odd ideas of the functions of a Post Office. A small proportion of correspondence is received with mixed stamps of various countries; stamps on the wrong side, both sides, or other irregular places on envelopes; puzzle or otherwise unclear addresses; incorrect postal rates; as well as other irregularities, and many of these are necessarily refused transmission. Conversely, the Post Office always endeavours to produce good clean franking impressions and to meet the next possible despatch for all clearly and correctly prepared items submitted to it. The great importance of this being done accurately and efficiently is recognised as although the service is primarily for communications, a substantial amount of revenue is raised from it and this is largely dependent on the reputation of the service.

The invasion and the Post Office

While HMS *Endurance* was at King Edward Cove on 15 and 16 March 1982, the Philatelic Officer from Stanley and the Deputy Post Master balanced the accounts, issued the 'Insect' stamp series, and made a substantial values despatch. On 17 March a new postal cancellor was introduced. Immediately prior to the invasion of King Edward Point on 3 April, the Deputy Post Master had taken action to prevent the Post Office cancellors, seals, and certain documents falling into enemy hands. These were dropped in deep water in lead-weighted bags at a memorised position. One bag was later recovered by the Royal Navy. After the surrender he used a ruse, distracting a guard's attention, to get about 10 minutes alone in the Post Office. This allowed the clearance of the outward mail, collection of most of the cash and mint stamps (including an unreleased series), and the destruction of personal cheques. The enemy found only the coins left. An enemy guard was used, unknowingly, to transfer the sealed mailbag with the despatch and other items to the Argentine ship *Bahía Paraíso*. This action effectively neutralised the Post Office and ensured

A Royal Air Force Hercules aircraft dropping mail to the garrison at King Edward Point, April 1983. (S.J. Martin.)

against abuse of its facilities. While the mails, civilian luggage and other items were in Argentine custody, they were thoroughly searched. Some things were stolen in the course of this, mainly papers of conceivable intelligence significance. What was left of the mails (the vast majority) was carried to the British Antarctic Survey Headquarters in Cambridge by the Deputy Post Master and thence the outward mails were despatched through the British Post Office with appropriate notes attached. The new issue stamps were returned to the Falkland Islands, after the liberation, for issue in South Georgia in the normal course of operations.

Since the Argentine forces on South Georgia were defeated and removed, the postal needs of the garrison on the island have been met by the British Forces Post Office. These have even included direct inward airmail deliveries from the Falkland Islands. The garrison established at King Edward Point has assisted the Falkland Islands Dependencies Post Office with two new issues, as related above, and in a number of other matters involving clearance of postal articles from South Georgia. A rubber cancellor was available for these purposes. No formal despatch was made by the garrison on behalf of the Dependencies Post Office and, owing to unfamiliarity with the procedures, some irregularities have resulted. The former Deputy Post Master returned to South Georgia as a member of the British Antarctic Survey inspection party on 19 September 1982 and, in the course of this, got the Post Office equipment, stationery, and other matters in order. On 21 September it was reopened when a registered and ordinary despatch was prepared which left the islands aboard HMS *Hecate* on 3 October with him, and the Post Office closed. It is not intended that the Post Office will reopen permanently until a civilian Deputy Post Master is available but the Commander of the garrison, who is also appointed Magistrate, will continue to provide some facilities in the interim. Damage sustained during the Argentine occupation included loss and destruction of records, much pilfering, loss of some instructional cachets and other equipment, as well as so much disorder as to suggest they threw the contents of drawers around the office. An officer of the Falkland Islands Post Office was able to spend some weeks on South Georgia in March and December 1983 and early 1984 during which the Post Office was re-opened and services re-established.

7

Physical sciences; the land, ocean and atmosphere

The land, sea and atmosphere of South Georgia are of particular interest because of the nature of the island and the region in which it lies. This chapter contains descriptions of various aspects of: the geological origin and structure, geomorphology, and glaciology of the land; oceanography of the adjacent seas; the climate, meteorology and upper atmosphere; and the work of the observatory.

Geology and the origin of the island

The first detailed investigation of the geology of South Georgia was made by the German International Polar Year Expedition in 1882–83. They explored the area in the vicinity of Royal Bay only and collected many specimens for subsequent analysis in Germany. Their results, with other details that were then available, led to the first hypothesis of geological relationships between South America and Antarctica through the islands of the Scotia Ridge (although Weddell in 1825 and Klutschak in 1881 had suggested this from other evidence).

Many later expeditions to South Georgia made geological observations and collections. However, it was not until the South Georgia Survey expeditions, led by Duncan Carse from 1951 to 1957, that attempts were made to complete a comprehensive geological reconnaissance of the whole island. Most of the latest information has been derived from a series of geological investi-

gations made by the British Antarctic Survey between 1970 and 1977. During these, landings were made throughout the island, its surrounding islets and rocks, and a detailed geological survey was completed.

South Georgia is largely composed of an approximately 8 km thick sequence of volcaniclastic sandstones and shales, termed the Cumberland Bay Formation. Individual sandstones and grit bands within this may be up to 5 m thick and the formation is intruded by various quartz- and feldspar-rich veins. It was derived from a series of active volcanic islands which were situated on the western side of the area which was to become South Georgia, during the Late Jurassic to Early Cretaceous ages (about 140–110 million years ago). The sandstones and shales were deposited by bottom-flowing currents laden with suspended sediments (turbidity currents) in a marine basin which formed along the Pacific margin of an ancient, large, southern hemisphere continent known as Gondwana. Gondwana subsequently disrupted and the sections drifted apart to form the present land masses of South America, Antarctica, India, southern Africa and Australia as well as components of South Georgia. The marine basin was bounded on its eastern side by continental rocks of what is now South America, and on its western side by the volcanic islands from which the Cumberland Bay Formation was derived.

Remnants of the ancient volcanic arc are still preserved on the small islands of the Hauge Reef, on the Pickersgill Islands and on Annenkov Island off the south-western coast of South Georgia. Here andesite, gabbro and granodiorite stocks, sills and plutons, aged 80–100 million years, are intruded into contemporaneous finely banded andesitic tuffs and mudstones, and non-stratified conglomerates and breccias, of the Annenkov Island Formation – another distinctive geological sequence of South Georgia – which is about 3 km thick. Similar rocks are also found on the mainland near Ducloz Head. The andesitic tuffs were deposited by the active volcanoes in shallow marine areas which the conglomerates, andesitic breccias and sandstones formed on their flanks.

As well as the volcaniclastic sediments of the Cumberland Bay Formation infilling this marine basin, sediments derived from the continent on its eastern side were also deposited within the basin. These now constitute the Sandebugten Formation, consisting of sandstones and shales which were also deposited by turbidity currents. This formation is less thickly bedded than the Cumberland Bay Formation and contains fragments of granites with metasediments typical of a continental margin. It occurs between the Barff Peninsula and Royal Bay. A major fault zone now separates the sediments of the Cumberland Bay Formation and the Sandebugten Formation. A similar sedimentary sequence, the Cooper Bay Formation, is present in the southern part of the island.

Sedimentary structures, including graded grits which give indications of directions of flow, are common within these formations and have been used to map the directions of the turbidity currents within the ancient marine basin. These include: slump folds, formed where soft sediments moved down the submarine slopes of the basin; 'tool marks', which include groove casts and prod and bounce marks resulting from pieces of shales or siltstone being ripped from the surface of the underlying beds and then dragged along the base of a flow (some have been found at the ends of the marks they originated); and current marks, which include flute casts, furrows and ridges, washout channels and ripple marks.

The marginal sea beneath which this occurred was not rich in animal or plant life and therefore the sediments have few fossils. With the exception of trace fossils (the tracks left by bottom-dwelling organisms) the formations are essentially un-fossiliferous. At four localities, however, relatively large numbers of fossils have been found. Of these the most common is the bivalve mollusc *Aucellina* which occurs as either single valves or in accumulations of disassociated valves as at Rosita Harbour, Prince Olav Harbour and Queen Maud Bay. These fossils have an age range of 100–110 million years. Three fragments of large heteromorph ammonites, eight small procerithid gastropods, a fragment of a serpulid worm (*Rotularia*) and some belemnite fragments have also been found in these formations as well as some microfossils. The latter, detected during examinations of thin rock sections, include radiolaria and echinoid fragments.

In contrast, the Annenkov Island Formation has a much more varied fossil fauna and flora, including ammonites, bivalves, foraminifera, cirripedia, with fish scales, bones and fragments of a badly crushed spinal column. Fronds of fossil cycads and fragments of fossil wood are also found throughout the island as well as some restricted small deposits of low grade lignite.

Although these three formations constitute the major part of South Georgia, the oldest and most varied rocks occur in the southern part of the island, south of the Ross Pass. They comprise the Larsen Harbour Formation and the Drygalski Fjord Complex. The former is composed of pillowed and massive lavas which are intruded by a sheeted dyke complex and small gabbroic plutons. At Smaaland Cove a composite gabbro-granite pluton intrudes the lavas. These rocks, collectively referred to as an ophiolite sequence, are characteristic of ocean floor rocks throughout the world. It is believed that they formed the floor of the marine basin on which the sediments described above were deposited and that the basin formed about 140 million years ago by a process similar to the sea-floor spreading which is active in the centres of large oceans today.

The oldest rocks are found within the Drygalski Fjord Complex which forms the Salvesen Range in the south of the island. These are gneisses and schists of the original continental rocks of Gondwana. They are intruded by a wide variety of granites, layered gabbros and basic dyke suites which are comagmatic with the rocks of the Larsen Harbour Formation of the marine basin.

South Georgia rocks are thus quite varied and

can be related to a classic island-arc back-arc or marginal basin system similar to the present marginal marine basin systems found along the western edge of the Pacific Ocean, for instance the Japan Sea. The marginal marine basin of South Georgia ceased to exist about 80 million years ago when it closed owing to movement of the volcanic arc towards the continental margin. This movement uplifted and folded the sediments deposited in the basin, resulting in the characteristic intense chevron folding displayed in many rock faces of the island.

Suites of rocks identical to those of South Georgia can be found in the southern part of South America. South Georgia formed part of the same system as present parts of Chilean Patagonia and Tierra del Fuego. It was originally situated less than 500 km south-east of Tierra del Fuego, prior to the disruption of Gondwana, after which it moved

Representative fossils from South Georgia: two ammonites and two bivalves. (M.R.A. Thomson.)

Kilianella sp

Aucellina sp

Neocomites sp

Inoceramus sp

eastwards along a transform fault system (the northern Scotia Ridge) to its present position during the formation of the Scotia Sea (the part of the Antarctic or Southern Ocean bounded by Burdwood Bank, Shag Rocks, South Georgia, South Sandwich Islands, South Orkney Islands and the 55° W meridian from 61° S).

Diagram showing the formation of South Georgia. Markings as in other Figure, below. (B.C. Storey.)

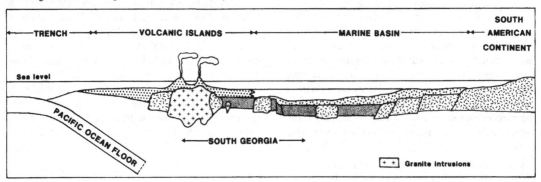

Tectonic activity

In contrast to the Antarctic Peninsula, the South Shetland Islands, South Sandwich Islands and the Cordillera of South America which all have active volcanoes, there are no traces of recent volcanism on South Georgia. The small areas of pillow lavas, massive lava and basalt columns that

The geological structure of South Georgia. (B.C. Storey.)

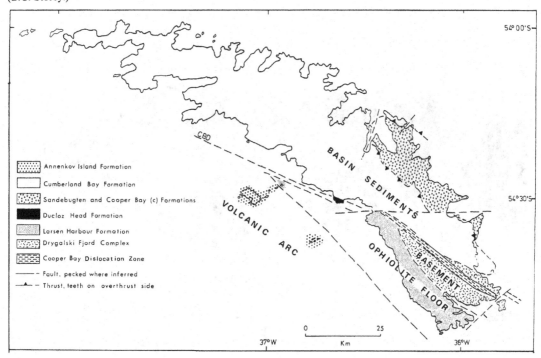

occur are all of great age. Large-scale seismic activity is consequently rare on the island though three perceptible earth tremors have been reported. James Weddell described the first of these in 1823. Seismometers have been operated at various times on the island. The latest series of observations commenced in 1971 and continued until the 1982 invasion. The observatory was part of a global system of stations which reported seismic events to a coordinating headquarters in Boulder, Colorado. The data obtained from South Georgia and other stations in the global system were used to determine where earthquakes and other events occurred in the world. South Georgia was the only station in a very large region and is situated on a major tectonic phenomenon, the Scotia Ridge. Hence the data from it were particularly significant. The instruments had been recording microseismic events virtually daily as well as the occasional large shock – usually from the South Sandwich Islands.

Minerals

One expedition, in 1912, was specifically to prospect for minerals of economic significance on the island. Several others have investigated this possibility, however no such minerals have been discovered. The geological structure renders it unlikely that any occur, with the possible exception of small deposits in the Drygalski Fjord Complex (which is extremely inaccessible due primarily to the mountains). Similarly, the surrounding continental shelf is not likely to hold reserves of hydrocarbons beneath it. There is, however, a deep sedimentary basin to the north of South Georgia, the thickness and extent of which have not yet been completely delineated. It is possible that hydrocarbons might occur beneath this, but it is unlikely that their

The western slopes of Spencer Peak near Maiviken, showing intense folding in the Cumberland Bay Formation and well developed scree slopes. (Author.)

extraction will be practicable. The only minerals exploited involve the local use of quarrying materials and stone for building.

Soils

The soils of South Georgia are largely podzols and other leached forms resulting from the effects of high precipitation and low temperatures. There are, however, substantial organic components in some soils near the shores where deep deposits of peat may also occur. There are four principal types of soil on the island: organic soils, meadow tundra soils, brown soils, and raw mineral soils. The organic soils include peat deposits as deep as 3.5 m in wet basin bogs and valley floors; the oldest basal layers exceed 9000 years of age with the maximum radiocarbon age of 9700 years for the base of a bog peat deposit. Other deposits occur beneath tussac areas with associated bryophyte communities, beneath tall turf-forming mosses on well drained but moist slopes, and more intermediate forms beneath dense stands of *Acaena magellanica*, which merge into meadow tundra soils. Meadow tundra soils have a shallow upper organic layer over a wet, brown-to-grey band of fine clay on a sandy or stony base. These occur in seepage slopes or in marshy areas. Brown soils occur on well drained slopes or on level ground beneath dry grassland. The mineral content of these soils is from 70% to 90%, in contrast to the organic soils (where it ranges from 10% to 30%). The mineral soils are principally glacially derived and vary from fine clay, silts and sands, to coarse gravel and stones. Vegetation is very sparse on them, partly in consequence of the frequent freezing and thawing cycles they experience and the resultant cryoturbic activity. Much of the mineral soil is of recent origin, being deposited at the margins of glaciers which are receding and at the lower ends of scree slopes (much of the South Georgian rock is very fissile, shattering easily to form large scree slopes). Nearly all the island's soils are acidic, the only alkaline ones being in small pockets on the Drygalski Fjord Complex.

Soils near colonies of birds or fur seals or around elephant seal wallows and similar places are highly enriched with nitrogen, phosphorus, and other elements. Vegetation growing on these areas is often restricted to tussac but is comparatively luxuriant except where severely trampled or eroded, but it is rarely of any great extent. Rapid humus accumulation due to low rates of decomposition frequently changes the character of the soils in these areas to a more organic type.

Periglacial features ranging from gelifluxion lobes to a variety of patterned ground (stone stripes, polygons, circles, and steps) are common on South Georgia; permafrost, however, is absent. Although most of the larger features are relict, much small-scale sorting and movement down slopes is evident. Many of the features are partly vegetated and, even in the most active areas, sorting rarely extends deeper into the soil than 20 cm. All patterned ground appears to be on glacial till or fluvioglacial deposits which are at least 1000 years old. Owing to the soil and climatic conditions of South Georgia, combined with the absence of disturbance of most of the island, it is an excellent location for the study of these phenomena.

Glaciology

South Georgia is presently in the grip of an ice-age such as have influenced many different parts of the Earth at various times and glaciers are some of the island's most prominent features. It provides an excellent location for the study of this and related phenomena. Approximately 56% of the area of the island is covered by permanent snow and ice and much of the rest of the island is too steep to permit these to remain. The climate of South Georgia is repeatedly affected by intense depressions coming in from westerly directions which bring heavy snowfalls to the island. The west and south-west flanks of the mountain ranges receive these first and are much more heavily glaciated than the opposite flanks. The prolonged and intense glacial activity has enormously modified the pre-glacial topography of the island and only remnants appear to have survived.

A total of 163 glaciers have been recorded recently on South Georgia of which 50 bear names. Four principal types have been described: those whose snouts are floating in the ocean or in bays; those which terminate on the coast; those which terminate inland; together with cirque glaciers and ice-caps. The five largest glaciers; the Brøgger, Neumayer, Nordenskjöld, Esmark, and Novosilski all have floating snouts and form over one quarter of the total ice cover of the island. Another six glaciers presently are of this type which, if glacial retreat continues, will merge into the next type. The most common type of glacier on South Georgia is that which terminates at the coast. These vary in size from 1 to 58 km². Between the shore line and each glacier is a strip of land, rocks, shingle or a cliff which, in several cases, is known to be getting progressively larger at present. Far fewer glaciers terminate inland and, in these cases, the area between the glacier snout and the shore is usually a smooth outwash plain. At elevations over 500 m, glaciers tend to form wherever snow can settle and accumulate. Thus, in shallow concavities in the mountains and in cirques carved during periods of more intense glacial activity, small glaciers exist. Some of these merge to form the occasional small ice-cap, and the heads of many larger glaciers are often composed of merged cirque glaciers.

Some other glacial structures worthy of note are the Kohl-Larsen Plateau, Mount Paterson, and the rock glaciers. The former is a remarkable ice field about 850 m high and 44 km² in area. It is situated at the heads of two large glaciers discharging on opposite sides of the island. Mount Paterson is a 2196 m high snow dome which, because of its altitude and the consequent low temperatures it experiences, may be expected to have an intact and deep ice record unaffected by seasonal melting. There is only one other comparable snow dome in a similar region of the world (Big Ben, on Heard Island). Two rock glaciers are known on South Georgia, one is ice-cored and the other possibly ice-cemented. Although composed of greater quantities of rock than ice the behaviour of these phenomena is essentially influenced by the ice they contain and resembles that of the true glaciers.

Glaciers probably first formed on South Georgia about 5 million years ago, before the advent

South Georgia showing the parts permanently covered by snow and ice (shaded). (Author, from Directorate of Overseas Survey map.)

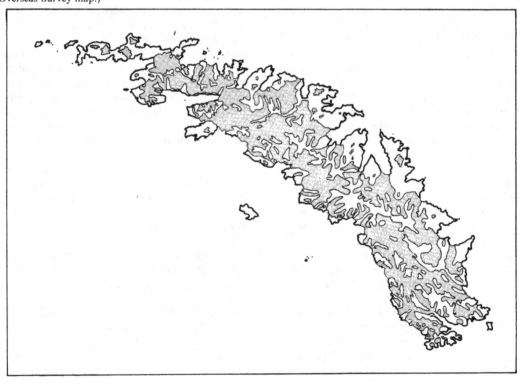

of the Pliocene epoch, as in continental Antarctica and parts of South America. Approximately 12 000–10 000 years ago the island was buried by a deep ice-cap, estimated to have been about 300 by 150 km (30 000 km² in area), 1100 m thick, and attaining an altitude of 900 m above the present sea level. The principal peaks of the island would then have towered above a smooth ice-cap surface as isolated nunataks. Glaciers extending from this ice-cap carved the fjords, now such a prominent part of the island's topography, and extended many of them well beyond the present shore lines, even in some cases to the edge of the adjacent submarine

shelf. A comparable ice-cap surrounded Shag Rocks at this time. Parts of the Antarctic continent presently resemble, on a much larger scale, South Georgia at that time.

The subsequent retreat of this ice-cap resulted in the deposition of many areas of moraine and other glacial features which have allowed interpretation of its extent. Much more recent glacial activity may be described from a major advance which probably reached its maximum at about the time of Captain James Cook's arrival (1775). This may explain his unenthusiastic descriptions of the island and was part of the 'Little Ice-Age' experien-

Neumayer Glacier in Cumberland West Bay. Note the ice-dammed lake and Stromness Bay in the background. The glacier front is about 50 m high and 3 km long. Photographed by HMS *Endurance* helicopter, November 1973. (Courtesy Royal Navy.)

ced throughout the world. Lesser advances probably reached their maxima around 1875 and 1925. Since the establishment of a permanent population on South Georgia, much more information has become available about its glaciers. These show many differences in their behaviour. Several expeditions have made glaciological investigations, but the foundation of a long-term study was laid by Jeremy Smith in 1957 during the International Geophysical Year. Climatic history and other information may be determined from glaciological studies and South Georgia has much to offer in this field.

Essentially continuous research on the glaciers of South Georgia has been conducted by the Falkland Islands Dependencies Survey and British Antarctic Survey since the International Geophysical Year in 1957. This presently forms part of the Survey's contribution to an International Hydrological Programme which began in 1965. Early results from this programme gave further stimulus to these studies which also used comparative data from American and continental Antarctic glaciers. The glaciers of South Georgia provide an interesting link between these two glacial areas as the island occupies a strategic position on the edge of the Antarctic regions and the glaciers below 1000 m altitude are mainly of a temperate alpine form rather than the frigid Antarctic one.

Harker Glacier (left) and Hamberg Glacier (right) entering Moraine Fjord. Photographed by HMS *Endurance* helicopter, November 1973. (Courtesy Royal Navy.)

Most of the research has been concerned with the present and past responses of the glaciers to the island's climate and thus to that of the region. Some glaciers have been monitored intensively. Hodges Glacier, in a small cirque near King Edward Cove, had a meteorological and micrometeorological station operating for 5 years and is regularly mapped for comparison with its state in 1957. Presently, however, owing to problems associated with measuring its melt-water drainage, research is concentrated on neighbouring Glacier Col.

The glaciers of South Georgia have been in general retreat approximately since 1925, but their fluctuations within this trend show significant differences from those elsewhere, which has particular importance for inter-hemispheric comparisons. In addition to the interpretations from the glaciers themselves, such research is greatly assisted by the meteorological record of Grytviken and King Edward Point from 1905. Studies of recent glacier recession indicate that those on the northeast coast have responded much more rapidly to climatic amelioration since the 'Little Ice-Age' than those on the south-west coast. This disparity is at first surprising because the south-west coast receives greater snowfalls so its glaciers flow faster and should therefore have responded earlier than those on the north-east. The influence on the climate of the north-east coast of the Allardyce Range appears to make it more sensitive to changes in atmospheric circulation – a matter which might prove generally applicable to other areas in the lee of mountain ranges.

Two particularly good examples of the present glacial recession are found in the vicinity of Husvik and Prince Olav Harbour. Behind Husvik, at the base of a small cirque glacier, is a series of annual morainic ridges formed during the retreat of the glacier. These may be counted back to about 1940 and vary in size and distance apart depending substantially on the climatic conditions in the appropriate year. Incidentally, they provide an excellent way of determining the time it takes various plant species to colonise recently exposed moraine. At the Prince Olav Harbour site, remarkably prominent annual bands are visible in the glacier surface and these may be allocated to particular years.

A rare and interesting series of observations

became possible at the Lyell Glacier from 6 September 1975 when an enormous fall of rock, estimated as 210 000 m³, from near the summit of 1887 m high Paulsen Peak landed on the glacier (fortunately at a time when there were glaciologists at the scientific station). The fall left seismographic records, an enormous cloud of dust, and large amounts of debris over the glacier's surface. This will greatly influence the mass balance of the glacier and a series of observations and photographs have been made of it as it moves towards the glacier snout. The rockfall represents about six times the normal debris load of the glacier and has substantially increased the surface area covered by debris which, because of the insulation provided, is expected to reduce its rate of ablation. This event provides an

Annual ridges on a glacier near Prince Olav Harbour. (Courtesy British Antarctic Survey, M.J. Skidmore.)

opportunity for testing many hypotheses of glacial activity through a fortuitous natural experiment.

Many glaciological investigations have been made by the British Antarctic Survey on glacier ice from the Antarctic Peninsula. These include detection and determination of trace elements (zinc, cadmium, lead, and copper), microparticles, radioactive strontium and caesium, beryllium isotopes, oxygen isotopes, carbon dioxide, chlorinated hydrocarbons, and other substances. It is expected that this research will be extended to South Georgia in due course and it will yield information about the age and chronology of the glaciers, amounts of radioactivity at various times, ancient climates, volcanic activity, atmospheric contamination, and other matters which leave preserved materials in chronological order in the ice record. Comparisons between these glaciological data and those from peat cores and stalactites, may be possible also, which will increase the significance of the information obtained. The results will be relevant to an area much larger than the South Georgian region alone.

Geomorphology

The principal geomorphological processes acting on the rocks of South Georgia have been glacial; although coastal, mass wasting, gelifluxion, fluvial, and even aeolian effects have contributed to the present form of the island. Although ice is a relatively weak solid which deforms under its own weight and slides down available valleys, it can move enormous amounts of rock over a very long period. A general topographical description of South Georgia was given in Chapter 1, here a brief account of the development of this is provided.

Several very early erosion surfaces have been recognised on South Georgia and it is considered that the concordant summit levels of the mountain ranges at altitudes of about 600–650 m and 1700–2000 m, are the remnants of two high-level peneplains the erosion of which commenced before the earliest glaciation of the island. Some of the most impressive landforms created by glacial erosion are the enormous cirques and troughs of the north-east flanks of the Allardyce Range. Other manifestations of these processes such as moraines, roches moutonnées, knob and tarn topography, glacially grooved and polished bedrock, and rock steps are abundant in the parts of the island free from ice. Glaciers can transport very large pieces of rock almost as easily as 'rock flour' and a consequence of this is some exceedingly irregular submarine floors in most of the fjords as debris separates from calving glaciers. The great variety of habitats for bottom-dwelling animals resulting from this effect is referred to in Chapter 8.

The rocks of South Georgia are nearly all highly fissile and undergo cycles of freezing and thawing many times in the course of a year. This has resulted in their becoming very fractured and in the development of vast quantities of talus or scree around peaks and in the valleys. Many rock faces are, in consequence, unstable so that great rockfalls occur and rock-climbing becomes a far more hazardous pursuit than usual. The scree slopes, a great nuisance when walking, are important sites for nesting of several burrowing bird species and, as they degrade further, give rise to the mineral soils of the island. This aspect of South Georgia's rocks also substantially accounts for the high debris loads of the glaciers.

Several raised beaches may be detected on South Georgia at heights of 2.0, 4.8, 6.0–7.0, and 20–50 m above sea level. As well as changes in the actual level of the ocean, isostatic uplift (resulting from the diminution of the effects of the very much greater masses of ice the island once supported) are responsible for these. Wave-cut platforms at the lowest level occur around many of the island's capes and promontories. These are particularly apparent in South Georgia and many other polar regions owing to the effects of ice-bearing tempestuous oceans, a factor which also influences living organisms on the shore lines.

Sea ice

Much of the southern parts of the Weddell Sea and Antarctic Ocean are covered by sea ice even during the summer. This becomes much more extensive during winter, and, in spring, these advances break up and drift northwards. The average limit of winter pack-ice is to the south of South Georgia although exceptional seasons occur when it may be far more extensive or, conversely, almost absent. In September 1980 nine tenths to total pack-ice extended to a position approximately 200 km to the north of the island, rendering it ice-bound and unapproachable. This was determined from satellite photographs – previous extreme pack-ice years occurred in 1959 (when it was over 100 km north of Cape Buller) and in 1924, but information about its extent is far less detailed. The extreme southerly voyage of James Weddell in April 1823 provides an example of a period with remarkably little sea ice.

Icebergs in the vicinity of South Georgia may be derived from the glaciers of the island, and from those and the ice shelves of the Antarctic continent.

The largest come only from the latter source and some may be almost half the size of the island. In 1978 one 65 by 38 km, which had previously been tracked by satellite for 15 years in the Weddell Sea, drifted from the South Shetland Islands vicinity. It passed between Shag Rocks and the Willis Islands on its way north. Such large tabular icebergs arrive with the prevailing winds and currents from the Weddell Sea in quantities which vary very greatly from year to year.

The region around South Georgia showing the Antarctic Convergence and the maximum extent of ice pack in September 1980 (Author, from Royal Navy and United States Navy diagrams.)

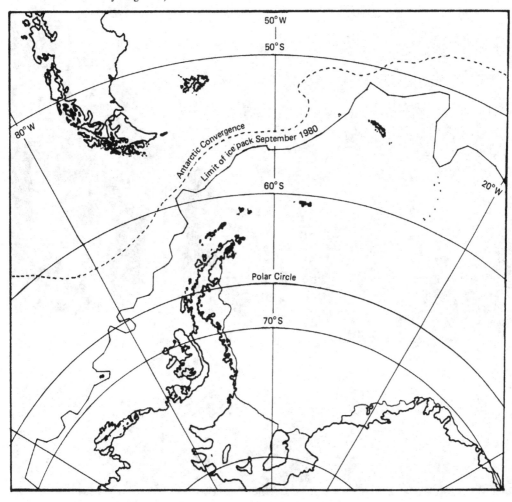

An iceberg, derived from the Nordenskjöld Glacier, in Cumberland Bay. Note the kelp bank in the foreground (Author.)

Smaller icebergs are common. Most of those found in the bays of South Georgia result from calving of local glaciers. As their size decreases they become progressively icebergs, bergy-bits, growlers, and brash ice. These are particularly common in spring when glacial advances from the preceding winter break off especially during the vernal equinoctial gales. The larger ones are sometimes grounded in shallow areas, where they may remain for perhaps several weeks. They may form a hazard to shipping as they are obstacles which can cause damage if not avoided, and larger ones may occasionally overturn and break up very violently. Growlers, in rough seas, are often difficult to see and are not apparent on radar.

Some bays may become frozen over and covered with drifting ice during winters. The frequent ocean swells and storms ensure this does not normally last for very long, as the ice is broken up and blown out to sea.

Oceanography

South Georgia is situated in the Antarctic or Southern Ocean which surrounds the continent of Antarctica. The low temperatures as well as several other features of this ocean greatly affect the climate and natural history of the island and of the region around it. The division between the Antarctic Ocean and the Atlantic, Indian, and Pacific Oceans is regarded in this work as the Antarctic Convergence, an oceanographical phenomenon which encircles the world roughly between latitudes 50° and 60° S. (Some other systems for defining the oceans in the southern regions are also used which differ from the above and Southern Ocean may be used synonymously for Antarctic Ocean.) At the longitudes of South Georgia, the Antarctic Convergence is situated approximately 350 km north of the island.

The Antarctic Convergence, essentially a surface phenomenon, results from meridional and vertical movements of three discrete water layers which differ in density owing to their temperature and salinity characteristics. The whole system may be compared with the discontinuity of an atmospheric polar front. North-flowing surface water of the Antarctic Ocean sinks at about the same latitude as that at which comparatively warm, south-flowing circumpolar water rises over very cold north-

flowing Antarctic bottom water. This resulting area of discontinuity in meridional temperature, salinity and chemistry has both surface and sub-surface expressions. At the surface, the convergence shows as an area where temperature changes of 2°C may occur within half a degree of latitude. The sub-surface expression shows a sudden change in the depth of the temperature minimum from 200–300 m in the Antarctic Ocean to 500–600 m in the oceans north of it.

The stability of the convergence is affected by local meanders and eddys which may grow into gyres perhaps 80 km in diameter. Nevertheless, the mean position is relatively constant, generally well within a 100 km span. It is sufficiently constant to be indicated on hydrographic charts. The flora and fauna of the oceans differ markedly across the convergence and, in consequence, birds which rely on the sea for their diet also differ. Krill, an organism of fundamental importance in Antarctic food-chains, does not occur north of the convergence. Islands south of it, in Antarctic waters, include South Georgia, the South Sandwich Islands, South Orkney Islands, South Shetland Islands, Bouvetøya, Heard Island, McDonald Islands, Balleny Islands, Scott Island, and Peter I Island. Iles Kerguelen lie virtually on the convergence while the remaining islands of the far southern regions are in other Oceans (Gough Island, Marion and Prince Edward Islands, Iles Crozet, Iles Amsterdam and Saint-Paul, Macquarie Island, and the islands south of New Zealand).

The ocean currents around South Georgia originate from the east-bound, peri-Antarctic Southern Ocean Current passing the west side of the Antarctic Peninsula and from the clockwise Weddell Sea Gyre (in which *Endurance* and *Deutschland* were caught.) These are set to the north of east and cause the marked extension, even beyond the fiftieth parallel, of the positions of the Antarctic Convergence, the pack-ice front, and the limit of icebergs which pass well to the north of South Georgia. The rate of these surface currents is about half a knot, where they are undeviated by local submarine topography. Some particularly important effects of the currents in the vicinity of South Georgia are local upwellings of nutrient-rich Antarctic bottom water which arise where the waters of the deeper currents impinge on the submarine shelf surrounding the island, the crescent shape of which lies almost at right angles to the currents. This causes much complexity in the movements of waters at various depths near the island and allows the development of a zone of particularly rich phytoplankton with associated food chains.

Tidal observations have been made for various periods at several different places on South Georgia since 1882. A chart datum was established at King Edward Point by the Discovery Expedition in 1927. HMS *Owen* operated a tide gauge in 1961 and an automatic recording apparatus was deployed there early in 1982, making observations every 18 minutes. It was intended to operate for a long period but stopped shortly after the invasion.

The tides in the enclosed bays of South Georgia are irregular, the usual range is about 1 m. They are greatly affected by meteorological conditions and are difficult to predict on a theoretical basis. Tidal surges are occasionally experienced; some of these have swept up to 100 m inland and have damaged buildings, for instance Duncan Carse's hut at Undine South Harbour in 1961.

Meteorology

The climate in the region of South Georgia is cold, wet and cloudy with strong winds, subject to rapid changes and without great seasonal variation. Conditions typical of the region are experienced on the south-western side and the extremities of the

The bathymetry of the South Georgia region.
(R.B. Heywood.)

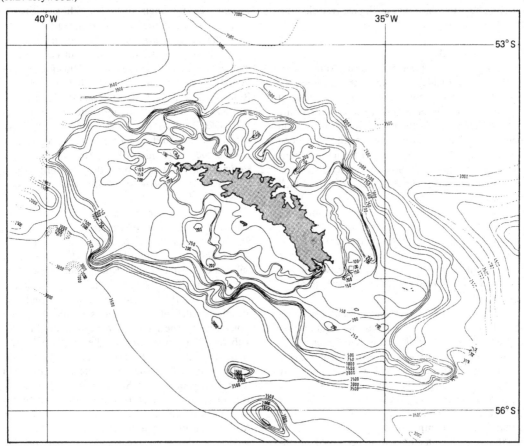

island. Those recorded at the observatory at King Edward Point and much of the north-eastern side are greatly affected by the mountain ranges and local topography. As a consequence of this and regular föhn wind action, observatory records show a climate slightly warmer than more exposed areas of the island where the climate is more representative of the region as a whole. Observations for periods longer than a year have been made at the Bay of Isles, Bird Island, Royal Bay, Maiviken and the whaling stations. Sporadic observations have also been recorded by various expeditions. Many of these demonstrate more rigorous conditions at places other than the observatory. South Georgia is in a position where it receives an almost continuous series of atmospheric depressions and frontal systems, many of which come through the Drake Passage and are often generated off the Antarctic Peninsula. In accord with apparent climatic trends on the Antarctic Peninsula during the last 30 years, the climate of South Georgia appears to be presently becoming slightly warmer, windier, and wetter. Such trends are reflected in the behaviour of glaciers and in the local distribution of some plants and animals.

Owing to its position, some 350 km south of the Antarctic Convergence, its relative proximity to the Antarctic Continent, and the very cold ocean currents which surround it, the island has a much colder climate than its latitude and geographical position might suggest. Sub-zero temperatures are recorded every month of the year. Temperatures in excess of 20°C have been recorded during föhn winds, though a screen temperature of 15°C is not uncommon on a warm windless summer day at King Edward Point. At Bird Island, where conditions are more representative of the region, air temperature rarely rises above 10°C. At the other extreme, it is uncommon to record screen temperature below −15°C near the sea at King Edward Point though in other parts of the island, especially inland and at greater altitudes, it gets very much colder. Although winter and summer seasons are clearly defined, the annual range of mean monthly temperatures at King Edward Point is only 7 degrees, largely owing to the proximity of the sea. Such average temperature data are quite inadequate to indicate the potential variations which may occur.

The prevailing winds of the Atlantic sector of

the Antarctic Ocean, channelled through the Drake Passage past Cape Horn, have long been commented on by mariners unfortunate to spend any length of time there. South Georgia receives these and is generally a very windy place. Gales are recorded in every month of the year although annual numbers have recently ranged from zero to thirty at King Edward Point. Those occurring near the time of the equinoxes are usually the most severe. Gusts approaching $50\,\mathrm{ms}^{-1}$ (100 knots) have been recorded at King Edward Point, a comparatively protected place. The strength of some winds may be enhanced by local orographic effects, and katabatic winds, resulting from cold air flowing down valleys, are commonly experienced. These may give rise to violent squalls and whirlwinds – 'williwaws', common in some harbours and which have blown several vessels aground. The terrain may also have converse effects leading to comparatively windless areas existing during periods which are very windy in the vicinity. Several mountain passes form distinct wind channels where high velocity air currents may commonly be encountered sufficiently strong to easily lift a heavily laden man.

Föhn winds are an outstanding feature of South Georgian weather. They are experienced in the region to the north-east of the main range. Their subjective effect is like that of the Chinook wind of Canada or the föhn wind of the Alps. They result from the arrival of an air mass at the windward side of South Georgia. After a long ocean crossing, this is heavily laden with water vapour. On reaching the island this is forced to rise over the steep mountain ranges up to about 3000 m. As it gains altitude it expands and cools, thus the water vapour it holds condenses into clouds and precipitates as rain and snow. Consequently its absolute humidity is reduced and comparatively dry air crosses the ranges to descend on the north-eastern side of the island. As it does so, its pressure (and thus its temperature) increases. The rate of warming is higher owing to the difference in thermodynamic properties of humid and dry air. By the time it reaches sea level again it may be many degrees warmer than it was prior to passing over the island. When a föhn wind begins to blow, the temperatures at King Edward Point may rise 10 degrees in as many minutes, any snow and ice present may melt quickly. As it descends the mountain sides and glaciers the air mass accelerates, often reaching gale force. Humidity drops rapidly also. Occasionally air from much greater altitudes is brought to sea level by lee wave effects during föhn winds and, when this occurs, it may cause local humidity to fall to 5% or less.

The average sea level atmospheric pressure at South Georgia for the period 1971 to 1980 was 996.5 millibars. Pressures from 1035.5 to an extreme low of 932.0 millibars (in February 1966) have recently been recorded and the rate of change in air pressure, rising or falling, is sometimes extremely rapid. During the period from 1976 to 1980 twenty-eight occasions were recorded when pressure changed by 10 millibars or more between observations three hours apart, 20 of these were of falling pressures. In December 1978 a 15.9 millibar fall occurred in such a period. These rapid changes, as well as the low average pressure, are a consequence of the position of South Georgia in the path of frequent depressions coming from the west.

Precipitation (rain, snow, etc.) occurs throughout the year although monthly variations may be vast. Dry periods are rare and not prolonged and a fortnight without precipitation begins to be a drought. It is greatest on the south-western side, for reasons discussed above. The rain gauge at King Edward Point has recently measured an annual rainfall or water equivalent of 180 cm. Although light rain and drizzle are the most common forms of rainfall, the occasional torrential downpour is also experienced. Snow depths of up to 1 m are commonly recorded at King Edward Point in winter, though it is very much deeper elsewhere on the island. Snow in the steeper valleys and slopes frequently tends to avalanche in slab and powder forms immediately after a heavy fall, and after being warmed by föhn winds followed by freezing which also leads to unstable conditions. Wet avalanches become common as the spring thaw develops. Virtually every other form of precipitation (sleet, hail, snow pellets, etc.) is regularly recorded at the observatory.

The atmosphere over the Antarctic Ocean generally, and thus around South Georgia in particular, is very cloudy. The mountainous topography of the island greatly influences the local distribution and types of clouds. Bird Island, a reasonably representative station for the region,

receives barely 10% of the theoretical maximum of possible sunlight. King Edward Point receives approximately 35% due to its lying beneath a 'radiation window' which forms in the lee of the mountains as a result of processes related to those giving rise to föhn winds. This exceptional amount of sunlight would be even greater were it not for the shadow of Mount Duse, which falls over the observatory for several weeks during winter (South Georgia is 14° latitude north of the Antarctic Circle and thus has no period of continuous night).

Cloud forms are varied though stratiform ones predominate. The mountains cause many unusual cloud effects, the most spectacular of which are stacks of lenticular clouds, resembling fleets of flying saucers. These are not uncommon and form in the ridges of standing waves in the atmosphere streaming over the Allardyce Range at heights of several kilometers. The clouds form almond shapes in these ridges and, at sunset, they become suffused with red and gold light. Long plumes of cloud as well as almost all other types are also seen.

Electrical storms associated with cumuliform clouds of great vertical extent are rare, as the frigid climate and uniform nature of the atmosphere generally preclude the local heating and instability required to generate large thunderstorm clouds. The occasional one that does occur may be spectacular, even bearing in mind the rarity of such events.

Dense fogs may occur on and over the oceans around South Georgia and these often fill certain bays. For topographic reasons they are less common near King Edward Cove. Much fog is of no great vertical extent and on climbing above it one may find a clear fine day. Offshore fogs, especially at times when much ice is about, greatly exacerbate dangers to local shipping.

The weather of South Georgia is notable for being subject to very great changes, both in time and location. A comparatively warm day can give way to a combination of low cloud, gales and snow in only 15 minutes, and the reverse may occur just as rapidly (perhaps even several changes in a day). Various combinations of weather can greatly affect travel on the island, making it treacherous for persons unfamiliar with and unprepared for these. A very good day in winter may have a brilliant blue sky, no wind and complete snow cover. Under these circumstances one may be comfortable in only light clothing and snow glasses. Whiteout may be experienced when illumination from a completely clouded sky is poor and uniform, the ground all snow covered and especially if fog is present. All sense of distance is, and sense of direction may be, lost under these circumstances. In these conditions it is not difficult to walk accidentally over snow cornices, cliffs, etc. Advice for trekking is to hope for good weather but be fully prepared for the worst. Similar remarks apply to operations involving ships, other vessels and aircraft. Precautions against powerful winds which are likely to change direction, heavy snowfalls, low temperatures and several other potential problems are essential.

Meteorological observations were initiated at Grytviken by C.A. Larsen assisted by E. Sörling on 17 January 1905. A set of instruments had been provided by Mr Walter Davis, formerly of the Ben Nevis Observatory in Scotland, who had become the Head of the newly established Argentine Meteorological Office in Buenos Aires. The observatory was moved to King Edward Point, about 1 km away, in August 1907 and has continued to function there. Erik Nordenhaag, a Swede, was the first meteorologist. A break in the records for a short period occurred after 3 April 1982, as a result of the Argentine attack on the station. Otherwise the records are the second longest series from the Antarctic and are an invaluable record from such a remote region. (The oldest continuous series is from Laurie Island in the South Orkney Islands which was initiated on 26 March 1903 by the Scottish National Antarctic Expedition and continued, at the invitation of the British Legation in Buenos Aires, by the Oficina Meteorologica Argentina independently from January 1905.) The Compañia Argentina de Pesca maintained the King Edward Point observatory, partly in cooperation with the

Orographic lenticular clouds over Cumberland Bay. (Author.)

Argentine Meteorological Office, in accordance with the requirements of its lease of the whaling station site, until 31 December 1949 (when it was relieved of this requirement). The Falkland Islands Dependencies Survey then operated it to 1952, when the Falkland Islands Meteorological Service took it over. The Service began to issue forecasts for the whaling fleet as well as continuing the records. Danny Borland was appointed senior meteorologist at South Georgia in 1950 and held the post to 1969. In that year responsibility for the observatory passed to the British Antarctic Survey, until the invasion in 1982. Subsequently, shortly after the invaders were removed, the British garrison continued collecting meteorological data.

Until April 1982, regular daily broadcasts of meteorological data were made from King Edward Point. These data were assembled at South Georgia

The original meteorological station at King Edward Point, erected in 1907. From a photograph by the first Pastor, Kristen Loken, about 1913. (Courtesy Norse Hvalfangstmuseum.)

from those of all other British Antarctic Survey stations, British and some other ships in the region, and cooperating foreign Antarctic stations. South Georgia was an official regional data collection centre for the World Meteorological Organization, which distributed these data on its global telecommunications system. Because of the isolated positions of the observatories participating in this collection routine, the data were of great value in global weather forecasting. After the invasion, data normally transmitted through South Georgia were transmitted by emergency links, established by friendly nations, to the world network. Arrangements for direct data transmission by satellite were well advanced at the time and should soon obviate the need for regional collection centres. Wireless transmission of meteorological data from South Georgia began in 1928; it was received at Stanley and passed to Brazil which had requested this in 1926 for forecasting purposes.

Summarised monthly tables, for observations over a period of 30 years at King Edward Point, are given in Appendix 4.

Atmospheric physics and geomagnetics

As well as meteorological data, the observatory at King Edward Point acquired data from several other physical phenomena until it was forcibly closed. Many of these observations may be re-established, although there will be prolonged interruptions in the records. The observations being made early in 1982 were geomagnetism, atmospheric turbidity, ozone concentrations, tidal records, and seismographic activity. Previously, programmes to measure solar radiation and regular soundings of the ionosphere were also operated. These observations are described in the following pages, with the exception of seismology and tidal observations which are referred to in the sections on geology and oceanography respectively.

The geomagnetic observations included measurement of the absolute values of three of the components of the Earth's magnetic field (the horizontal field, vertical field and the local declination), daily variations in these and magnetic micropulsations. The absolute values and declination are regularly determined with instruments not unlike those used a century ago by the German International Polar Year Expedition at Royal Bay.

Daily variations are recorded on La Cour vario-meters, a type that has long been used throughout the world, and the standardised data obtained from them are easily intercomparable. The observations contribute to an estimation of the variation of the world-wide magnetic field at the Earth's surface. This varies not only on a daily and seasonal basis but also the alignment direction progresses gradual-ly. The causes lie deep within the Earth's core and long continuous records from as wide a geographic distribution as possible are important for in-vestigating these. The importance of remote stations is again great. A direct practical result of these observations is in the indications of magnetic deviation of the north point on navigation charts. In 1980 the magnetic deviation was 8°43′ W with very small drift and the angle of dip was 37° at the King Edward Point observatory.

The three La Cour instruments measure vari-ations in the horizontal and vertical geomagnetic fields as well as those in the declination. They consist of a very small magnet suspended by a very fine quartz fibre. A small mirror is attached to the magnet from which a beam of light is reflected through a series of prisms and lenses on to a revolving drum covered with a strip of photo-graphic paper. A variation in any of these three aspects of the Earth's magnetic field yields a move-ment of the light beam which is recorded by the paper. The paper is changed daily and after development it reveals the time, date and amount of the variation. Regular checks are made of the calibration of the La Cour instruments against the absolute ones.

Rubidium vapour magnetometers were used to detect micropulsations in the Earth's magnetic

The instruments of the meteorological observatory at King Edward Point in 1981. Note also the lenticular clouds. (P.R. Stark.)

field. These rely upon detection of changes in the atomic state of rubidium caused by small, rapid field changes. Such micropulsations result from effects of perturbations in the stream of subatomic particles reaching the earth from the sun (the solar wind). These induce small brief changes in the Earth's magnetic field, which the instruments detect and measure. Such interactions between the solar wind and magnetic field are responsible for magnetic storms, which may severely disrupt radio communications throughout the world. They are associated with sun spot activity and aurora. A coded index of hourly records of micropulsations was regularly transmitted to the British Antarctic Survey head-

quarters in Cambridge and thence to the world data centre for this information in Paris. Predictions of solar and magnetic events are made and distributed from these data.

Another effect of the solar wind on the upper atmosphere is the aurora – the northern and southern lights. Unfortunately these are not usually visible from South Georgia. They occur most commonly along 'Auroral Zones' centred on the north and south magnetic poles. The south magnetic pole lies almost diametrically on the other side of Antarctica from South Georgia and the closest parts of their highest active regions are far south of the island.

Calibrating the Quartz Horizontal Magnetometer at the geophysics observatory, King Edward Point, December 1980. (Courtesy British Antarctic Survey, C.J. Gilbert.)

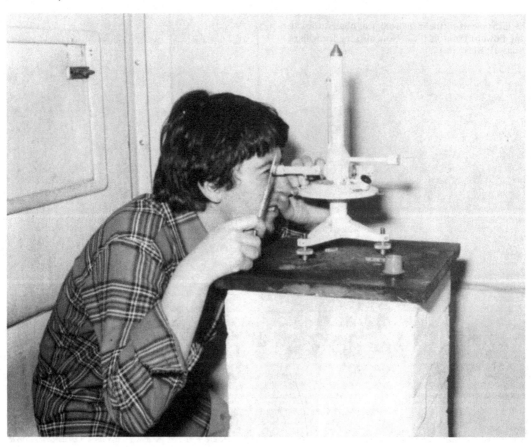

Regular monitoring of the concentration of ozone in the upper atmosphere was also conducted by the observatory (ozone is a variety of molecular oxygen consisting of three atoms rather than the normal two). The ozone layer begins in the stratosphere and is found at its greatest density at about 22 km above sea level. It is responsible for absorbing the greater part of the ultraviolet radiation from the sun. Interest in the ozone layer has been particularly strong since the discovery that certain chemicals in common use (notably fluorocarbons) may deplete the layer with potentially serious consequences. Although evidence for this is equivocal, world-wide data about ozone concentrations are important to detect any significant changes beyond seasonal variations. The King Edward Point observatory is one of over 200 ground ozone monitoring stations, most of which are in the northern hemisphere. A Dobson ozone spectrophotometer is used to determine the concentrations and has been in use since 1973. Since 1977 some determinations have been made from artificial satellites and the ground-based observations are important to calibrate and verify these.

Measurements of solar radiation were made with four instruments deployed between 1973 and 1981. These determined the total radiation (sun and sky), diffuse radiation (sky alone), reflected radiation (surface albedo) and the nett flux (difference between incident and reflected radiation). The data from this programme would permit calculation of a localised energy balance between the lower atmosphere and the ground as well as giving information about the amount of solar radiation available for plant growth.

Another aspect of the radiation programme on South Georgia is concerned with periodic monitoring of atmospheric turbidity. The solar radiation incident on the upper atmosphere is depleted by both atmospheric gases and minute suspended particles, so the final intensity of radiation at ground level is below the theoretical incident values. Measurements of the amount of attenuation of solar radiation are made with an Ångstrom pyroheliometer in which a narrow beam of sunlight is directed onto platinum thermocouples through filters. Information about the concentration and sizes of atmospheric dust particles may be calculated from these results. The measurements can be made only on clear sunny days and, as these are not common, the frequency of observations is irregular (very few are possible from late autumn to early spring owing to the shadow of Mount Duse over the observatory). Interesting variations in the total dust concentrations over South Georgia have been observed. Compared to much of the rest of the world, the air over the island is, as might be expected, remarkably free from artificial pollutants.

From January 1970 to September 1978, regular soundings of the ionosphere were made from Kind Edward Point. The location was a particularly valuable one as, apart from its remoteness, the local geography assisted greatly in shielding the instruments from external interference and the earth's magnetic field is particularly weak over the regions to the south of the Atlantic Ocean (about 30 rather than 35 micro-Tesla). South Georgia lies generally to the south-west of this geomagnetic field anomaly and is thus in a good position to monitor changes in it due to magnetic storms and any unusual ionospheric effects. Number 10 Union Radio Mark II ionosonde equipped with a 33 m aerial was installed to make observations every 15 minutes. This work caused much interference on base to radio, internal telephones, gramophone, tape recorders and other apparatus. As it took 4 out of every 15 minutes to make each observation, it caused local adverse (mainly good-natured) criticism from all save physicists. Records (ionograms) were produced on a photographic film which were reduced to a numerical form for transmission. They were derived from three layers of the ionosphere; D-region (70–90 km), E-region (90–180 km) and F-region (180–700 km). The ionosphere greatly affects propagation of radio waves and at some frequencies it can act as a reflector, thus allowing communications over very large distances. Data from South Georgia were relayed to the World Data Centres for the Ionosphere which provide forecasts of world-wide conditions for radio communications.

Future atmospheric research

There are several possibilities for future atmospheric research at South Georgia should the scientific station resume normal operations. Some potential projects in which the British Antarctic Survey may collaborate with other scientific organisations are: a southern hemisphere interplanetary scintillation programme to investigate the solar wind, the use of balloon-borne radio-sondes for upper atmospheric studies (including ozone observations, a radar examination of the troposphere and stratosphere, and the study of plasma released from satellites). The first of these would involve apparatus operating jointly in the United Kingdom at a latitude of about 54° N and South Georgia at 54° S. A suitable telescope will be required for the last project, which may permit various astronomical observations to be made. Only one such telescope has, to date, been deployed on the island, by the German International Polar Year Expedition in 1882–83.

8

The natural history of
South Georgia

The natural history of South Georgia and
its surrounding oceans is one of the most
fascinating aspects of the island. The
species composition, life histories, origins,
interdependence, feeding, communities, and
other aspects of its biota have been
investigated since Captain Cook first
brought naturalists to the island and
particularly since the German International
Polar Year Expedition a century ago. Many
animals and plants have been introduced to
South Georgia by Man and some now
make a substantial contribution to its
biology. These isolated communities
provide interesting opportunities for
research in biogeography, ecology, genetics,
adaption, and the effects of human
activities.

Some biological research has been
undertaken by nearly all expeditions to
South Georgia and many of them were
principally for that purpose. The Falkland
Islands Dependencies Survey and its
successor, the British Antarctic Survey,
have been responsible for a greater amount
than all other expeditions (owing to the
length of time they have worked on the
island). As well as these organisations and
expeditions, several individuals – most being
employees of the whaling companies – have
independently made biological observations
and collections.

Biogeography and ecology

South Georgia is extremely isolated from the
rest of the world not only by its remote location but
also by that substantial biological frontier, the
Antarctic Convergence. Its climate and location
have resulted in its flora and fauna being, to a large
extent, part of the transition which extends from the
Antarctic continent and Peninsula, South Shetland,
South Orkney, and South Sandwich Islands, to
South Georgia thence the Falkland Islands and
Gough Island, and ultimately Tierra del Fuego,
Patagonia, and Tristan da Cunha. The species
on the island also have many peri-Antarctic and
South American relationships although some are
cosmopolitan or even bipolar in distribution. They
include relict species from those of widespread pre-
glacial distribution and post-glacial immigrants,
both of which have had sufficient time for some to
have evolved endemic species, as well as recent
introductions from the time of arrival of Man on
South Georgia.

Three principal divisions of the biology of
South Georgia with greatly different characteristics
may be distinguished: terrestrial and freshwater,
marine, and that of the sea birds and seals. The
terrestrial and freshwater species are comparatively
few and adapted to a very rigorous environment.
Their food chains are simple, as might be expected
on an isolated Antarctic island. In contrast, the
marine community is quantitatively one of the most
abundant although its species composition is restricted. The abundance is due to the nutrient-rich

water, with its high concentration of dissolved oxygen (owing to its low temperature), as well as the relatively constant conditions, especially temperature which never falls below − 1.8 °C. The marine food chains are also short, however, the phytoplankton-feeding krill is the most important constituent of many of them. The sea birds and seals take advantage of both of these systems, relying on the island as a breeding site, and the oceans as a food source. It is interesting to contemplate that South Georgia is one of the most remote places where humans, with appropriate scientific knowledge and basic equipment, could live off the land, for long periods.

The islands and continents surrounding Antarctica and the Antarctic Convergence. (Author.)

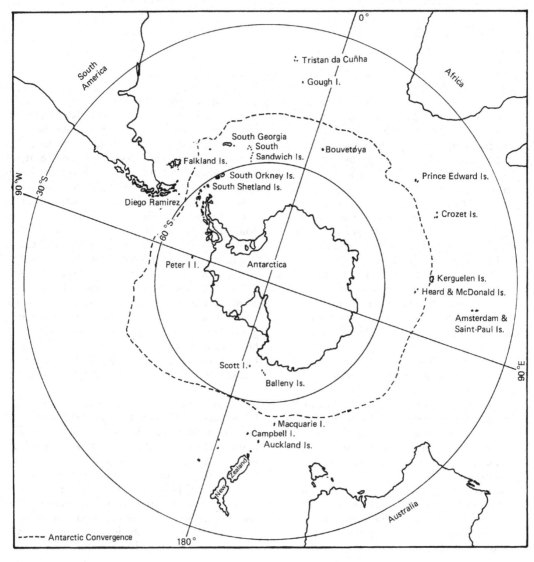

The names of species

Animal and plant names in this chapter and its appendices are given in both their Linnean and, where practicable, vernacular forms. A particular problem with vernacular names is that, in many cases, the species is by no means common and the scientific name is generally better known and in wider use. Both types of name are subject to revision and, with vernacular names, many variants occur as there is no standard system of application. United States and British vernacular names are also frequently different. With regard to the various groups, most of the whales, seals, and terrestrial mammals have vernacular names which are in common use, short, stable, and specific. However, the birds often have several vernacular names some of which are descriptive and of several words; those preferred by ornithologists at South Georgia are given in the text and some others in the appendices. Vernacular names of plants are, with some exceptions, not often used. The Linnean names used in this work are from the most recent revisions known to the author and some differ from those in much earlier literature; they are used in the binomial form only. Norwegian and Spanish vernacular names for many South Georgian animals

and some plants exist; no account of these is taken in this work.

Botany and the indigenous flora

The vegetation of South Georgia gives the impression of being dominated by higher plants. In fact there are only 26 native species of vascular plants (5 grasses, 4 rushes, 1 sedge, 9 small dicotyledon herbs, 6 ferns, and 1 clubmoss) of which only 5 achieve widespread dominance. Of the macroflora, mosses (about 125 species), liverworts (about 85 species), and lichens (about 150 species) have the greatest species diversity. In addition about 50 species of macrofungi (toadstools) and perhaps 10 of macroalgae have been recorded. Species lists for the microflora (including algae, fungi, and bacteria) are at present very incomplete. Futhermore about 35 flowering plant and several lichen and macrofungi species have been introduced to the island, mainly around the whaling stations and King Edward Point; some of these have become successfully naturalised. Numerous other flowering plants have occurred briefly as transient introductions. Accounts of most of the macroflora groups have been published and a list of the native vascular flora is given in Appendix 5.

Principal marine food chains and the influence of Man upon them in the South Georgian region in the 20th century. (Author.)

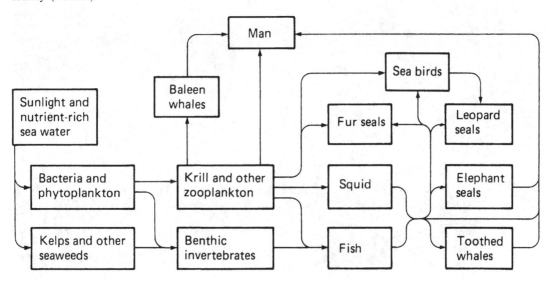

The paucity of South Georgia's flora is caused primarily by the isolation of the island rather than climatic factors alone, although the cold summers are a critical factor. The success of many adventive weedy vascular plants of European origin together with transplant experiments with species native to the Falkland Islands indicate that many other species could be expected to survive if introduced. The native flora are principally related to those of the Falkland Islands, Tierra del Fuego, and Patagonia although some species have general southern-hemisphere affinities and a few have bi-polar or world-wide distribution. The ancient flora of South Georgia was sufficiently tropical to in-clude cycads but the fossil record is poor. What vegetation existed prior to the last major glacial epoch, which ended about 10 000 years ago, is uncertain but it is probable that some species of mosses and lichens survived on nunataks which served as biological refugia. Colonisation of the land

by plants during the comparatively recent retreat of the glaciers was greatly dependent on the fortuitous arrival of propagules. Some interesting results from experiments using microsporomorpha traps for sampling air-borne particles include the detection of pollen from tropical plants in the winds of South Georgia. Tree trunks and similar objects are occasionally found on the island's beaches; these probably originate in southern South America. All hypotheses concerning the arrival of propagules at remote islands involve transport (mainly by birds and wind) over long distances, prolonged periods of establishment and much un-certainty.

The most notable and the largest plant species on South Georgia is the tussac or tussock grass *Parodiochloa flabellata* (known as *Poa flabellata* until a recent revision of the tussac grasses of the Southern Hemisphere). It occurs mainly on raised beaches and other low coastal areas but it may also

Tussac, *Parodiochloa flabellata*, near sea level at Tortula Cove, March 1980. The scale object is 186 centimetres high. (Author.)

grow on steep slopes rising from the shores. Its rate of growth is quite rapid but the rate of decomposition of its leaves is slow and this, combined with erosion of the peaty soil by seals and penguins between plants, may result in the formation of pedestals which can raise the uppermost crowns of leaves higher than a person. Its flowers begin developing early in summer and remain dormant during the winter. They emerge early in spring, often before the plants are completely free from snow, and develop rapidly to set seed by early summer, much earlier than other plants. Tussac leaf bases are rich in carbohydrate and have supported shipwrecked mariners in the Falkland Islands. It is also the staple diet of the introduced reindeer, especially in winter, and is the reason for their survival on South Georgia, as well as that of the rats and mice.

Of the other grasses, *Festuca contracta* (the tufted fescue or land tussac) is dominant in a dry grassland which is common in areas from Fortuna Bay to Royal Bay and occurs less extensively elsewhere. *Phleum alpinum*, the alpine cat's tail, is notable in being bipolar and found in many Arctic and northern alpine regions as well as southern South America. The vascular plant with the most southerly distribution of any species, *Deschampsia antarctica*, the Antarctic hair grass, occurs on South Georgia as a pioneer plant on recent glacial soils and at high altitudes as well as in dense swards near the sea. The rushes and sedge are short plants seldom more than 15 cm high, which typically frequent damp to very wet boggy areas close to the shores. The fruit of one of them, *Rostkovia magellanica*, the brown rush, is a large noticeable brown capsule.

The woody stemmed *Acaena magellanica* and *A. tenera* (burnets) are very common. Their seeds have small barbed appendages and are sometimes seen attached to birds (as well as to clothing). The

The grass *Festuca contracta*, about 15 cm high. (D.W.H. Walton.)

Acaena magellanica, the flower head is about 15 cm high. (D.W.H. Walton.)

two species often hybridise when they are growing in close proximity and the hybrid forms a clearly recognisable zone which is known only from South Georgia. *Acaena magellanica* has two forms of flower heads, hermaphrodite and female. Their ripening times are such as to ensure a high degree of cross fertilisation. Although usually in reasonably well-vegetated regions, either species may also be a pioneer plant in exposed areas, particularly the smaller *A. tenera*.

Two species of the genus *Colobanthus* (pearlworts) have a compact cushion-like growth form that renders them less exposed to the harsh environment. *C. quitensis* is one of the two flowering plants which occur on the Antarctic continent. The remaining flowering plants are small forbs which

occur in mossy flushes and other wet areas and three have flowers with prominent but small petals. *Galium antarcticum*, Antarctic bedstraw, has small creamy-white flowers and is the only scented plant on the island, producing a delicate patschouli-like odour. *Ranunculus biternatus*, the Antarctic buttercup, has small yellow-green flowers and forms a fruit of distinctive red clusters of achenes. Owing to the absence of alternatives, most South Georgian plants are dependent on wind or, rarely, water for cross fertilisation although it is possible beetles may be involved in some cases. Thus petaloid flowers are exceptional and most species are able to propagate efficiently by vegetative processes.

Five of the six species of ferns usually occupy crevices and ledges in low altitude, north-facing

Ranunculus biternatus, Antarctic buttercup, the flower and cluster of fruits are each about 1 cm wide. (D.W.H. Walton.)

rock faces or scree slopes in areas which receive regular drainage from above. Several can, however, tolerate short periods of virtual dehydration. *Cystopteris fragilis* is remarkable as it also occurs on Spitzbergen, Mount Kilimanjaro and several other cold places. *Ophioglossum crotalophoroides* is a distinctive but rare adder's tongue fern found in wet flushes which produces distinctive spherical rhizomes. *Lycopodium magellanicum*, the Magellanic club moss, occurs on dry stony ground and is often locally abundant. It is the only species in the island's flora which attains a distinctive autumnal colouring when the scale-like leaves change from a yellow-green to a brilliant orange.

Mosses are very widely distributed and are found in virtually every plant community on the island. Many occur in wet areas where the reddish-brown species *Tortula robusta* is often dominant. The bog moss genus *Sphagnum*, which is typical of northern hemisphere wetlands, is represented by *S. fimbriatum* on South Georgia, which occurs in only a few sites although it appears to be spreading quite rapidly. *Polytrichum* and other tall turf-forming species develop banks of peat which may be as deep as 2.5 m. Many mosses, such as species of *Andreaea*, *Grimmia*, and *Racomitrium*, growing on exposed rock faces and fellfield areas (sometimes at high altitude), have a short compact cushion growth form. Liverworts are commonly associated with the mosses in wetter areas. The closely appressed thalli of *Marchantia* and *Schistochila* species sometimes form a firm mat in flush areas.

Lichens are the most widespread and diverse plant group on the island with the encrusting forms

Colobanthus quitensis, Antarctic pearlwort, about 2 cm across. (D.W.H. Walton.)

The ferns *Cystopteris fragilis* (foreground) and
Polystichum mohrioides (background) about 8 cm high.
(D.W.H. Walton.)

predominating. They are found from exposed rocky shores deluged by sea spray – for instance the prominent yellow and orange species of *Caloplaca* and *Xanthoria* – to high on exposed rocks in the mountains (many species of *Lecidea, Lecanora,* and *Rhizocarpon*). The drier grass and moss communities often contain many species of *Cladonia* including the reindeer 'moss' *C. rangiferina* which is so common in the Arctic. The large yellow and grey thalli of *Pseudocyphellaria* species are also typical of these areas while rock faces may be festooned with the bushy growth of *Usnea* species. Seven introduced lichens are known on South Georgia, all growing on timber imported by sealers, whalers or by expeditions.

Several types of algae are common on soil, the largest and most common is *Prasiola crispa* which grows in damp areas, especially near seal wallows and abandoned penguin colonies. On wet soils and

in some pools, globular gelatinous colonies of *Nostoc* are found and numerous filamentous species abound in the more eutrophic lakes and pools. Cryoalgae may develop and concentrate in melting snow banks and glacier surfaces, where aggregations of their cells impart a curious red tinge to the snow and ice. Endolithic algae occupy one of the strangest habitats of terrestrial algae, within minute fissures in light-transmitting rocks.

The marine macroalgae are a far more varied group with many more species. Three of these are large and spectacular, *Macrocystis pyrifera, Lessonia antarctica,* and *Durvillea antarctica* (the giant kelps), which grow around rocky shores of South Georgia where the first may reach 40 m or more in length. They provide a favourable habitat for many species of young fish and invertebrates which yield food for numbers of seabirds, especially during winter. Since these seaweeds grow in quite shallow

Ophioglossum crotalophoroides, adder's tongue fern, about 3 cm across. (R.I. Lewis Smith.)

waters and rarely in depths greater than 20 m, they provide warnings of hazards to navigation by indicating rocks and shallows. The establishment of an industry to extract agar and other alginate products from them has been considered. The marine microalgae form the phytoplankton and include diatoms, dinoflagellates, and many other unicellular species. The three water masses which make up the oceans surrounding South Georgia each have a distinct flora and a total of almost 250 species are known. These plants are the food of krill and hence are the beginning of Antarctica's marine-based food chains. Their concentrations increase as much as 36 times in areas off South Georgia where the water masses mix and deep nutrient-rich currents are diverted by local submarine topography.

Many fungi occur and several produce prominent fruiting bodies. These are usually in the form of small toadstools, particularly species of *Galerina* and *Omphalina*. Bright red disc-shaped fruiting bodies of *Scutellinea* species are quite common on wet peat or mud in many parts of the island, especially where reindeer occur. Many fungi cause rust-like infections in the leaves of vascular plants and a few myxomycophyta are similarly parasitic. Fungal parasites of the lichens are also recorded. Many species of soil fungi and bacteria occur but a systematic examination of these has yet to be made.

There are many different plant community types or ecosystems on South Georgia. The most distinctive include littoral (intertidal) and sub-littoral zones, moss banks, bogs, grasslands, fell-fields, and rock faces. Much recent botanical research has been concentrated on five of these in which designated sites are being intensively investigated. The marine plant communities, dominated by the giant kelps, are very varied ecosystems for reasons similar to those outlined below in the

Mosses of the genera *Polytrichum* and *Tortula*. (Author.)

description of the marine invertebrates. The freshwater ecosystem often includes floating fringes of mosses of the genera *Drepanocladus* and *Calliergon* with *Juncus, Callitriche, Ranunculus* and other herbs extending into it. There is no emergent vegetation growing from the beds of the water bodies owing to the action of thick ice in winter. Some aquatic mosses grow at depths of 30 m and mats of algae form at lower levels in the deepest lakes. Flushes and streams often have their margins lined by floating areas of pale yellow-green *Pohlia* and *Philonotis* species. The freshwater community merges into the bogs and mires where *Rostkovia* or *Juncus* become dominant, usually with a dense carpet of mosses and liverworts, many of which are also found in the aquatic habitats. The deepest peat deposits on the island occur beneath some of these bogs; ages of around 9700 years have been deter-

mined by radiocarbon dating for peat from about 3.5 m deep.

Moss banks are notable and many are of great age, up to 8500 years; locally they form a major component of the peat deposits of the island. *Polytrichum* and *Chorisodontium* are their main moss components while *Juncus* species, various liverworts, and lichens are also usually associated.

Acaena herbfields may have virtual complete cover by the deciduous *A. magellanica* during the summer with an understorey of the moss *Tortula robusta*. They form dense communities matted by the woody stems of *Acaena* on stable slopes and sheltered areas. The microbial activity in their litter and soil is intense and the decomposition of cellulose and other plant products is more rapid than in any other South Georgian ecosystem.

Two principal types of grassland cover much

Lichens, mainly *Cladonia rangiferina*, about 10 cm deep in *Festuca* grassland at Hestesletten. (*C. rangiferina*, reindeer 'moss' also occurs in the northern hemisphere; it is now rare on South Georgia in areas where reindeer graze.) (R.I. Lewis Smith.)

South Georgian lowland. Tussac grassland, dominated by *Parodiochloa flabellata* occurs in nearly all coastal areas and provides the major habitat for many bird species as well as the introduced rats and mice. It is also inhabited at various seasons by seals and penguins; their excreta may greatly enrich the soil and produce a deep green and luxuriant growth of the grass. *Festuca* grassland, dominated by *F. contracta*, occurs on dry loamy soils and appears light brown owing to the large standing dead component of the individual plants. These grassland communities include several other grasses, *Acaena* species, mosses, and lichens (especially *Cladonia* species).

Fellfields occur on exposed slopes and plateaux, screes, and rock faces as well as recently deglaciated areas and moraines. Their vegetation comprises scattered lichens and mosses, and (where there is an immature mineral soil) pioneer plants of some of the grasses, especially *Phleum alpinum* and *Deschampsia antarctica*, *Acaena tenera* and *Colobanthus quitensis* cushions may also occur.

In many places plants are slowly colonising new areas exposed by presently retreating glaciers on South Georgia. Pioneer species are becoming established on recent scree, moraines, glacial outwash debris, and similar areas. As the vegetation develops, it accumulates organic matter and forms 'islands' which provide foci for more rapid colonisation. The present general amelioration of the climate is being reflected in an increase of the amount of vegetation cover of the island. However, an analysis of pollen and spores in deep deposits of peat has indicated little change in the composition of the flora since the end of the last major glaciation.

Introduced vascular plants

Approximately 60 species of alien vascular plants are recorded from South Georgia, of which some 35 are presently represented on the island. The majority of these are of European provenance although some have a world-wide distribution and were accidentally introduced by Man during the whaling era. Only two species occur in areas well away from sites of human habitation. The most abundant and widespread of these is the weedy grass *Poa annua*, a hardy and adaptable species of world-wide distribution which even occurs as the first pioneer plant on newly exposed glacial moraines. *Cerastium fontanum*, a chickweed, is widespread but sparsely distributed, mainly within a few kilometres of the whaling stations. Of the remaining species, with the exception of one at the former German station at Royal Bay, none is found farther than about 3 km from the whaling stations; most occur within them. Some of those previously recorded grew close to steam pipes or in similar artificial environments and died out shortly after the end of the whaling era. About half a dozen species are represented by only one or two long established plants – individuals of which have been known for over 25 years. The number and composition of the alien species is distinct for each whaling station and King Edward Point, only one third occur in half these sites.

The alien vascular flora have been divided into four categories: transient aliens, persistent aliens, naturalised aliens of restricted distribution, and naturalised aliens of widespread distribution. A list of plants in these categories is given in Appendix 6, all but a few of the transient aliens have been recorded over the last 5 years.

Only three alien species; *Poa annua*, *Cerastium fontanum* and *Taraxacum officinale* (dandelion) commonly set fertile seed although several others may do in favourable seasons and many propagate vegetatively. Introduced seed of some species appear to lie dormant for many years in the cold soils of South Georgia; these sometimes germinate after disturbance or in particularly favourable conditions. The study of these plants has provided interesting data about their adaption to harsh environments and their ability to become distributed to very remote areas.

Other plants have been deliberately introduced to the island as ornamentals, for food, and for experimental purposes. These species have not been able to survive a winter in the open and, in any event, are subject to careful control to ensure they do not become established. A greenhouse at King Edward Point, generally under botanical management, provides crops of several salad species which are a most welcome addition to the diet of the scientific station, especially during winter. The whaling stations similarly kept greenhouse and ornamental plants during their occupation. Even a century ago the German International Polar Year Expedition had some success with salad vegetables, although all their other crops failed.

Terrestrial and freshwater invertebrates

The free-living terrestrial and freshwater invertebrates of South Georgia are found in virtually every available habitat on the island although their species diversity is not great. The parasitic fauna, principally of bird hosts, are also invertebrates and are described in this section. Free-living invertebrates include about 40 species of insects, 10 crustaceans, other arthropods, annelids, tardigrades, rotifers, gastrotriches, protozoans, and one species each of mollusc, platyhelminth, and coelenterate. Approximately one third of these are endemic and it is probable that there are several species, especially of the smaller phyla, awaiting discovery and identification.

The insects represent a particularly interesting group which comprises 7 Coleoptera (beetles), 13 Diptera (flies), 2 Hemiptera (bugs), 1 Thysanopteran (thrips), 1 Hymenopteran (wasp), and 16 species of Collembola (springtails) among the free-living species. Most of the insects are found in the coastal lowlands although some beetles and flies inhabit the *Festuca* grassland further inland, while certain species of Collembola are found almost wherever moss growth occurs. Only a few of the insect groups normally capable of flight can do so on South Georgia, an adaption common to those inhabiting windy, isolated islands.

One insect species is remarkable as it is a dytiscid water beetle, *Lancetes clausii*, which is active in many lakes and ponds. Under favourable conditions, the adult may be seen swimming beneath layers of ice on the water surface. Its larvae, which pupate in the soil surrounding these water bodies, and the adult are carnivorous and presumably feed on arthropods, annelids, and nematodes. Two other beetles, *Perimylops antarcticus* and *Hydromedion sparsutum*, are members of the rare family Perimylopidae found only in South Georgia, the Falkland Islands, and southern South America. They are the most widespread of the island's beetles and are sometimes found walking over the snow surface during winter in areas several kilometres inland. Their body temperatures are then substantially higher that the substrate (with which they make minimal contact on their extended legs) and

their dark colour may allow them to absorb infrared heat from the sun. The other beetles are rarely found far from the coast and the two staphalinids are common inhabitants of bird and rat nests.

The flies are also principally coastal in distribution and are especially common around elephant seal wallows. One species, *Parochlus steinenii*, often forms swarms which fly over tussock plants on sunny days. Another, *Antrops truncipennis*, has markedly reduced wings and is flightless. It is common on beaches, particularly near penguin colonies and rotting kelps. The larvae of the large fly *Paractora trichosterna* are principally responsible for the later degradation of carrion and may be found in rotting seals, birds, and reindeer. The single hymenopteran belongs to a family of small parasitic wasps, its larvae are thought to be egg parasites of the beetles.

The Collembola are all Arthropleona except for one sminthurid. They are widespread and sometimes occur in enormous local concentrations. The species *Cryptopygus antarcticus* is common and is also found as far south as 68° on the Antarctic Peninsula.

As well as the indigenous species of insects there are several introduced ones recorded from the whaling stations and King Edward Point. The ubiquitous German cockroach (*Blatella germanica*) once inhabited Grytviken but died out after the station's closure. It is usual to find an occasional exotic living insect arriving with cargo from ships especially with vegetables from South America but, even if they escape the entomologists, these rarely live very long. A group of insects noticeable by its absence from South Georgia is the biting flies which infest almost all other tundra regions of the Earth, frequently rendering living conditions very uncomfortable.

The mite fauna of South Georgia is comparatively rich and contains members of five orders which exploit virtually every terrestrial habitat and niche on the island. Two species of *Rhagidia* are predatory in habit, comparatively large, orange, and fast moving. These hunt lesser mites and Collembola. A recently discovered new species, *Hyadesia maxima*, is remarkably hardy; it inhabits the littoral zone in large numbers living in crevices and fissures in rocks on the sea shore and withstands heat, cold, desiccation, submergence, ice abrasion, and deprivation of oxygen. Its food is mainly filamentous green algae. The mite fauna of soil and plant litter shows large seasonal fluctuations in populations and developmental stages at different depths in the profile during the year. These reflect their differing life cycles and ecology.

Four species of spider are known from South Georgia although only one, which inhabits drier grassland and scree, is abundant. They, with *Lancetes*, two other beetles, with mites of the genera *Rhagidia* and *Gamasellus* are the only carnivorous terrestrial arthropods on South Georgia. The diet of the other species consists largely of fungi and algae although some, notably the beetles, feed on higher plants such as the grasses *Festuca contracta* and *Parodiochloa flabellata* and leaves of *Acaena* species.

Amongst the remaining terrestrial animals, a small land snail has recently been discovered. It is a new species yet to be named and is the only Antarctic land mollusc. It is known from crevices in sheltered rock faces in only one location. An earthworm, belonging to the southern hemisphere family Megascolecidae, is common in wet vegetated areas and under *Acaena* mats. It is known to winter in peat bogs below the ice level in small groups and, presumably, as eggs elsewhere. Many enchytraeid worms, tardigrades, and nematodes are found in moss, peat, and similar situations. Quite large nematodes are also common beneath reindeer dung.

The native arthropods have several strategies for overwintering; many do so as inactive stages, eggs or pupae, but others survive as larvae or adults. During winter those seeking cover in plants or litter and the soil fauna are well protected by the snow layer from the extremes of temperature. Their main survival problem is enduring temperatures near zero for several months and being able to continue their development and reproduction during the brief, cold summers. Another particularly interesting adaption involves their capacity for extensive supercooling and the presence of polyhydric alcohols and sugars in arthropod body fluids. Such compounds increase the capacity for cold survival by maintaining the body fluids in the liquid phase at temperatures well below their normal freezing point. Some other species, usually in their larval stages, are capable of survival even after being frozen solid. The rate of growth and increase in numbers when

favourable conditions occur again may be very great. Comparative studies of the cold-tolerance of arthropods from South Georgia, Signy Island in the South Orkney Islands, and the Antarctic Peninsula have yielded interesting data on their adaptions to increasingly rigorous environments.

Protozoa are common throughout the terrestrial environments of South Georgia but most have not yet been investigated. The testate amoebae have recently been the subject of research and 29 species are now known. The sizes and seasonal fluctuations of the population of 18 species of these in 4 contrasting habitats, were examined for over a year and showed a noticeable spring bloom with daily rates of increase averaging 5% with a maximum of about 20%. They have the capacity to exploit, very rapidly, the short periods when comparatively high temperatures occur in their habitats at any time throughout the year.

Invertebrates inhabiting freshwater include 10 species of crustaceans, 16 rotifers, and at least one each of annelids, tardigrades, nematodes, gastrotriches, coelenterates, and many protozoans as well as the aquatic beetle previously mentioned. Most of the species are widespread in the Falkland Islands and southern South America while a few extend to the South Orkney Islands and some to the Antarctic Peninsula. Many of the rotifers are found specifically in association with aquatic mosses. The freshwater fauna of South Georgia is remarkably depauperate and appears to provide no basis for higher food chains. Thus the practicality of successfully introducing fish to these areas, as has occasionally been suggested, is remote. The larger lakes and pools do not freeze to their entire depths during winter although thick ice may cover them for almost half the year; rivers and streams are usually flowing beneath an ice cover during winter. Although shallow ponds and tarns are completely frozen, many of their inhabitants survive this, often as eggs, and their populations re-establish rapidly during summer.

The parasitic invertebrate fauna is mainly associated with the birds although reindeer, rats, and seals also support some species. The ectoparasites of the birds include 38 Mallophaga (biting lice), an Anopluran (sucking louse), and two flea species as well as many mite species. Some occur on virtually every species of bird, including the penguins. The elephant seal has some interesting ecto-parasites, one species of which can withstand prolonged immersion (although it is air breathing) and another which lives in the seal's nasal sinuses. The South Georgian reindeer are remarkably fortunate in being free from the large numbers of parasitic flies and other species which torment reindeer in their habitats in the northern hemisphere. They and the rats have only small numbers of ectoparasites. Endoparasites of several phyla occur in most of the island's vertebrates but little is yet known of their life cycles. The relationships and ecology of South Georgian parasitic fauna follow closely those in similar situations elsewhere, owing to the specialised nature of the parasitic environment.

Marine invertebrates

Marine invertebrates may be divided into planktonic (open sea) and benthic (bottom-dwelling) species. The planktonic invertebrates are characterised by very large populations of small numbers of species, a classic response to an extreme environment. In contrast, the benthic invertebrates have a very high species diversity, a response to the great variety of habitats available to them. Many of the marine invertebrates, especially the planktonic ones, are more to be regarded as creatures of the Antarctic or Southern Ocean rather than of South Georgia but several are of great importance; much of the island's biology, as well as human exploitation, is associated with them.

One planktonic species, *Euphausia superba*, the Antarctic krill, accounts for about half the biomass of the zooplankton in the ocean around South Georgia. This occurs at certain seasons near the island in swarms which may have a total mass of 10 000 000 tonnes and cover several square kilometres. Antarctic krill are small, shrimp-like crustacea, about 5 cm long when adult and occur in all parts of the Antarctic Ocean, although their high concentrations are localised. Their diet is principally phytoplankton which they obtain by filtering sea water through specialised 'baskets' developed on their thoracic appendages. The movement of water masses near South Georgia and the great increases in quantities of phytoplankton resulting have already been described. The abundant phytoplankton allows the development of such quantities of krill and thus the associated food

chains of the region. Most other marine organisms (including some squid, fish, seals, and baleen whales as well as many sea birds) are directly dependent on krill for their food and the majority of the remainder feed on krill feeders. This, with the probability of its greatly increasing commercial value, demonstrates the great importance krill has in the Antarctic marine ecosystem.

Of the other planktonic invertebrates, the most numerous are the copepods, and several other crustacea are common. Ctenophores are also abundant and occasional very large ones may be found, perhaps 25 cm long by 15 cm in diameter. Large (up to 1 m in diameter) cnidarian medusae (jelly fish) are occasionally noticed. Squid of several species occur around the island; these form much of the food of many sea birds, some seals, and, especially the larger squid, of the sperm whales.

The benthic invertebrate fauna of South Georgia includes members of the majority of phyla; a single trawl haul taken in a bay may be teeming with different species. This is largely due to the very wide range of habitats available to them in the kelp beds, in areas of glacial deposition where there is a high diversity of substrates from fine material to rough hard rocks, boulders, and stones, and to the isolation of the Antarctic regions. This has led to a great variety of adaptions and the evolution of many endemic species. Interestingly, such isolation and extreme conditions are reflected in the absence of some groups which are very common elsewhere. For example, the larger decapods (crabs, lobsters, etc.) are virtually absent and Cirripedia (barnacles), so common throughout the rest of the world, are restricted to unusual habitats such as areas subjected to strong currents at depths of 500 – 1000 m, on the skin of whales, and on guard hairs of the fur of seals.

The seas around South Georgia, and Antarctica generally, are characterised by great clarity in the surface water. This has led to the development of communities living on the sea bed composed of attached filter-feeders such as tunicates, sponges, hydroids, tube-worms, and molluscs with associated mobile predatory groups such as echinoderms and crustacea which browse over the organic 'carpet'. The 'carpet' may have a biomass of up to $10 \, \mathrm{kg \, m^{-2}}$.

The benthic community includes several

Euphausia superba, the principal component of krill.
(I. Everson.)

noticeable members. The king crab or centolla is very rare but occasionally found in Cumberland Bay. It is one of the few decapods of South Georgia and is unfortunately far too uncommon to be passed to the cook. Nemertines reach lengths of over 1 m and are substantial worms. A priapulid, *Priapulus caudatus*, occurs which has a bipolar distribution (being found also in the Baltic Sea). Isopods and pycnogonids are common and relatively huge animals compared to their relatives elsewhere; to a large extent, these fill the niche normally occupied by the absent decapods. Large numbers of delicious small bivalves of the genus *Gaimardia* are found attached to *Macrocystis* in the kelp beds and many serpulid worms occur on kelp fronds.

South Georgia's foreshores are poorly colonised owing to their exposure to sub-zero temperatures, ice abrasion in winter, and lack of strong regular tidal changes. However a zonal

Benthic invertebrates in Cumberland Bay. (D. Sanders)

succession of communities of seaweeds and invertebrates develops in some areas which includes small mussel-like bivalves such as *Kidderia* species and even some mites described with the terrestrial invertebrates. A limpet, *Nacella concinna*, may occasionally be seen browsing on algal-coated rocks near the tideline also but its shells are more frequently encountered on land near the nests of gulls.

Fish

In the waters south of the Antarctic Convergence, almost all the fish encountered elsewhere are absent and only 4 of the 41 species recorded near South Georgia also live north of the convergence. A single group, the Nototheniiformes, predominates. This is subdivided into four major families: Nototheniidae, Channichthyidae, Harpagiferidae, and Bathydraconidae (Antarctic cod, ice fish, dragon fish, and plunder fish). These fish are generally sluggish, bottom-dwelling species with large, often rounded, heads and rather slender tapering bodies. The Nototheniidae are scaled but some of the Bathydraconidae and all of the others have a completely naked skin well-equipped with mucous glands which make them very slimy. Lantern fish form part of the fauna of the bathypelagic zone, from 200 to 2000 m deep near the island. *Notothenia rossii*, one of the more common species, may reach almost 1 m in length and has been recorded as living up to 15 years (from examination of its otoliths). *Dissostichus eleginoides* is another common species also taken commercially which may be 2 m long. The largest fish is *Raja georgiana*, an endemic ray which may weigh 14 kg. Even a flat-fish, *Mancopsetta maculata* is found in deep coastal waters near the island. The shark *Lamna nasus* is known from only one specimen taken to the north-east of the island. A systematic list of the fish recorded from the South Georgia region is given in Appendix 7.

Most Antarctic fish have reduced haemoglobin in their blood. This is thought to be an evolutionary response to the high oxygen-carrying capacity of very cold water and of haemoglobin giving blood an increased viscosity at low temperatures. The most extreme case is found in the ice fish, such as *Chaenocephalus aceratus*, a large pike-like fish common at South Georgia. These have completely colourless blood and no functional red blood cells. Another adaption to Antarctic conditions is the tendency for the fish to produce large yolky eggs from which comparatively advanced larvae emerge.

Krill is an important part of the diet of many South Georgian fish, notably of adult *Notothenia* species. They also take tunicates, amphipods, and small pelagic fish. The larvae feed principally on copepods and amphipods. The bulk of the food of

Notothenia rossii, Antarctic cod, from Cumberland Bay. (Courtesy British Antarctic Survey, C.J. Gilbert.)

Parachaenichthyes georgianus, Crocodile fish, from Cumberland Bay. (Courtesy British Antarctic Survey, C.J. Gilbert.)

many bottom-dwelling fish is crustaceans such as *Serolis* and *Gammarus* as well as smaller fish, molluscs, and worms. Interestingly, many of the fish support large numbers of parasites. Their ecto-parasites are mainly highly specialised leeches and are specific to the group of fish and thus are as endemic to Antarctica as the Nototheniiformes.

Several attempts at establishing commercial fisheries near South Georgia have not met with success although abundant quantities have been caught for local consumption, as discovered by Fanning in 1800. However, the large concentrations of bottom-dwelling fish which are found over the continental shelf around South Georgia have at-tracted several eastern European fishing fleets from about 1970. The principal fish caught are three species of ice fish (*Chaenocephalus aceratus*, *Para-chaenichthyes georgianus*, and *Champsocephalus gunnari*) and the Antarctic cod (*Notothenia rossii*). During the 1970/71 season, nearly half a million tonnes of fish were caught around the island but in recent years this has fallen to 10 000–20 000 tonnes. There is much additional evidence to indicate that this fishing is above the maximum sustainable yield of the population, that fish populations have fallen substantially and their age distributions have been greatly reduced. Foreign fishing vessels are commonly seen near the coast of South Georgia; in 1980, at least one was seen during every month of the year. They occasionally call at King Edward Cove to take fresh water and, with the Magistrate's permission, at other parts of South Georgia to tranship cargo. The local fish are fairly easily caught with a hook and shiny sinker and the author can testify to their excellent taste.

Juvenile lampreys of the species *Geotria austr-alis*, primitive parasitic fish, also occur around the island and near the Antarctic Convergence. They are an important constituent of the diet of several albatross species. The adult lampreys are found in South American rivers; they and the juveniles feed on the blood of many vertebrate species.

There are no freshwater fish on South Geor-gia; suggestion that some should be introduced experimentally has been made on several occasions but only one attempt has been reported. In late 1964 ten rainbow trout were introduced into Gull Lake by the Japanese company which was opera-ting Grytviken whaling station. In the following summer no trace of them could be found and none have subsequently been recorded. It is unlikely that freshwater fish could survive a South Georgian winter and, even in summer, food to support them would be very scarce. In any event Gull Lake, which is principally an artificial reservoir, would have been a far more unfavourable place than most lakes on the island for such an introduction.

Indigenous birds

One of the outstanding features of South Georgia is the bird life. Nicolas-Pierre Guyot referred to them in 1756 and their great numbers were first commented upon by Captain Cook and others aboard *Resolution* in 1775. Virtually every subsequent account has shown its writer was greatly impressed by the island's avifauna which includes some of the great albatrosses and several species of penguins. To date, 30 breeding and 27 non-breeding species have been recorded as naturally occurring on the island, and in addition at least 9 species have been introduced by human agency. The breeding species represent members of 11 families. As is to be expected from an isolated oceanic island, the

Anthus antarcticus, the South Georgia pipit, Elsehul February 1977. (T.S. McCann.)

avifauna is dominated by sea birds but 5 terrestrial and fresh-water species are resident. Of these, two – the South Georgia pipit (*Anthus antarcticus*) and South Georgia pintail (*Anas georgica*) – occur only on the island. The majority of the remaining species have a distribution south of the Antarctic Convergence.

Most scientific investigations of South Georgia have been concerned with the birds of the island and many have been specifically for ornithological purposes. From 1958 to 1964 and since 1972 there have been summer ornithological investigations at Bird Island (named by Captain Cook owing to the great number of birds present), and studies during winter were made in 1963 and 1983. The majority of the birds of South Georgia, as with much other Antarctic wildlife, have virtually no fear of humans and one may approach most species quite closely. The only significant exploitation of them has been the collection of penguin eggs, under licence, by whalers. References to the use of penguins during the first sealing epoch exist but are somewhat tenuous. A list of the bird species recorded from the island is given in Appendix 8.

The smallest terrestrial bird, the endemic *Anthus antarcticus* (South Georgia pipit), is slightly smaller than a sparrow. It breeds on Bird Island where 150–200 pairs occur, on some other small islands and in relatively rat-free areas elsewhere. It has a distinctive twittering call and its flight is somewhat like that of a wagtail. Its diet consists largely of invertebrates, mainly insects and spiders in summer and those associated with tideline debris in winter. Two waterfowl species occur, both with South American affinities. The endemic South Georgia pintail (*Anas georgica*) feeds predominantly on algae and thus frequents ponds in summer and sheltered bays in winter. It is a comparatively small, very mobile bird, the drake of which calls with a shrill repeated whistle while the duck has a far softer note. The speckled teal (*Anas flavirostris*) was first found breeding in 1967 and occurs only in the Cumberland Bay area. This may reflect its feeding preference for aquatic invertebrates from glacial ponds and lakes.

The sheathbill (*Chionis alba*) is dependent on sea birds and seals for its food. It is an enthusiastic, omnivorous scavenger around rookeries and seal beaches near which it nests in rock crevices, under

tussac etc. In winter, most migrate to the Falkland Islands and South America, although many remain near king penguin rookeries where food is available throughout the year. Despite these migratory flights, the sheathbill is noticeably reluctant to fly when on land. *Catharacta lönnbergi* (brown skua) is also a scavenger at seal, penguin, and albatross colonies; it feeds on afterbirths, eggs, and corpses. In some areas it also takes small petrels at night and is a predator of the rat. Though breeding throughout South Georgia, it is usually found where petrels or seals are numerous. It too migrates for winter, sometimes to the northern hemisphere. During its breeding season, from November to January, pairs defend a territory where they may resolutely attack other skuas or any person who ventures into it.

The remaining 25 species of bird known to breed on South Georgia are dependent directly on the oceans for their food and the great productivity of the waters around the island largely accounts for their enormous numbers. Much recent study has concerned the nature of the factors which separate them ecologically and the descriptions which follow give details of some of these.

Three species are inshore feeders: Dominican gulls (*Larus dominicanus*), blue eyed shags (*Phalacrocorax atriceps*), and Antarctic terns (*Sterna vittata*). These occur throughout the coastal regions of South Georgia and avoid direct competition by feeding on shoreline and inshore invertebrates (mainly limpets), large fish at depths to 30 m, and small fish near the surface respectively. Similarly with the remaining 22 bird species, which are pelagic feeders, each has a distinct feeding habit. The very high concentrations of nesting species on places like Bird Island are possible because of this and demonstrate the rarity of areas suitable for breeding sites for the sea birds of the region.

Penguins are a very distinctive flightless group of birds, which are excellent swimmers. They are adapted for exploiting food resources well below the ocean surface, in contrast to the remaining groups which are virtually restricted to taking food at the surface, but over wider areas. Eight species of penguin are recorded on the island; there are four main breeding species, one which breeds only occasionally, and three others.

The king penguin (*Aptenodytes patagonicus*) stands about 1 m high and has distinctive golden-

Adult and juvenile *Phalacrocorax atriceps*, blue-eyed
shag, with a nest in a tussac plant. (Courtesy British
Antarctic Survey; C.J. Gilbert.)

A colony of *Aptenodytes patagonicus*, king penguin, at Right Whale Bay in February 1980. (Courtesy British Antarctic Survey; C.J. Gilbert.)

yellow plumage on its neck and throat. Its call is remarkable – a deep trumpeting sound which is particularly impressive when made by a colony of several thousands. The breeding cycle is unusual in that it takes from 14 to 16 months to rear a chick. The most successful penguins may raise two chicks every three years but the majority are probably successful only in alternate years. There are large colonies at the Bay of Isles, St Andrews Bay, and Royal Bay together with many smaller ones elsewhere. A total of 57 000 adults and chicks at 31 breeding sites is known. The main colonies are continuously occupied by adults and chicks and each has over 10 000 breeding pairs. Their diet consists principally of squid and fish.

The smaller macaroni penguin (*Eudyptes chrysolophus*) is far more numerous, occurring in immense numbers particularly on the Willis Islands. The breeding population on South Georgia may exceed 5 000 000 pairs and many non-breeding birds are also present. They feed almost exclusively on krill and probably forage well out to sea. Gentoo penguins (*Pygoscelis papua*) nest in small colonies throughout the island. They take adult krill and fish during their November to March breeding season. Chinstrap penguins (*P. antarctica*) are less common, being at the northern extent of their range on South Georgia. There are several breeding colonies on the island and its population is thought to be increasing. The rockhopper penguin (*Eudyptes*

A portion of a colony of *Eudyptes chrysolophus*, macaroni penguin, at Bird Island. (P.A. Prince.)

chrysocome) is at its southern limit but has been recorded with eggs on a few occasions. It is very common on the Falkland Islands. Emperor, Adelie, and Magellanic penguins (*Aptenodytes forsteri, Pygoscelis adeliae,* and *Spheniscus magellanicus*) are the other species recorded as visitors to the island.

There are about 8 600 breeding wandering albatrosses (*Diomedea exulans*) on South Georgia, mainly confined to Bird Island and several other small islands. This species is amongst the largest of the albatrosses and may have a wingspan of 3.5 m. Eggs, laid in late December, hatch in late March and the chicks are reared during the winter, finally fledging in November or December. Birds which have successfully raised a chick cannot then breed again that season and so most of the population breeds every second year. Wandering albatrosses regularly travel around the southern regions of the world; birds ringed at South Georgia have been recovered in Australia and vice versa. Their diet consists principally of squid and fish. The wandering albatross as well as several others and the giant petrels cannot take to the air without first gaining substantial speed. Their nests commonly have a 'run way' adjacent and a take-off from water may be a laborious affair. In flight, however, they are magnificently graceful.

The light-mantled sooty albatross (*Phoebetria palpebrata*) is a remarkably handsome bird which nests solitarily or in small colonies on rocky bluffs around the island. The total population is about 10 000. Its food is mainly squid together with krill,

Courting of the wandering albatross, *Diomedea exulans*, at Bird Island. (W.N. Bonner.)

other crustaceans, fish, and carrion. Its call is distinctive, a two-note wild drawn-out scream, and can be heard regularly during the nesting period.

The two other breeding albatrosses, black-browed and grey-headed albatrosses (*Diomedea melanophrys* and *D. chrysostoma*) are closely related. They breed in colonies, sometimes mixed, and there are approximately 60 000 pairs of each, breeding mainly in the north-west of the island. During January to May, when chicks are on the nests, the black-browed albatross is mainly a krill feeder while grey-headed albatrosses take squid with lamprey and other fish. The former is an annual breeder and the latter biennial.

The blue petrel (*Halobaena caerulea*) and dove

prion (*Pachyptila desolata*) have different breeding times, but nest in similar situations – burrows in tussac. The latter is probably the most common breeding species on South Georgia. South Georgia diving petrels and common diving petrels (*Pelecanoides georgicus* and *P. urinatrix*) probably have separate feeding sites, one being inshore and the other in the ocean. South Georgia diving petrels nest in burrows on high scree slopes throughout the island thus, to some extent, escaping predation by rats. The common diving petrel is numerous on Bird Island and other rat-free areas where it nests in steep tussac slopes.

Two giant petrel species (*Macronectes giganteus* and *M. halli*) have only recently been taxono-

A male wandering albatross, *Diomedea exulans*, incubating an egg at Bird Island. (Courtesy British Antarctic Survey; C.J. Gilbert.)

mically separated since they are morphologically very similar. The latter breeds some 6 weeks earlier than the former whose chicks may be present on the nest till early May. Most of the *M. halli* population is on the north-west of the island whilst *M. giganteus* occurs throughout it. They are large powerful birds, mainly carrion eaters and scavengers, but either may take live food and have been aptly described as Antarctic vultures. Both the adults and chicks are particularly efficient at using a potent defence mechanism – the ability to project a stinking oily vomit some distance with considerable accuracy against any supposed enemy.

Of the remaining breeding species, only white chinned petrels (*Procellaria aequinoctialis*) and Wilson's storm petrels (*Oceanites oceanicus*) are common. The former is large and powerful enough to resist rat predation; it forages over large distances, catching squid and krill, and is the petrel most likely to be seen on land during daylight hours as most of the smaller ones are nocturnal. Wilson's storm petrel is a common breeder in coarse scree and rock crevices. It is a very small bird which may often be seen feeding among kelp beds and in many bays. Despite its size, it migrates to the northern hemisphere for the austral winter. The two other storm petrels are also small birds, black-bellied storm petrels (*Fregetta tropica*) are not uncommon

A nesting pair of *Phoebetria palpebrata*, the light-mantled sooty albatross, at Hope Point. (Courtesy British Antarctic Survey; C.J. Gilbert.)

while grey-backed storm petrels (*Garrodia nereis*) are rare and local. Fairy prions (*Pachyptila turtur*) are only a very recent addition to the avifauna of the island, they were discovered in 1977 and found breeding in 1979. Cape pigeons (*Daption capense*) are now most frequently seen at sea but once occurred in huge flocks around the whaling stations when these operated, feeding on floating refuse. They breed in small numbers on the island. The snow petrel (*Pagodroma nivea*) is often seen at sea as well as in the mountains, where it breeds in only small numbers. All the petrels are notable as they have a distinctive musk odour which, in some species, is unpleasantly strong and very persistent, especially when clothing is contaminated with it. It has been suggested this assists them in locating their nesting sites.

Macronectes giganteus, giant petrel, adult and chick. (Courtesy British Antarctic Survey; C.J. Gilbert.)

Vagrant and introduced birds

The 27 species of birds which have been recorded at South Georgia as naturally occurring non-breeding species, include several sea bird species from other regions of the Antarctic as well as more northerly regions in the Atlantic Ocean, together with birds from South America and the Falkland Islands blown over by storms and prevailing winds, and one other. Sightings of herons, egrets, ducks, sandpipers, and others are most commonly reported after periods of several stormy days. These birds usually do not survive for very long although the occasional one has been seen for several months near King Edward Point. An interesting new arrival is the cattle egret (*Bubulcus ibis*) which has only recently become established in South America from Africa. Several small flocks

have arrived over the past few years with strong winds from South America. The most interesting vagrant to arrive is the little stint (*Calidris minuta*). This was first recorded at South Georgia in 1977 and is unknown from South America. It breeds in Siberia and winters in Africa and elsewhere, and thus it probably arrived in South Georgia from the east against the prevailing winds. A list of the vagrant species recorded to 1982 is given in Appendix 9.

Several species of birds have been introduced to South Georgia by human agency but only one of these was released with the intention that it should become established on the island – the others remained substantially domestic. In 1882 the German International Polar Year Expedition made the first introduction of a domestic bird, at Royal Bay; this was of two geese, their fate is related in Chapter 3. Domestic fowls, ducks, and geese have variously been kept at the whaling stations and King Edward

Point. Hens, especially, were kept for egg production and are still maintained as a breeding population at King Edward Point with occasional imports of fertile eggs from the Falkland Islands. Pigeons were once kept at Prince Olav Harbour, Leith, and Grytviken. At the former station they were used as message carriers for a short period before the advent of radio. At Grytviken they bred in a warm loft above the bakery and occasionally provided squabs. The last caretaker, Ragnor Thorsen, supplied them with food until his departure in 1971 after which they soon died out. Sparrows have occasionally arrived with a ship from the Falkland Islands and might be regarded as having introduced themselves by human agency; they survive on the island only briefly. Canaries, parrots, budgerigars and other caged birds have also been present at various times.

The only bird species introduced with the aim of establishing it as a game bird was the Upland

Oceanites oceanicus, Wilson's storm petrel, feeding on the surface of King Edward Cove. (R.I. Lewis Smith.)

Goose (*Chloephaga picta*) a native of the Falkland Islands and southern South America. On 21 March 1911, 17 were released near Sappho Point by the Magistrate and they were specifically mentioned in a preservation ordinance in 1912. Several years later they were reported to be thriving when several families of geese with goslings were seen near Maiviken. However, by 1927, they were assumed to have died out. The second introduction was made by another Magistrate, at Sandebugten on 2 February 1958. Five ganders and six geese were released but failed to become established. The Magistrate hoped to repeat the experiment but counsel from the Government Naturalist prevailed and further introductions of species alien to the island were discouraged.

Indigenous mammals

The only native mammals found at South Georgia belong to the marine environment and include two orders of whales and two of seals.

Several species of terrestrial mammals have been introduced over the last two centuries, however, and are described later.

Six species of seal have been recorded at South Georgia. Two occur in great numbers, the Antarctic fur seal (*Arctocephalus gazella*) and the southern elephant seal (*Mirounga leonina*). Two species are far less common, the leopard seal (*Hydrurga leptonyx*) and the Weddell seal (*Leptonychotes weddellii*). Two species are only rare visitors, the crabeater seal (*Lobodon carcinophagus*) and the sub-Antarctic fur seal (*Arctocephalus tropicalis*). Of the latter two, the crabeater seal normally frequents the edge of the pack-ice far to the south of South Georgia and reaches the northern limits of its distribution at the island while the sub-Antarctic fur seal is at its most southern limit and is probably a vagrant from Gough Island which is its nearest breeding site.

The Antarctic fur seal once bred in enormous numbers all around South Georgia. Now, after

Pagodroma nivea, snow petrel. (R.I. Lewis Smith.)

Part of a fur seal, *Arctocephalus gazella*, rookery on
Bird Island in 1976–77. Note the light-coloured animal
in the foreground. (P.A. Prince.)

A 'pod' of female elephant seals, *Mirounga leonina*, in tussac near King Edward Cove. (Courtesy British Antarctic Survey; C.J. Gilbert.)

many years of protection, it is making a remarkable recovery from virtual extermination during the first sealing epoch (see Chapter 3). The main colonies are on beaches at the north-western part of the island and their populations are presently increasing at about 10% per year. This remarkably rapid rate is probably assisted by the relative abundance of krill, the main constituent of their diet, as a consequence of the reduction in the baleen whale populations. Recent estimates suggest over 300 000 fur seal cows are breeding annually on South Georgia and at least this number of immature animals is present. Breeding populations of the species are distributed from the South Shetland, South Orkney, and South Sandwich Islands to Bouvetøya, Heard Island and McDonald Islands, as well as South Georgia. Increasing numbers of non-breeding seals are seen in sites along the west coast of the Antarctic Peninsula.

Fur seals are active, agile creatures which are polygamous and gregarious during the breeding season. This begins in early November when adult bulls compete to hold territories on breeding beaches. Those which are successful are usually from 8 to 10 years old and defend harems of up to 15 cows. Pups, conceived the previous season, are born singly in early December and weaned about 4 months later. Sexual maturity is reached at 3 to 4 years of age in cows and 6 or 7 in bulls and they may live to 23 and 13 years respectively.

Outside the breeding season the whole population is pelagic. The staple diet of the species is krill at South Georgia but fish, squid and birds are also occasionally taken. Their fur, which has two types of hair, is grey to brownish; when prepared, it is of very fine quality. Occasional animals lack pigment in the guard hairs and appear white or straw-coloured. Adult bulls may be up to 2 m long and weigh 200 kg, cows up to 1.5 m long and 50 kg while pups at birth are about 66 cm and only 5–6 kg.

A challenge between male elephant seals, *Mirounga leonina*. (W.N. Bonner.)

The southern elephant seal or sea elephant is the largest of all seal species. Its vernacular name is derived from its size, and the large erectile proboscis of the adult male. The largest bulls may reach a length of 6 m and weigh 4.5 tonnes of which 40% is skin and blubber. Cows are less than a quarter of this size. Following considerable overexploitation in the first sealing epoch, a scientifically managed plan was successfully followed for their exploitation during the second epoch of sealing (from 1909 to 1964). Their population on South Georgia is now approximately 300 000 and is relatively stable.

Elephant seals are notably inactive on land except when fighting. Their breeding season commences when the first bulls haul out of the ocean in late August followed, in September and October, by the pregnant cows. The harems formed are much larger than those of the fur seals, perhaps 100 cows may be controlled by a 'beachmaster', although most are smaller. Many bachelor bulls lurk in the surrounding areas and challenges frequently develop into spectacular fights. These occur on the beaches throughout the breeding season and cause much pup mortality (many pups die after being squashed by bulls). Pups are born soon after their mothers haul out. Their milk diet is exceedingly rich in fat and their weight increases from about 44 kg at birth to about 180 kg at the conclusion of lactation some 23 days later, when they are deserted by their mothers. They then moult and enter the sea before the end of the year. Prior to this, the adults have returned to the sea. They haul out again in late summer to moult while spending much time in large 'pods' in foetid wallows.

The principal constituent of their diet at South Georgia is squid with some fish. By absorbing energy from their blubber they are capable of remaining continuously on land without feeding during the breeding season. Their maximum life span is about 20 years for both sexes. For most of the year they are pelagic and an animal tagged at South Georgia has been detected in South Africa 4725 km away. The species occurs in three main breeding stocks; South Georgia and the other islands of the Scotia Ridge with southern South America and the Falkland Islands; Kerguelen, Heard and McDonald Islands; and Macquarie Island with the islands south of New Zealand.

The leopard seal is essentially a solitary species and individuals are often seen around the coast of South Georgia especially near penguin colonies. The female reaches a length of up to 4 m – considerably larger than the male. Normally it breeds on the edge of the pack-ice well south of South Georgia but there are several records of breeding on the island. Leopard seals have impressive recurved tricuspid molars and their diet includes fish, krill, birds (especially penguins), seal pups, and almost anything else they can kill. Divers consider them potentially dangerous and they have been known to savage inflatable boats in King Edward Cove. The species is peri-Antarctic in its distribution.

The Weddell seal is also a peri-Antarctic species which is at the northern limit of its range at South Georgia and is not often seen. There is a long established breeding colony at Larsen Harbour at the southern end of the island and the seals are regularly seen in several coves in the vicinity. Its diet is principally fish with squid and other invertebrates. The South Georgian population is less than 100.

South Georgia was once the centre of the world's whaling industry but excessive exploitation, principally from factory ships, with inadequate controls have drastically reduced the numbers of whales throughout the Antarctic regions as well as much of the rest of the world. It has been estimated that populations of the great whales are presently only about 10% of those at the beginning of the century. International acceptance of improved conservation measures may permit their eventual recovery but this will take many decades. The natural history of whales is essentially an oceanic matter rather than one particularly associated with South Georgia, which provided a land base for their capture and processing for 60 years. Up to the early part of this century, whales abounded in the island's fjords as well as in the surrounding ocean. The species encountered included; blue (*Balaenoptera musculus*), sei (*B. borealis*), fin (*B. physalis*), minke (*B. acutorostrata*), humpback (*Megaptera novaeangliae*), southern right (*Eubalaena australis*), and sperm (*Physeter catodon*). Fin and sei whales were those most commonly killed towards the end of the whaling era while humpbacks were the first to become rare. The southern right whale is a species which tends to remain in one location more than other whales do; they have recently been seen several times in certain fjords of South Georgia – the first recorded for many years and perhaps early signs of the recovery of their numbers. The largest whale ever recorded was taken at South Georgia around 1912 when a female blue whale which measured 33.58 m long (107 Norwegian fot) was hauled up at Grytviken. Another, taken at

Prince Olav Harbour in 1931 measured 29.48 m and was estimated to weigh 177 tonnes excluding blood and gut contents.

All except sperm whales are krill feeders. They swim through swarms of krill with their mouths open, filtering the crustaceans through baleen plates, dermal processes of their upper jaws; the 'whale-bone' of commerce. These whales have well-

A Weddell seal, *Leptonychotes weddellii*, on a beach strewn with glacier ice, Moraine Fjord. (Courtesy British Antarctic Survey; C.J. Gilbert.)

defined migratory cycles corresponding with seasonal fluctuations in krill abundance around South Georgia. Sperm whales belong to the toothed whales and feed mainly on squid, which they may take from great depths. Their bulbous heads with reservoirs of oil-like spermaceti serve to control their buoyancy.

The killer whale (*Orcinus orca*) is sometimes seen around South Georgia. Though not hunted, it was often encountered during whaling operations, feeding on dead whales. It is an intelligent predatory carnivore which sometimes hunts in packs and may attack much larger whales. Other lesser cetaceans recorded near the island include pilot whales (*Globicephala melaena*), Commerson's dolphin (*Cephalorhynchus commersonii*), Peale's dolphin (*Lagenorhynchus australis*), and the rare cruciger dolphin (*L. cruciger*) which is sometimes seen with fin whale groups.

Introduced mammals

There are two introduced species of mammal which have had a major effect on the native flora and fauna of South Georgia. These are the brown rat (*Rattus norvegicus*) and reindeer (*Rangifer tarandus*). One other introduced mammal, the house mouse (*Mus musculus*), has a very localised breeding population on the island. The rats and mice were probably introduced during the first sealing epoch, perhaps almost two centuries ago, although the earliest records of them are 1877 and 1913 respectively. The reindeer and majority of the rest of the introductions took place after the establishment of a permanent human population on the island.

A female sperm whale, *Physeter catodon*, on the flensing plan at Leith Harbour in 1913, about 17 m long. (From Salvesen, 1914.)

The introduced mammal which has had the most profound effect on the island is the rat. Rats are well known almost everywhere visited by man, including other islands around Antarctica. Their effects on South Georgia are described later. Two other species which have severely affected many similar far southern islands are cats and rabbits. Both have been introduced to South Georgia several times but neither has established successful wild breeding populations. This is most probably due to the climate as these species, although able to endure mild winters, are not suitably adapted to prolonged periods of severe weather, deep snow cover, frozen soil, and shortage of food, such as occur in a particularly severe South Georgia winter. On Îles Kerguelen and Crozet, Macquarie Island and Marion Island, all similar to South Georgia but with slightly warmer winters, these species have become well established with serious consequences.

The alien mammals which have been introduced to South Georgia are (roughly in order of their long-term impact on the island): rats, reindeer, mice, cats, sheep, cattle, pigs, goats, ponies, horses, rabbits, dogs, silver foxes, and monkeys (as well as other pets). Many of these species were imported for food although some were working animals and a few were pets. Perhaps it is reasonable to argue on a zoological basis that Man should be included in the list but it is difficult to decide whether he should come before or after rats.

The present rat population is probably the result of many separate introductions. They were first noted by Klutschak in 1877 who recorded the name Rat Hafen for Prince Olav Harbour. C.A. Larsen noted that rats arrived at Grytviken before the whaling station. Although only the brown rat is established in South Georgia, the ship rat (*Rattus rattus*) was probably introduced several times contemporaneously but did not survive. It is, however, the latter species which occurs on several other far southern islands. Rats are especially widespread and abundant on the north-west coasts of South Georgia and populations in different parts of the island, separated by glaciers and fjords, may be descended from separate introductions. Rats live mainly in areas with dense stands of tussac which provide food and shelter for them. Tussac is rich in carbohydrates, especially in the leaf bases, and provides a major part of the rat's diet throughout the

year. The tussac stools are ideal habitats in which rats can excavate nests which they line with dry litter, insulating them very efficiently. During winter they make tunnels under the snow which may interconnect and cover distances of several hundred metres. Separate faecal chambers are also constructed. After the thaw these rat runs are revealed as distinctive trails through the vegetation. The rat is among the most omnivorous of creatures and its diet on South Georgia reflects this. Tussac rats, those living away from the whaling stations, feed on insects (mainly beetles and their larvae), forage on the sea shore for a variety of invertebrates, eat carrion when this is available, and are major predators of ground-nesting birds and their eggs. The effect of rats on the South Georgia pipit (*Anthus antarcticus*) is particularly severe and the pipit is almost absent whenever rats occur. It is notable that Bird Island, the Willis Islands, Annenkov Island, the islands of the Bay of Isles, and some others with the best developed avifauna are free from rats and one wonders what the rest of South Georgia was like before their arrival. Rat tracks are occasionally seen for several kilometres over snow in winter and the possibility exists that, during a winter with severe pack-ice, they might gain access to some of these islands. The abandoned whaling stations have substantial rat populations which were very much larger during the early days of their operations. These rats are still feeding on remaining dried and other foodstuffs as well as meat meal and other products of the factories.

Several times in the past it has been suggested that the rats should be eliminated from South Georgia. Regrettably this is virtually impossible. Observations by Sörling in 1905 caused Lönnberg to postulate a new sub-species for the South Georgia rat. As the resources at the whaling stations, from the point of view of the rat, decline and as further recruitment from outside South Georgia is now very unlikely, it is conceivable that a South Georgian strain will become more well defined in the long term. Studies show that no significant changes in them have as yet occurred.

Three introductions of reindeer (*Rangifer tarandus*) all from Norway, have been made to South Georgia by the whaling companies (to provide sport and fresh meat). Only two of these now have any representation. The date of the first introduction, which now forms the southern herd, has been variously given as 1909 and 1911; it was made at Ocean Harbour by C.A. Larsen and his brother L.E. Larsen. The deer came from Numedal in central southern Norway. The original numbers introduced are also quoted differently in various records. It is probable that three stags and seven hinds were introduced early in November 1911. They were provided with fodder on landing to assist them in becoming established. This may be the origin of several isolated patches of an introduced sedge (*Carex nigra*) and rush (*Juncus filiformis*) in the vicinity. These reindeers' descendants have spread throughout the Barff Peninsula and south through the St Andrews Bay area to the northern side of the Ross Glacier in Royal Bay. The most recent extension of the population, to beyond Doris Bay, probably commenced about 1959. This area now contains a population which is, to a large

extent, independent of that from which it originated. However, as glacial recession becomes greater, more regular contact may be established between them. A second introduction of two stags and three hinds was made in 1912, probably on 15 March, at Leith Harbour in Stromness Bay by Christian Salvesen's Company. These survived and had increased to 20 animals by 1918, when all were killed by an avalanche. A third introduction of three stags and four hinds was made at Husvik in 1925. This formed the basis of the present northern population. It is interesting to note that the two present populations of reindeer are genetically distinct and, because of natural physical barriers, have not associated on South Georgia.

As the introductions became established and their breeding cycle quickly accommodated to the seasons of the southern hemisphere, the populations increased. Some controversy resulted when

Reindeer grazing on a sward of *Poa annua*.
(P.R. Stark.)

the Magistrate informed the managers of the whaling stations that his permission would be needed for hunting them although the original application for the protection of the reindeer came from Captain C.A. Larsen. A small amount of venison is regularly obtained by controlled shooting from the southern herd. It forms a welcome addition to the diet of the settlement at King Edward Point as well as the British Antarctic Survey ships and the other stations. Killing of reindeer is however, subject to strict licensing by the Magistrate under the fauna conservation ordinances. The numbers, sex, location and date on which they may be taken are matters decided for and specified on each licence to ensure no serious effects on the populations result.

Several studies of the reindeer and their effects on South Georgia have been made, principally in 1928, in 1954 and an intensive study of their population dynamics, breeding behaviour, diet and health was made from 1971 to 1976. A series of experimental sites was established during the 1973/74 summer throughout the majority of the reindeer ranges. These consisted of 19 fenced constructions which prevented the reindeer from gaining access to the vegetation within them. They were erected over a selection of the principal vegetation types and control areas were established near them with free access for the reindeer. From regular examinations of these sites as well as other studies it has been possible to examine the effects of reindeer on many types of the island's vegetation and to determine their feeding preferences. Among the flowering plants the greatest effect has been on *Acaena magellanica* which is greatly reduced by overgrazing throughout their range. In contrast, the introduced grass *Poa annua*, which is resistant to grazing and trampling, is more abundant within their range than elsewhere. The majority of the reindeer's diet is tussac due largely to its high carbohydrate content. This grass is especially important in sustaining them through winter because it remains green and is accessible even under moderate snow cover. Its abundance is one of the major controls on their numbers. Although tussac is moderately resistant to grazing, extensive stands of it have been destroyed, particularly on coastal slopes where this is resulting in some serious erosion. Unlike in the northern hemisphere, where the populations of reindeer often rely to a large

extent on lichens as a food, these are unimportant in the diet of the South Georgian animals. Most of the larger lichens have been almost eradicated from their range and this, combined with the very slow growth rate of most South Georgia species, has resulted in much greater damage to them than to other constitutents of the reindeer's diet. During periods of severe weather or deep snow in winter, the reindeer sometimes eat kelp washed up along the shoreline. Although relatively rich in protein, these seaweeds have a low food value.

The island's reindeer population is now about 2000, accurate estimates are difficult owing to the nature of the terrain. Aerial photographs have provided the best estimates but these have only recently been available. Combined with details from various other sources, these have enabled determination of population trends since the introductions of the animals. The southern population is presently near the limit of its size as there is little possiblity presently available for extending its range. It has substantially decreased in numbers over the last 20 years. The northern population, in contrast, has easy access to many areas (some of which its predecessors used from 1912 to 1918) to which it may spread if its population increases. The reindeer have no natural predators on the island apart from Man.

One export of reindeer from South Georgia has been made. It was an unsuccessful attempt to transfer some to Isla Navarino in the Magellanic Archipelago at the request of the Government of Chile. Eight animals were captured at Corral Bay in December 1971 but only one survived the exceedingly rough sea voyage aboard HMS *Endurance*. The introduction of reindeer from South Georgia to the Falkland Islands has been considered several times but no action has yet resulted.

The most recent discovery of an introduced mammal on South Georgia is that of mice (*Mus musculus*), found by a geological field party at Shallop Cove on the southern side of the western end of the island in 1976. A previous report of mice found at Rosita Harbour was made by seamen in 1913 but not confirmed. It was suggested at the time that they were small rats (as was the first report of the later discovery). The area where they are now known to occur is small and isolated. It has several remains from visits of sealers made in the previous century which provides a likely explanation of their origin. An investigation of the South Georgian mice was made during the 1977/78 summer. It revealed that they show some adaption to the rigour of the island's environment by being larger animals with greater deposits of brown fat than is usual for the species. Their genetical relationships were also investigated and were confirmed to belong to a well known type.

The survival of the mice is dependent on the tussac which, as with the rat, provides food and shelter. Tussac seed forms a major part of their diet, while much of the rest consists of terrestrial arthropods, especially beetles. The mice live in a number of discrete territorial groups which occupy from two to six individual tussac plants with interconnecting burrow systems. The South Georgian mice are believed to be the most remote and environmentally stressed population of the species.

Cats and dogs have been brought to South Georgia on many occasions as pets and for other reasons. James Weddell referred to what was probably the first dog on the island in 1823 and its attack on an albatross. Both dogs and cats have, on occasions, proved rather a nuisance at the whaling stations and 'dog shoots' have been necessary. Sledge dogs have been landed on South Georgia on their way to the Antarctic continent by several expeditions, where they undoubtedly enjoyed and benefited from the vast amounts of whale meat included in their diet while they were there. The last resident sledge dog was a husky from Halley Bay which died early in 1973. Cats have lived at all the stations where, to a certain extent, they assisted in controlling the rats. They have bred for many generations on the island but only when partly supported by Man. About two dozen lived near King Edward Point in 1970 and hunted as far as Maiviken and Hestesletten. Most were culled, and in 1982 there were only two left on the island, remote from each other, and it is likely that they will soon die out unless others are marooned from visiting vessels.

Domestic animals imported for food have been sheep, cattle, goats, and pigs. The earliest introduction of the first three was made in 1882 by the German International Polar Year Expedition at Royal Bay. The goats, which had kids with them, provided fresh milk as well as meat. They and the sheep flourished while grazing on local tussac grassland during the summer but spent the winter in stables. The whaling stations were equipped with animal houses, mainly for pigs, although sheep, cattle and more rarely, goats were also maintained. The pigs were fed partly on products from whaling operations and were allowed to roam freely in the early days. Slaughterhouses had facilities for preparing pressed and smoked ham as well as sausages. With the latter, a combination of pork and whale meat had a good reputation. Sheep and cattle were regularly brought from the Falkland Islands and sometimes from South America, and their hides were sometimes exported. The Norwegian name for King Edward Point, Sauodden (Sheep Point) reveals where they were once allowed to graze. Several fence runs were established at the whaling stations to allow sheep and cattle to be grazed during summer. An exceptional sheep once lived for several years in the wild near Sheep Point at Prince Olav Harbour. Most of these species, however, were maintained in the animal houses which were provided with heating. Despite the apparent unsuitability of South Georgia for sheep raising several applications have been received for grazing rights. One, at the Dartmouth Point area, was let for a year from 1 January 1911 for £2 and 10 shillings. No success in these ventures resulted however as the sheep proved unsuitable for stock raising – and it would have been unlikely in any event. Sheep were kept by a whaling company on Mutton Island (later called Grass Island) in Stromness Bay in the 1920s. They were reported as eating virtually all the tussac and thereby causing much erosion prior to their removal. The tussac has since largely regrown.

A stud horse and three mares were left on Hestesletten in 1905 by the South Georgia Exploration Company. Another horse was introduced by the Manager of the Ocean Harbour whaling station, but the possibilities of riding it must have been quite restricted. Nothing is recorded about the eventual fate of these animals and there is nothing to suggest any breeding by the first introduction. In 1912 some Manchurian ponies were left by the Filchner expedition aboard *Deutschland* and they too grazed at Hestesletten. R.C. Murphy later reported that they produced some colts but no certain information exists about their eventual fate (two 'horse hides' were, however, exported in 1917).

Rabbits were first reported as being introduced in 1872 by sealers from Tristan da Cunha but information is lacking about this. C.A. Larsen made another introduction after founding Grytviken and they became established at Hope Point. These were reported by Captain Hodges as thriving about a year later but subsequently succumbed to predation by skuas (*Catharacta lönnbergi*). Several other introductions were attempted; a small population existed on Jason Islet in 1931 and the name Kanin Point (Kanin is rabbit in Norwegian) near Husvik, suggests their presence there.

The breeding of fur foxes, to be fed on whale meat and other products was first considered in 1920. The experiment was tried at Grytviken from 1939 when some foxes were unsuccessfully kept in captivity. They died out before 1947 as their breeding was inadequate, probably owing to the unsuitable accommodation. Monkeys and other small mammals were occasionally introduced as pets – mainly from South America. Fabled animals have also been reported, including a fox in 1775 by Johann Forster and an otter in 1906 by Captain Hodges. The introduction of several other mammals has been proposed at various times including; mink, musquash, bighorn goats, Alaska sheep, and the South American fur seal (*Arctocephalus australis*), but wiser counsel has prevailed.

Human effects on the island's biology

Of all the Antarctic regions, South Georgia has suffered the most intense human activity and exploitation for the longest period. For much of its history it was by far the most industrialised and densely populated place south of the Antarctic Convergence. Although this exploitation has direct-

Cattle grazing at King Edward Point. Photographed by the first Pastor, Kristen Loken, about 1913. Note the severely damaged tussac and also the meteorological station in the left background. (Courtesy Norse Hvalfangstmuseum.)

ly involved only seals, whales, and fish, all of which had their populations severely reduced, there have been various other consequences. These include the introduction of alien animals and plants, massive local industrial pollution, acute and chronic oil spillage, and large civil engineering works. Several of these effects of human activity are discussed in the following paragraphs.

Much of the early history of South Georgia involved the exploitation of the Antarctic fur seal which became so rare that it was thought to be extinct near the beginning of this century. The recent recovery of its populations has been remarkably rapid; existing numbers might reasonably be compared to those before sealing began. The dis-

tribution of fur seal breeding beaches is strongly concentrated at the western end of the island, and thus quite different from that prior to exploitation. The seals are, however, spreading from this area and can be expected to recolonise all their former breeding beaches, perhaps by the end of the century. The elephant seal suffered less severely but nevertheless was greatly reduced in numbers during the last century. It recovered sufficiently to support the scientifically managed second sealing epoch which, but for the failure of whaling, might profitably have continued to the present.

The severe reduction in numbers of whales around South Georgia was a more recent event, although southern right whales and humpback whales

A stallion, mare and foal of the Manchurian ponies from the Filchner Expedition, near Brown Mountain. Photographed by Kristen Loken about 1913. (Courtesy Norse Hvalfangstmuseum.)

had become very scarce by the mid 1920s, as described in Chapter 5. There is considerable evidence that whale stocks, near the island and generally, are now increasing; it will, however, take many years before they reach a level where scientifically controlled exploitation becomes practicable.

The commercial fisheries around South Georgia are presently used mainly by eastern European fleets. There is strong evidence that these are being exploited above their maximum sustainable yield. Owing to reduced catches, many of the fishing vessels are now operating elsewhere. The length of the reproductive cycle and fecundity of the fish concerned are both favourable for a comparatively rapid recovery of their populations to a level where a scientifically controlled fishery would be practicable. Questions about fishing limits, territorial waters, and control of the industry are unresolved and will also strongly affect future commercial fishery.

Krill exploitation has not yet occurred on a large scale, and the state of scientific knowledge about this resource is improving rapidly with investigations being undertaken by the British Antarctic Survey and several other organisations. This may permit initiation of rational use of the resource prior to large-scale exploitation. If so, this will be in marked and very favourable contrast to previous commercial activities in the region.

The animal introduced by man which has had the greatest effect on South Georgia is the rat. It has been present for perhaps two centuries, as discussed earlier in this chapter. There is some evidence that it is slowly adapting to conditions on the island. The reindeer have had substantial consequences for some vegetation in their ranges. The effects of the introduced mice are minimal. None of these three introduced animals has had serious effects on the distribution and populations of most native animals and plants (although the South Georgia pipit has suffered to some extent). Only one species of alien plant, *Poa annua*, is widely distributed on the island and this occurs in dense concentrations only where other plants are grazed by reindeer. Any effects of the other alien plants are much less owing to their being almost completely confined to the localities of the whaling stations. Some are, however, spreading slowly and these are the subject of scientific observation. Future deliberate introductions of animals and plants are now unlikely owing to the present regulations and controls.

At the maximum, seven of the island's bays had whaling factories operating in them. The pollution resulting from the industry was enormous. Vast quantities of very big rotting corpses, blood and other body fluids, liquid and solid discharges from cookers, putrefaction tank residues, various chemicals and other substances were all released in the bays. The odour of a whaling station was described very unfavourably by many visitors. After regulations requiring full utilisation of whales were introduced, this was reduced but enormous quantities of some of these materials continued to be discharged. Local atmospheric pollution, especially when coal was used as fuel, was minor compared to the above but nevertheless intense in local areas; about six large furnaces operated when a factory was working (mainly to produce steam). Disposal of rubbish, sewage and so forth from concentrated human populations of up to 2000 men was also largely done in the bays where the whaling stations operated. Mineral oil has been released in the vicinities of the whaling stations from spillages, flushing of tanks, and in a variety of other ways. Some deteriorating tanks still hold oil and small seepages are chronic although rapidly dispersed on reaching the sea.

The present results of this pollution, less than 20 years after the closure of the last whaling station, are quite minor. Whale bones lie around many beaches near the abandoned whaling stations, bleached white and greatly eroded by the sea. The whaling station bays now have a full complement of animals and plants whereas previously they were biologically depauperate and almost anaerobic. Some aromatic hydrocarbons and other chemicals have been measured in marine benthic invertebrates of King Edward Cove in concentrations several times greater than those in comparable Antarctic locations unaffected by industry. The sediments of the cove contain similar chemicals which reflect the whaling station's operations. The high concentration of dissolved oxygen in sea water around South Georgia may be largely responsible for the recovery of the bays from the effects of the industry and no evidence exists that its effects are now significantly deleterious to the environment.

The construction of the whaling stations and

the settlement at King Edward Point involved much levelling, river diversion, and similar works on the restricted flat areas they occupied. Other civil engineering involved the construction of dams and aqueducts for water supply and hydroelectric-power generation. In addition structures erected included factories, accommodation, stores, and other buildings – up to 60 substantial structures in a few hectares. Since the whaling stations have closed many of these buildings have collapsed and others deteriorated generally, this is proceeding more and more rapidly. Prince Olav Harbour, abandoned in 1931, is largely derelict and the majority of buildings have collapsed to ground level. Rusting, rotting, the consequenes of the severe climate and similar processes are reducing the remains even further. Slowly but ineluctably the stations and their effects are being erased.

Because of all this, the present circumstances of South Georgia are particularly interesting. Parts of the island have sustained very great human effects, especially by the standards of the region, and the populations of some animals of the surrounding waters have been drastically reduced. Yet, fewer than 20 years after the cessation of industrial activities, recovery from this is very substantial – indicating the great resiliency of the natural environment.

Conservation of wildlife

The conservation of the flora and fauna of South Georgia to ensure preservation of the species and of the island's specialised ecosystems, as well as to permit rational exploitation of some of these, have been matters of concern to the Government for many years and legislation has been progressively introduced and improved with increases in scientific knowledge. The first protection ordinances for the Falkland Islands and Dependencies applied to seals and provided for licensing of their hunting and the establishment of reserves to control their exploitation. These were followed by comparable ordinances regulating whaling. In 1909 an ordinance for the preservation of the penguins in the Dependencies was passed following a specific recommendation. The Governor, Mr (later Sir) W.L. Allardyce, at the time of the beginning of the whaling era, was a strong proponent of regulations to control the exploitation of whales and to ensure

the maximum utilisation of those taken. He pursued a policy to endeavour to establish whaling as a permanent industry rather than as a rapid source of a large revenue. Many of the regulations were, however, relaxed owing to the requirements of the First World War. In 1912 an 'Ordinance to provide for the preservation of certain Wild Animals and Birds in South Georgia' was passed which provided protection for listed species but allowed permission to be given for them to be taken for scientific purposes.

These and other Ordinances were amended as necessary until, in 1975, they were consolidated into 'The Falkland Islands Dependencies Conservation Ordinance'. This was based partly on the 'Agreed Measures for the Conservation of Antarctic Flora and Fauna' drawn up under the Antarctic Treaty with the assistance of the Scientific Committee on Antarctic Research, and which came into force in the same year. Under the new law the main principles are that all killing and exploitation on the island and in its territorial waters should be subject to a permit system, that import and export of living organisms require licences, and that all access to the island is controlled. Three types of areas are established by proclamation under the Ordinance. 'Specially Protected Areas' are designated to preserve their ecological systems unchanged for later comparisons and it is the aim of the administration to keep everyone out of them. 'Sites of Special Scientific Interest' are designated to prevent scientific investigations being jeopardised by disturbance and permits to enter them are only issued for compelling scientific reasons which cannot be served elsewhere. The former presently covers Cooper Island and the latter Bird Island and Annenkov Island. In contrast 'Areas of Special Tourist Interest' are selected places which are representative of wildlife and scenic beauty where the effect of tourist activity may be systematically assessed, and have been declared for the area between Cumberland East Bay and Cumberland West Bay near Grytviken, and at the Bay of Isles. It is now prohibited to land in South Georgia for mountaineering or other recreational pursuits except in the latter areas, unless granted a special permit to visit other places, in addition to undergoing the normal entry formalities for the Falkland Islands Dependencies. This Ordinance provided an excellent and comprehensive basis for scientific control and conservation of the flora and fauna of the Dependencies.

The regulation of whaling around South Georgia was first made under laws licensing the whaling stations which limited their numbers as well as the numbers of whale catchers each might use. Concern about decreasing whale stocks and the preservation of the industry had been expressed by scientific observers since before 1920, and subsequently this became more urgent. The first licensing laws were also responsible for raising substantial revenue for the Government and, with various amendments, remain in force (although whaling from the island finished in 1965). As whales are pelagic and the territorial waters extend for only three nautical miles from the shores, factory ships were unregulated while on the high seas. British, Norwegian and other regulations and agreements were produced as factory ships became both more efficient and more numerous, exacerbating the problems of whale conservation. However, after the Second World War, delegates from 19 countries with substantial whaling industries met in Washington from 20 November 1945. Their deliberations resulted in the formation of the International Whaling Commission which assumed the functions of producing agreed measures for rational exploitation of whales. The Commission's decisions have often been subject to controversy but are the only regulation of pelagic whaling. It remains a voluntary association of nations with whaling interests and can produce recommendations for its members which are of necessity often a compromise between conflicting interests. Several countries not associated with it also undertake whaling.

Although the Falkland Islands Dependencies are north of the limit of the Antarctic Treaty area, the 'Convention on the Conservation of Antarctic Marine Living Resources', signed in Canberra on 21 May 1980, recognised that the geographical parallel of 60° S is of little biological significance and that a more appropriate biological frontier is the Antarctic Convergence. The Convention was the culmination of two and a half years' work which began after the Antarctic Treaty Consultative Meeting in 1977. The final negotiations were attended by observers from the European Economic Community, United Nations Food and Agriculture Organization, International Whaling Commission,

International Union for the Conservation of Nature and Natural Resources, Scientific Committee on Antarctic Research, and Scientific Committee on Oceanographic Research. The first article states 'This Convention applies to the Antarctic marine living resources of the area south of 60° South latitude and to the Antarctic marine living resources of the area between that latitude and the Antarctic Convergence, which form part of the Antarctic marine ecosystem'. The Convention concerns 'the populations of fin fish, molluscs, crustaceans and all other species of living organisms, including birds, found south of the Antarctic Convergence' and established a 'Commission for the Conservation of Antarctic Marine Living Resources' which is to acquire data, conduct research, identify conservation needs and formulate appropriate measures, and to publish information about these matters. The Commission has its headquarters in Hobart, Australia. A Scientific Committee to provide a forum of consultation and cooperation concerning the collection, study, and exchange of information as well as other activities as the Commission directs, was also established under the Convention. These provisions, with those of the Falkland Islands Dependencies Conservation Ordinance of 1975, may well mark the commencement of a virtually ideal legal structure for biological conservation of South Georgia and its surrounding oceans as well as providing an example for many other regions. It also marks the first extension of provisions of the Antarctic Treaty to the Falkland Islands Dependencies.

The most urgent conservation matter presently concerning the South Georgian region is probably that of the fisheries. There is much evidence to indicate that these have been overexploited during the last 10 years, mainly by eastern European fishing fleets. Scientific investigations of the region's fisheries potential is presently being conducted by several nations, some of which are not fishing commercially in the region. Much of the present British Antarctic Survey's biological research is concerned with krill, fish, and other resources of the Antarctic Ocean in and around the Falkland Islands Dependencies and British Antarctic Territory. Several international and other national research programmes are also cooperating in these projects. By such means it is hoped to obtain sufficient knowledge for future exploitation of krill, fish, and other animals to be done on a scientific basis.

9

Military actions, events on South Georgia in 1982 and the island's future

South Georgia came into prominence in 1982 due to the armed invasion by Argentina. There had been a long-standing pretension by Argentina to sovereignty over the Falkland Islands, which were also invaded at that time. During the first half of this century, this pretension was extended to include South Georgia and other parts of the Falkland Islands Dependencies. Considerable discussion of the status of the Falkland Islands and Dependencies and its interpretation in conformity with international law has taken place, largely following the events of 1982. Although much of this is inappropriate to include in this work, a brief discussion of the problem as it affects South Georgia is relevant. This and an account of the hostilities follow. An economic survey was made after these events which considered the future for South Georgia, thus a discussion of this is included at the end of this chapter.

Argentine pretensions of sovereignty over South Georgia

The Argentine claim to sovereignty over South Georgia is difficult to interpret in isolation; to a large extent, it relies on the association of South Georgia, as part of the Falkland Islands Dependencies, with the Falkland Islands. This argument maintains that the Government of the Falkland

Islands is without legal status and that, as the islands are part of Argentina under illegal British occupation, so also are what the British Government has designated as their Dependencies. The whole are grouped in the territory designated 'Islas del Atlantico Austral' and supposedly governed from Ushuaia in Tierra del Fuego. Although the first indications of Argentine claims to the Falkland Islands came early in the previous century, her earliest interest in South Georgia was conceivably demonstrated in 1892 when a Romanian, Jullu Popper, received authorisation from the Argentine Government to take possession of the island. The matter was brought to the attention of the British Foreign Office which was informed by the Admiralty that it was not expedient to send a man-of-war to the region. This action subsequently proved appropriate as Popper did not proceed to South Georgia or any of the other places in Antarctica where he had proposed to begin whaling and sealing. (Popper was a very well-travelled adventurer who had established a virtually independent state in northern Tierra del Fuego which mined gold, produced coins and issued its own postage stamps.) In 1898 the Argentine Government showed interest in prospecting for coal on South Georgia and a reward for its discovery was offered. The Governor of the Falkland Islands made some inquiries on learning of this, as related in Chapter 4.

Argentina has argued that the first manifestation of her sovereignty over South Georgia was in

November 1904 when the Norwegian Captain C.A. Larsen established a whaling station at Grytviken for the Compañia Argentina de Pesca. This private company, registered on 29 February 1904 with a head office in Buenos Aires, operated the station from its foundation to 1960 after which it was sold to Albion Star Ltd, registered in the Falkland Islands (with a head office at Grytviken). When the first cargo of whale oil arrived in Buenos Aires from South Georgia, aboard *Rolf* in March 1905, the Argentine Government had to decide what customs duty to charge; and, after consideration, classed it and charged it as a product from outside Argentina. On 2 November 1905 an application to the British Legation in Buenos Aires was made by the president of the Compañia Argentina de Pesca and the Director of Armaments of the Argentine Navy (Marina de Guerra), Captain Núñez, for a lease of the site of the whaling station at Grytviken. Núñez was the technical advisor and an important shareholder in the company and the application was made following receipt of a protest from the South Georgia Exploration Company of Chile, a group which had purchased a pastoral and mining lease for the island from the Governor at the Falkland Islands.

Captain C.A. Larsen, the founder of the Compañia Argentina de Pesca and of Grytviken, had no doubt about South Georgia's status. When applying to the Governor, through the island's Magistrate, for the grant of British citizenship he wrote: 'I have given up my Norwegian citizens rights and have resided here since I started whaling in this colony on the 16 November 1904 and have no reason to be of any other citizenship than British, as I have had and intend to have my residence here still for a long time'. The application was made at King Edward Point on 6 January 1910 and granted.

The funding of a meteorological station (which was established in January 1905 by C.A. Larsen) from July 1907 by the Servicio Meteorologico Nacional of Argentina has also been claimed as an expression of sovereignty over South Georgia by Argentina. In 1904 the original instruments were provided by Mr Walter Davis, Director of the Servicio, and the Servicio received reports from the station for a long period. From 1907 it paid the salary of an observer but subsequently the whaling company appointed and paid a Norwegian

observer. The meteorological station was operated under the requirements of the lease of the site of the whaling station until the end of 1949, although the requirement was deleted when the lease was renewed in 1948. Similar requirements were incorporated in the leases of the other whaling stations of South Georgia and all of them furnished details to the Magistrate. No question of sovereignty was raised until 2 years after these meteorological observations were no longer required by the leases, and then only in one instance. After these changes had occurred and the Falkland Islands Dependencies Survey took over the meteorological observations from 1 January 1950, a protest was received by the British Secretary of State for Foreign Affairs from the Argentine Ambassador in London on 29 January 1952 which included an assertion of sovereignty said to be represented by the station and complained about the alleged interruption of scientific activities.

There have been occasional small hydrographic surveys of parts of South Georgia and vicinity by Argentina, as well as by several other nations. The first of these was in 1905 by the Argentine naval ship, *Guárdia Nacionál* on charter to the Compañia Argentina de Pesca. Argentine charts are available for the island which incorporate these surveys and many from other sources. These bear a substantially different series of place-names (many derived from Argentine national heroes and dates) from those on British charts, and are occasionally changed. (The author was interested to observe that British charts were used by the Argentine Navy during the invasion in 1982 as their own were not regarded as sufficiently accurate.) Likewise Argentina publishes navigational details for the island and vicinity which, although appearing correct, are of an inaccurate and outdated nature.

The first contemporary and definite indication that Argentina had conceived a pretension to sovereignty over South Georgia came in 1927 when it was represented by a delegate of the Argentine post office to the International Postal Bureau at Berne, Switzerland, that Argentine territorial jurisdiction extended over the island, with an announcement that 'Argentine territorial jurisdiction extends in fact and in right over the continental area, the territorial sea and the islands of Tierra del Fuego, the archipelagos of Estados, Año Nuevo, South

Georgia and the polar lands not yet delimited'. The International Postal Bureau and its successors have continued to take no regard of this however, as has the International Telecommunications Union. Although the matter was discussed in the Argentine press during subsequent years the first statement by a President of Argentina asserting sovereignty over South Georgia was made as late as 1938, also as an adjunct to a postal matter. On 22 September of that year, when ratifying the Postal Convention signed by the Argentine representatives at Cairo in 1934, he added an expressed categorical reservation asserting the Argentine claim to the Falkland Islands and Dependencies as an inalienable right.

These matters marked the begining of a campaign by Argentina directed to creating a semblance or fiction of sovereignty over South Georgia which included reinterpretation of history and related actions. She is the only state ever to have initiated direct military action to assert a claim to Antarctic territory and has now done so thrice.

Other matters raised by Argentina to support her assertion of sovereignty over South Georgia include a concept of continental propinquity and the application of the Treaty of Tordesillas of 1494. The first of these has been used to support Argentine sovereignty over the Falkland Islands and Dependencies as well as Argentine and Chilean claims to the Antarctic Peninsula and parts of the Antarctic continent. Geological events have been quoted in support of the assertion. Such a propinquity concept has no standing in international law (which is fortunate, considering the problems it could cause in many other places if applied). The Treaty of Tordesillas allocated between Portugal and Spain much newly discovered territory to the west of the Atlantic Ocean and elsewhere. It was made following a Papal decree intended to resolve a dispute between these nations over their discoveries and, in its final form, roughly followed the meridian 46° W. Certain lands, then undiscovered, were later divided in accordance with it. The application of the Treaty to Antarctic regions and its force over nations other than its signatories, Spain and Portugal, are highly dubious. The newly independent Spanish colonies in South America in the early part of the nineteenth century, generally corresponded to the former Spanish administrative units. This, combined with the Treaty of Tordesillas, has been used by Argentina to rationalise her pretensions of sovereignty over, among other regions, South Georgia. This too is highly conjectural and, owing to the position of South Georgia and the division of the Treaty, it is difficult to understand why, if it has any force, it does not impose Portugese sovereignty over the island (a matter of interest to Brazil).

The basis of British sovereignty over the island

The British claim to sovereignty over South Georgia relies principally on discovery, followed by occupation and effective government. The first sighting of the island was probably made by a London merchant and another was later made from a Spanish ship. Claims have also been made that the island was seen in 1501 or 1502 from a Portuguese ship on which Amerigo Vespucci, a Florentine, was pilot; his writings were said to have described land resembling South Georgia. This suggestion is unfounded and arose through confusion about the origins of manuscripts (Christie, 1950). Captain James Cook, in 1775, made the first landing, prepared charts, described and formally took possession of South Georgia in the name of His **Britannic Majesty George III on 17 January** of that year, and left indubitable records of this, described in Chapter 2.

British government of South Georgia developed slowly, principally owing to the location and nature of the place. A provision for the government of the Falkland Islands and their Dependencies was made by Letters Patent of 23 June 1843, although the Dependencies were not specifically defined at that time. The Colonial Office Year Book first mentioned South Georgia among the Dependencies in 1887. The Government received several inquiries about leases and land purchases on the island towards the end of the nineteenth century. The most significant action in this connection was the advertisement of a mining and general lease of South Georgia in 1900. As related earlier, several inquiries resulted and eventually a Chilean company was granted a lease in 1905. The problems which developed when it was discovered that a whaling company had established a station on South Georgia without proper authorisation have already been described in Chapters 4 and 5. This was resolved when the whaling company made a suitable application; the previous neglect of this was believed to be an endeavour by the company to avoid paying fees for the lease and similar matters. An inspection was made of South Georgia from HMS *Sappho* from 31 January to 5 February 1906 by Captain Hodges as a result of these incidents.

The Letters Patent of 1843 were, as circumstances required, modified in 1876, 1892, 1908, and 1917. The Dependencies of the Falkland Islands were first specifically defined in those of 1908 as 'South Georgia, the South Orkneys, the South Shetlands, and the Sandwich Islands, and the territory known as Graham's Land, situated in the South Atlantic Ocean to the south of the 50th parallel of south latitude, and lying between the 20th and 80th degrees of west longitude' and this was published in the Falkland Islands Gazette at the direction of the Governor. A copy of the Gazette was transmitted to the Argentine Foreign Ministry by the British Minister in Buenos Aires on 20 February 1909, this was acknowledged on 18 March and no protest or other action resulted at the time. In 1948, however, Argentina conceived an argument that the 1908 Letters Patent had no validity as they claimed parts of Patagonia and Tierra del Fuego in addition to the places specified in them. (The reasoning which gave rise to this fatuity would also have made the Falkland Islands part of the Dependencies.) The 1917 Letters Patent redefined the Dependencies to incorporate parts of the Antarctic continent (Coats Land and polar parts of the sector) and the revised sector then delimited included only the Dependencies of the Falkland Islands and so stated.

In response to the establishment of the whaling industry and a permanent population on South Georgia, a resident Magistrate was appointed and an administrative post established on the island in 1909. The island was subsequently administered peacefully from King Edward Point until 3 April 1982 when the settlement was attacked by Argentina and the Magistrate, a civilian, with the rest of the population taken as prisoners of war.

During the period from 1923 to 1939, New Zealand, France, Australia, and Norway acquired sectors of Antarctia by processes similar to those applied by the United Kingdom. These were mutually recognised and distinct, and comparable to divisions made in the Arctic. By a similar procedure, Chile in 1940 claimed a sector which partly overlapped the prior British claim of the Falkland

Islands Dependencies. In 1943, Argentina claimed a sector by a note to the British Government dated 15 February. This claim, which overlapped prior British claims and partly overlapped the prior Chilean claim, was enlarged in 1946 and 1947. Various problems resulted which were not all solved diplomatically. Argentina was the first state to resort to military force to assert her pretensions. Her armed services, based at a nearby Argentine station, opened fire with a machine gun over an unarmed British scientific party at the Falkland Islands Dependencies Survey station at Hope Bay in February 1952. The British station, originally established in 1945, was burnt down accidentally in 1949 with the loss of two lives, and was to be rebuilt. The Argentines then sought to prevent further landings and expelled previously landed British personnel at gunpoint. The matter was resolved when the Governor arrived aboard HMS *Burghead Bay*. The Argentines removed and admonished the commanding officer of their forces involved in the incident.

Fortunately many problems of sovereignty over the regions of the Antarctic south of the 60° parallel are now relieved owing to the Antarctic Treaty of which 16 nations, including Argentina, Brazil, Britain, Chile, Norway, the Soviet Union, and the United States of America are full members (Peru, Spain, Uruguay, and 12 other states are acceding parties). The provisions of the treaty include that no action taken during its operation will affect previous claims or stand as future claims, to sovereignty over the area it covers. This effectively puts the disputes in abeyance for the foreseeable future in the Treaty area. The Treaty also requires that Antarctica shall be used for peaceful purposes only and prohibits the establishment of military bases. Unfortunately interpretation of and adherence to these and some other provisions of the Treaty are apparently very flexible (in the case of some member states). A recent disregard of the Treaty has been the proclamation of a war zone which included part of the Treaty area south of Southern Thule, in the South Sandwich Islands, on 29 April 1982, by Argentina.

As a result of the Treaty, the United Kingdom promulgated an Order in Council in 1962 which established and separated British Antarctic Territory as the region, previously part of the Falkland

Islands Dependencies, over which the Antarctic Treaty operated; and provided for its government. Thus the Dependencies came to contain only South Georgia with Shag Rocks, Black Rock and Clerke Rocks, and the South Sandwich Islands.

International Court of Justice

Prior to the Antarctic Treaty, several attempts were made by the United Kingdom to resolve the problem of sovereignty over South Georgia and other parts of the Dependencies, as then defined. The appropriate place for testing the conflicting claims was deemed to be the International Court of Justice, established under the aegis of the League of Nations for the peaceful solution of international disputes, and now a component body of the United Nations Organisation (of which Argentina is one of the original members). There are several precedents for such action and the Court has rendered judgement on analogous cases. On 17 December 1947 the British Government invited the Argentine Government, should they be disposed to challenge the British title to sovereignty over the Falkland Islands Dependencies, to refer the matter to the International Court of Justice, whose jurisdiction the British Government bound themselves in advance to accept, and with whose decision they would be bound under the Charter of the United Nations Organisation to comply. The Argentine Government replied on 28 January 1948 and rejected the British offer to accept the arbitration of the International Court of Justice, a decision hardly consistent with a belief that its claim was demonstrable.

Following the failure of further approaches by the British Government in 1951, 1953, and 1954 to Argentina for a judicial solution to the dispute before the Court or before an independent arbitral tribunal summoned solely for that object, which were all rejected, a unilateral application was lodged with the Court by the British Government on 4 May 1955. The Registrar of the Court communicated with Argentina and received a reply that the Argentine Government reaffirmed its refusal, in the most express way, with regard to the jurisdiction of the Court in the matter. The reply continued to assert that Argentina had 'unquestioned rights and titles derived from and based on legitimate methods of acquiring territorial domaine, and effective, notori-

ous and peaceful possession'; but yet she found herself unable to defend this before an impartial international tribunal.

Relations with Argentina

Despite these problems, Argentine ships have regularly visited the Falkland Islands and (during the whaling era) the Dependencies, without raising difficulties about the sovereignty dispute. Normal harbour, customs, immigration and other procedures have been complied with. On the contrary, the entry of British and Falkland Islands ships to Argentine ports has not been so easy, as demonstrated by the incident of the *Polar Maid* in 1950. On this occasion the ship, a whaling company's oil tanker, was held to ransom in Argentina for payment of tax claimed to be due for landing oil in South Georgia. Owing to the value of the ship the company paid to obtain her release. In 1971 a joint declaration was signed by the British and Argentine Governments which resulted in closer contact between the Falkland Islands and Argentina by removing several obstacles by mutual agreement. Although this did not apply to the Falkland Islands Dependencies, some of its provisions (particularly those concerning postal communications) benefited the inhabitants of South Georgia. The policy adopted following the joint declaration allowed the pursuit of more normal relationships between two, geographically close, sovereign states. During this period occasional difficulties concerning the Argentine pretensions to sovereignty were experienced in the Falkland Islands and, to a lesser extent, in South Georgia. The latter were mainly brief incursions of Argentine ships into the Dependencies' territorial waters. An Argentine naval station was, however, clandestinely established on Southern Thule, South Sandwich Islands, a remote part of the Falkland Islands Dependencies, in November 1976. This later functioned partly as a meteorological station and was the subject of a British diplomatic protest on 19 January 1977 and subsequently. At talks in Lima in February 1978 Britain proposed an arrangement to provide for British and Argentine scientific activities to be undertaken in the Dependencies. Little progress was made at the first meetings of two working groups and no arrangement resulted.

The events of 1982 were the first major

disruption of this situation and corresponded with a period of greater than usual instability in the government of Argentina.

Salvage operation

Since the closure of the last four stations operated by whaling companies on South Georgia, there have been various interests shown in them. Materials have been purchased from them by the Government of the Falkland Islands for their own use and for the Dependencies, by the British Antarctic Survey and others. Large quantities of smaller items have also been removed illegally by the complements of private yachts, a whaling ship, the fishing fleets of certain eastern European countries, and others. The Far Eastern Shipping Company of the Soviet Union made inquiries in 1970 to obtain a lease of Grytviken or Husvik whaling station for used as a bunkering depot and transfer station for cargo and personnel. Following a negative response from the British Government the matter was not pursued.

Albion Star held the leases of Husvik and Grytviken but, since the mid 1970s, had not maintained payments necessary to continue them. The leases were sold to Christian Salvesen Ltd, the company owning those for Stromness, Leith Harbour, and Prince Olav Harbour. This company, which had been involved with South Georgia since 1909, thus gained the leases for the whaling stations situated at the five best harbours of the island. It retains these against the eventuality of restarting whaling or the commencement of other commercial operations at South Georgia.

In 1978 a Buenos Aires businessman, Sr Constantino Davidoff, contacted Christian Salvesen and Company with a view to salvaging materials from their whaling stations on South Georgia. Sr Davidoff had been successfully involved in this type of trade for a number of years in several parts of the world. Negotiations were protracted, partly owing to others proceeding at the time (concerning Salvesen's acquisition of Husvik and Grytviken). The Governor of the Falkland Islands and Dependencies maintained an interest in the dealings and much of the contact with and inquiries from Davidoff were passed through the British Embassy in Buenos Aires. The Governor was not enthusiastic about the matter, but declined to impede what was believed to be a normal commercial transaction.

On 19 September 1979, a contract was signed between Salvesen's and Davidoff. For the payment of £10 000, Davidoff was to receive the right to exercise an option to purchase certain materials under defined conditions from Salvesen's whaling stations in Stromness Bay and at Grytviken. If Davidoff were to decide to proceed with the salvage, the price of all he could take before a specified date was to be £105 000. The contract required Davidoff to adhere to all terms and conditions of the leases held by Salvesen's and the British Embassy, as did the contract, made quite clear the legal requirements for the operation (which included entry and other formalities for South Georgia).

Davidoff endeavoured to obtain more information about the condition of the abandoned whaling stations from sources other than Salvesen, whose information was somewhat out of date, having not inspected their property on the island since 1966. He also made inquiries through the Embassy to secure a passage on a British Antarctic Survey ship or on HMS *Endurance* in order to visit the stations. Accommodation on and time available for operations of these ships are usually fully committed and, in any event, they do not carry passengers but only persons directly connected with their work, so Davidoff was unable to obtain a passage. The Master of a private yacht which arrived at King Edward Point late in 1980 reported that, while in Buenos Aires, he had been contacted by Davidoff to arrange a charter of the yacht to visit South Georgia. The charter was declined. Owing to these problems Davidoff asked for an extension of the time allowed to exercise the option and time to complete the contract. Salvesen extended both by a year, and later by a year again. During this period some 'false starts' occurred (when details of intended visits were received from Davidoff by the Embassy), but nothing actually proceeded. For the venture, Davidoff and an Argentine bank established a company (Empresa Georgias del Sur S.A.), registered in Buenos Aires.

Argentine military involvement

In December 1981, the Argentine naval icebreaker *Almirante Irizar* went into radio silence for several days. During this period she illegally entered

Stromness Bay, South Georgia, and endeavoured to remain undetected. This was not the first time that Argentine and other foreign ships have entered South Georgian waters unlawfully. Davidoff later admitted that he was on board and that he made an inspection and photographed the abandoned whaling stations in Stromness Bay; as, presumably, did accompanying officers of the Argentine armed services present at the time. Upon being questioned about this by the British Embassy in Buenos Aires and being reminded again of the laws in force requiring ships to arrive through the port of entry (King Edward Cove), he claimed that as it was only for a few hours he considered this procedure unnecessary. Other evidence indicates that the ship was present for a considerably longer period. It thus became apparent that Davidoff had become closely associated with the Argentine armed forces, and that he and they showed no regard for the laws of the Falkland Islands Dependencies.

Argentine military aircraft have occasionally flown over South Georgia since about 1980 without reference to the local administration as the law and established international practices require. These flights became more frequent from the winter of 1981. Three were made early in 1982, one of which, on 11 March, was comprehensive (flying from Bird Island to Royal Bay along the northern coast).

The visit of *Caiman*

The next development was the detection of a foreign yacht, *Caiman*, 22 m long and 40 registered tons, illegally present at Leith Harbour on 13 February 1982. She was flying the flag of Panama and her complement was two Argentines, an Italian resident of Argentina, and a Peruvian. She had sailed from Argentina to Prince Olav Harbour, South Georgia, and had been looting some items while claiming to be sheltering from a storm – for 7 days. The meteorological station at King Edward Point cannot support this claim. The yacht was boarded and ordered by the Magistrate to proceed to King Edward Point for investigation. Her Master claimed he was acting on behalf of a bank financially involved with Davidoff's salvage contract but he was unable to produce any written evidence whatsoever in support of this. This information and other matters were investigated through the British Embassy in Buenos Aires and, rather surprisingly,

Davidoff disclaimed the yacht's connection with his operation. He also suggested her Master was trying to secure materials for his own gain from South Georgia and threatened to institute legal proceedings against him on his return to South America.

The complement of the yacht was able to obtain information about South Georgia and was in regular contact with Argentina. Most of what they discovered, apart from conditions of the whaling stations, could probably have been determined from published literature of the British Antarctic Survey and elsewhere. Local relations with the owner, master and crew of the yacht were generally amicable, and she departed for Argentina after approximately 4 weeks at South Georgia. Information was passed in both directions and the Magistrate was informed that the expenses estimated to be involved in the salvage operation by Davidoff were: to obtain salvage rights, £ 500 000, to effect the salvage, £ 2 000 000; sale of the salvage, £ 7 000 000; thus total profit, £ 4 500 000 and total turnover £ 9 500 000. These estimates seem excessive, even bearing in mind that Davidoff was to operate from the nearest mainland country, that he had arranged Argentine naval assistance, and that many items of machinery etc., of little use now in their countries of origin owing to obsolescence would still be quite saleable and useful in Argentina or elsewhere in South America.

The ship *Bahia Buen Suceso*

Information was received at South Georgia in late February from Davidoff through the Embassy, that he intended to commence operations in the near future and had arranged transport in a fleet auxiliary ship of the Argentine Navy. The procedures for correctly receiving them were outlined; these apply to any visiting foreign naval ship. It was considered strange, however, for the salvage work to begin almost at the start of winter.

HMS *Endurance*, commanded by Captain N.J. Barker (who was awarded a CBE for action in subsequent events), was at South Georgia on 15 and 16 March to embark the members of a Joint Services Expedition who had been on the island since 12 December 1981, and to assist with some British Antarctic Survey operations. An inspection of Leith Harbour was made by helicopter and supplies at the Survey's refuge there were replenished. HMS

Endurance departed for Port Stanley on the evening of the 16 March.

A four-man British Antarctic Survey field party was despatched by launch from King Edward Point to Carlita Bay and thence overland to Leith Harbour on 18 March to relocate and generally improve the Survey's refuge there. The refuge had been ransacked and looted by persons aboard some of the visiting yachts earlier in the year. The party arrived at Leith Harbour on the evening of 19 March to find an Argentine ship, *Bahía Buen Suceso* in the harbour, in the process of unloading cargo. The refuge had been occupied and its contents taken over. Shots were being fired and freshly killed reindeer (a protected species) were seen. Approximately 100 persons had landed, the Argentine flag was flying ashore, and many other irregularities were apparent. News of this was conveyed to the Magistrate, Mr S. Martin (who had assumed duties a month previously), at King Edward Point by radio before the field party made camp in the abandoned Customs House and spent the night as best they could. The ship had arrived to deliver men and equipment for Davidoff's salvage operation and preparations for a long stay ashore by Davidoff's employees were being made. It was later found from some of these men that the ship had arrived on 18 March directly from Buenos Aires and had maintained radio silence during the voyage.

The Magistrate contacted the Governor, Mr (later Sir) Rex Hunt on the evening of 19 March, to inform him and to request advice. He was given a message to be conveyed to the captain of *Bahía Buen Suceso*. This was passed by radio to the field party

Bahía Buen Suceso in Leith Harbour photographed by a member of the British Antarctic Survey field party 19 March 1982. (R Banner.)

early on the following morning and thence to Captain Briatore by a member of the field party, Mr T. Edwards. It stated that the salvage party had landed illegally, that they must go back on board the ship and report to the Magistrate at King Edward Point, that the Argentine flag must be removed, that the British Antarctic Survey's depot must not be interfered with, that no military personnel are allowed to land on South Georgia, and that no firearms are to be taken ashore. The field party left Leith to return to King Edward Point shortly afterwards as they were unable to attend to the refuge and gunfire had continued in the area. They obtained many photographs of the Argentine activities which included a series of the Argentine flag being removed about a quarter of an hour after they had conveyed the Governor's order that this be done. None of the other directions was complied with.

In order to support the island's administration and resolve these irregularities, HMS *Endurance* was despatched from Port Stanley to return to South Georgia. Meanwhile the British Antarctic Survey had established an observation post on 21 March, at the request of the Governor. This was at a high pass near Jason Peak which overlooked all the whaling stations in Stromness Bay and allowed the activities at Leith Harbour to be kept under surveillance. The Argentine ship to shore communications were also monitored from the observation post by a Spanish-speaking member of the Survey until *Bahia Buen Suceso* departed on 22 March. Later, following their arrival aboard HMS *Endurance* on 24 March, the Royal Marines continued the surveillance and briefly extended it to a position within Leith Harbour whaling station. A plan, arranged while HMS *Endurance* was on the way to South Georgia, to arrest the Argentines at Leith Harbour immediately on her arrival at the island was abandoned at short notice owing to diplomatic negotiations taking place between London and Buenos Aires.

Arrival of *Bahia Paraiso*

It was hoped that a correct and peaceful settlement of the problem could be achieved. Several courses of action were proposed to promote this, some of which were conveyed to the Argentine Government by the British Embassy in Buenos Aires. In an effort to ease the tension, Britain made it clear that if the salvage party requested proper authorisation this would be given retrospectively. Argentine reaction was to make public announcements, many of which were broadcast, strongly and uncompromisingly asserting her pretended sovereignty over South Georgia and stating that her citizens on the island, who were claimed to be acting correctly in accord with the salvage contract, would be protected from any action by the British authorities. Approximately half the Argentine navy had put to sea on exercises at this time and all attempts at negotiations were unsuccessful. *Bahia Buen Suceso* departed very early on 22 March leaving the salvage team of 39 men at Leith Harbour (at least 10 of whom were seen working on the water front, from the observation post). On 25 March another Argentine naval ship, the ice-strengthened *Bahia Paraiso*, arrived there. She carried Argentine special forces who reinforced the occupation of that part of the island. Large amounts of supplies and more equipment for the salvage operations as well as military requirements were unloaded. The Argentine flag was raised again accompanied by highly nationalistic oaths and anthems which caused concern among some of the salvage workers, five of whom were not Argentine citizens. A few of these stated, when being deported later, that they had expressed a desire to leave the island when this started.

The *Bahia Paraiso* had been on a voyage supplying Argentine stations in the Antarctic and was flying the pennant of the senior naval officer of Argentina's Antarctic squadron. The additional materials for the salvage operations, numbers of weapons and a large number of armed personnel aboard her might be interpreted as indicating that preparations for the invasion of South Georgia were made well before Davidoff's party landed on the island. *Bahia Paraiso* left Buenos Aires on her maiden voyage on 26 December 1981 and called at several Argentine military stations and the Argentine scientific station on the Antarctic Peninsula, then Ushuaia in Tierra del Fuego. On 18 March she left Ushuaia for the Argentine army station on the Peninsula, the Argentine naval station at the South Orkney Islands, and the Argentine naval station at Southern Thule where she is believed to have collected 32 special forces troops to attack King Edward Point.

Much delicate diplomatic negotiation continued between London and Buenos Aires to attempt to resolve the problem. On 31 March, in one further endeavour to reach a peaceful settlement, the British Government proposed that a senior official sould be sent to Buenos Aires for discussion of the incident but this was rebuffed on the next day when the Argentine foreign minister told the British Ambassador that the diplomatic channel, as a means of settling the dispute, was now closed. Meanwhile an Argentine naval frigate, *Guerrico*, had joined *Bahia Paraiso* at Leith Harbour (another, *Granville*, was in the vicinity). Four more Argentine naval ships (two destroyers, a submarine, and a tanker) were reported to be making way towards South Georgia. The tanker experienced engine trouble, thus neither she nor the destroyers, which relied on her for fuel, reached the island.

French and Russian vessels

A French yacht, *Cinq Gars Pour*, with a complement of three arrived at King Edward Cove from Buenos Aires on 14 March. After several days there her skipper requested permission to visit other parts of the island. He and the others aboard were to be granted a permit to go anywhere except the areas protected for scientific purposes and Stromness Bay where Leith Harbour is situated, as provided in the Conservation Ordinance of 1975 of which he had received a copy. In disregard of this the yacht sailed directly to Leith Harbour early in the morning of the 22 March without reference to anyone at King Edward Point. She was reported by the British Antarctic Survey's observation post and witnessed assisting the Argentine personnel. One of her crew was a cinematographer who photographed some of the Argentine activities for sale to the press and television. From this and later information from those on board many details were obtained of events at Leith Harbour. The yacht sailed directly to Buenos Aires shortly after the attack on King Edward Point. The accounts her complement subsequently produced of their journey to and arrival at South Georgia show evidence of embellishment.

A Russian deep-sea tug, *Storki*, called at King Edward Cove on 30 March arriving early in the morning after travelling almost 1500 km from near Elephant Island in the South Shetland Islands. This was ostensibly a normal visit to take water at Grytviken where she remained 36 hours.

Emergency and invasion

An emergency situation developed at the Falkland Islands which led to the departure of HMS *Endurance* for Port Stanley at 21:00 on 31 March 1982 under orders from the Ministry of Defence, London. Her course went undetected past *Bahia Paraiso* stationed outside Cumberland Bay, thence around the south-eastern end of the island. She left a platoon of Royal Marines commanded by Lt. Keith Mills at King Edward Point who, with assistance from some of the Survey's personnel, began preparations for the defence of the area. On the evening of 1 April, the Governor of the Falkland Islands and Dependencies made a broadcast announcement of impending Argentine attack and subsequently proclaimed a State of Emergency on the Falkland Islands. On 2 April, early in the morning, Argentine forces attacked Stanley in overwhelming strength, took the Governor prisoner, and occupied the Falkland Islands.

On the same day, at approximately 10:00, the Argentine naval ship *Bahia Paraiso*, entered Cumberland Bay and was seen from King Edward Point. At the time the weather was extremely windy with blowing snow; she was thus unable to launch either boats or helicopters and departed after about an hour. Subsequently her Master passed a message for the Magistrate to expect an important communication in the morning. The Royal Marines used that day to improve their defensive positions. Civilian parties at two of the four occupied field stations were augmented and informed about events by men from King Edward Point. Most other civilians evacuated the scientific station to take refuge in the church behind the abandoned Grytviken whaling station. News of events, as far as was compatible with security, was passed by radio, which was being monitored by the enemy, to the other two field stations on South Georgia, to the other British Antarctic Survey stations, and to RRS *Bransfield* on her way towards the South Orkney Islands. Action was taken that afternoon to prevent Government seals, cryptographic materials, and some other items from reaching enemy hands.

The weather on 3 April 1982 was clear and windless, in great contrast to that of the previous day. At first light, the two Royal Marines at the observation post were brought to King Edward Point under circumstances which earned an award. At 10:00 the two Argentine warships, *Bahia Paraiso* and *Guerrico*, entered Cumberland East Bay and enemy helicopters started flying around the area. The Magistrate was contacted by radio from *Bahia Paraiso* shortly afterwards with the signal: 'Following our successful operation in the Malvinas Islands the ex-governor has surrendered the islands and dependencies to Argentina. We suggest you adopt a similar course of action to prevent any further loss of life'. The Magistrate advised that a ceasefire had been declared in the Falkland Islands only, that it might take some time, under the circumstances, to determine the appropriate course of action in the Dependencies, and he offered to discuss the matter with the ship's captain in order to effect a peaceful solution. The Magistrate was then ordered to have the personnel of King Edward Point stand in a prominent place and informed that a helicopter, from the ship, would land. The Magistrate refused permission for the helicopter to land and repeated his offer of discussion with the Captain. No response resulted. He then asked if it was the intention to take South Georgia by force; this was answered affirmatively. The Magistrate then advised the Captain of *Bahia Paraiso* that there was a military presence on the island and that it would be defended.

By this time the enemy frigate, *Guerrico*, was lying off King Edward Point with her guns directed at the radio room but holding fire. The Magistrate abandoned the radio and authority was passed from the civil administration to the officer commanding the Royal Marines in order to defend the island. *Guerrico* moved a short distance further off. During these events the civilians at Grytviken had been monitoring the radio communications and, towards the end of these, had taken shelter on the floor of the Library at the rear of the church just before firing commenced. The Magistrate's radio transmissions were made on a British Antarctic Survey frequency at high power and were monitored by several ships and land stations. The *Bahia Paraiso* transmitted on 'VHF' and was received only locally.

Argentine troops, landed from a helicopter at the King Edward Point jetty, then opened fire on the Royal Marines. *Guerrico* returned to the vicinity of King Edward Point and opened fire on the scientific station as she passed. When she reached an appropriate position, the Royal Marines directed several Karl-Gustav anti-tank rockets into her starboard side. Owing to the position of Hobart Rocks the damaged frigate had to continue on course before being able to turn, after which she presented her port side to receive more rockets. She was also hit by well over 1000 rounds of machine-gun fire. The rockets had penetrated her below the waterline, damaged her guns, and hit her Exocet missile equipment. Two enemy helicopters were hit by gunfire; one crashed across King Edward Cove and the other, trailing much smoke, eventually managed to land near Grytviken. The survivors of the enemy troops who had landed earlier were confined in the vicinity of the Post Office by defending fire. Others had landed near the whaling station and made their way towards King Edward Point. The severely damaged frigate began to bombard the vicinity of the scientific station with 100 mm shells, from a position well out to sea and was starting to get the range of the defensive positions. Shortly after the shelling commenced the officer commanding the Royal Marines surrendered his forces. The battle had lasted for almost two hours. Enemy forces had probably lost at least 15 men killed (Argentine sources have reported only 3), with a similar number wounded; the defending forces sustained one wounded man. The officer commanding the Royal Marines was subsequently awarded the Distinguished Service Cross for the defence of South Georgia.

The civilians, all men of the British Antarctic Survey, remained at Grytviken during the conflict. At approximately 14:00 ten Argentine troops passed behind the whaling station and were approached by the author who advised them of the presence of the civilians within the church; they passed this information on by radio. Shortly afterwards the remaining functional enemy helicopter made a close inspection of the vicinity. At 15:00 the officer commanding the Argentine forces, Captain Alfredo Astiz, accompanied by the officer commanding the Royal Marines arrived with about 20 Argentine troops carrying machine guns, who surrounded the church. The author, acting for the Magistrate, surrendered the civilians one by one as ordered. They were searched and marched at gunpoint to the scientific station. On the way, the Magistrate was captured. He had taken refuge in a ground depression after leaving the radio and passing control to the officer commanding the Royal Marines, as he had had insufficient time to reach the civilian refuge at Grytviken before firing commenced.

By the time the 13 civilians had arrived at King Edward Point, nearly all the Royal Marines had been removed to *Bahia Paraiso*. The civilians were held at gunpoint around the empty flag pole until, as a result of several requests, they were eventually allowed less than 20 minutes, under individual guard, to gather personal property, scientific data, and similar items. In consequence they lost much personal property, notably small valuable items which were stolen by the Argentines when they later looted the scientific station. Most of the civilians were embarked on *Bahia Paraiso* immediately after-

wards. The author had the opportunity to ensure the security of some dangerous substances and, by a ruse, collected several items of value and intelligence interest as well as the last despatch. These were smuggled, unknowingly, aboard the ship by a guard. The Magistrate, commander of the Royal Marines, and author were then embarked and the ship departed for Leith Harbour and thence directly into Argentine waters off Rio Grande, Tierra del Fuego.

The Argentine forces were purely military and had no interest in the scientific station as such. All meteorological, geophysical, and other data being recorded at the observatory, some of which were continuous from 1905, as well as other scientific programmes in many fields of biology, glaciology, and other sciences had to be abandoned as a result of the attack. Laboratory animals and plants had to be left to die and no opportunity was given to switch

The Argentine Puma helicopter shot down by Royal Marines during the battle at King Edward Point 3 April 1982. Photographed in September 1982. (Author.)

off most apparatus and equipment.

As a result of these events, Argentine military forces occupied King Edward Point and Leith Harbour. British field stations remained at four other sites; Schlieper Bay, Lyell Glacier, St Andrews Bay, and Bird Island, with two, four, five, and four persons respectively. All were employees of the British Antarctic Survey except for two women who were at St Andrews Bay, making a wildlife film for a television company. These stations were in radio contact with each other and, by a relay link, with the British Antarctic Survey Headquarters in Cambridge, England, Some intelligence concerning enemy movements was secured by them, especially by some of the Lyell Glacier party who observed King Edward Point from a peak overlooking it. Otherwise they remained at or near their stations, performing normal duties in as far as was possible during the enemy occupation of King Edward Point and Leith Harbour. It had been the enemy's intention to take all civilians prisoner and remove them from South Georgia. The loss of one and damage to another helicopter during the battle at King Edward Point frustrated this.

United Nations Organisation

On 3 April 1982, the same day as the attack on King Edward Point, the United Nations Security Council met in emergency session. It passed a Resolution, number 502, which required Argentina to cease hostilities and immediately withdraw her forces from the Falkland Islands and Dependencies. Argentina ignored this mandatory directive, regardless of the conditions she accepted upon acceding to the United Nations Charter in 1945. Some of these include the specific obligation that the settlement of international disputes must be by peaceful means and a prohibition of use of force in such matters. Great Britain, after a number of attempts through intermediaries to secure an acceptable negotiated settlement, acted under Article 51 of the United Nations Charter to expel the invasion and restore the legal government of the Falkland Islands and Dependencies.

HMS *Endurance*

HMS *Endurance* received news of the Argentine invasion force soon after she left South Georgia on 31 March. She was impotent to defend the Falkland Islands, or even herself alone, against the majority of the Argentine navy and changed course to return to South Georgia when the strength and intention of the enemy became apparent. Her new course took her back around the southern end of the island to Royal Bay. From there she was able to establish an observation post high on the Barff Peninsula using her helicopters. The post overlooked Cumberland East Bay, King Edward Cove, the scientific station and Grytviken; it was manned just in time to witness the course of the battle on 3 April.

HMS *Endurance* was under orders from the Ministry of Defence, London, not to involve herself in the battle. Her weapons may not have been sufficient against both the enemy's ships (one of which had prisoners aboard later in the day). A third enemy ship, the frigate *Granville* armed with Exocet missiles, was known to be searching for her at the time. Thus, as her most useful task was to obtain intelligence about the enemy, she avoided them and concealed herself amid the icebergs to the south of the island and in some of its bays. Two visits to a field station were made secretly during this period by one of her helicopters which landed briefly during the second one. HMS *Endurance* remained in the region until 5 April and then moved north to meet a forward section of the Royal Naval task force on course for South Georgia and to replenish food and fuel. Argentine news broadcast that she had been sunk during the invasion of South Georgia.

Defeat of the Argentine forces

In order to expel the invaders and restore the legal government of the Falkland Islands and Dependencies a Royal Naval task force was despatched to the region. HMS *Conqueror*, a nuclear submarine, first made an inspection of the area from South Georgia westwards. Then HMS *Endurance*, accompanied by HMS *Plymouth* and *Antrim* with RFA *Tidespring*, arrived off South Georgia on 21 April when Mr Peter Stark of the British Antarctic Survey, who had a good local knowledge of the island, was taken aboard her from St Andrews Bay to provide local intelligence and a Royal Naval officer replaced him to monitor radio transmissions. An unsuccessful attempt to reconnoitre enemy positions was made on the 21 April by a helicopter-borne patrol, landed on the Fortuna Glacier (against local advice). The attempt failed because of the location and weather of the landing place. On the next day boat-borne observation parties had more success but still experienced many difficulties, mainly from the weather.

Information was obtained that an enemy submarine was patrolling the vicinity and the ships temporarily withdrew. On 24 April an Argentine 707 aircraft spotted HMS *Endurance* in Hound Bay and early on 25 April HMS *Brilliant* joined her, HMS *Antrim* and *Plymouth* near the island. A helicopter patrol detected the submarine, *Santa Fé*, which was attacked and rendered unable to dive. *Santa Fé* had previously left King Edward Cove and altered course to return there on being attacked. The Royal Naval helicopters came under mortar and machine-gun fire as they approached the shore

The Argentine submarine *Santa Fé* captured on 25 April 1982. Photographed beached off Hestesletten in October 1982. (Author.)

and *Santa Fé* was eventually able to moor alongside the King Edward Point jetty.

Later in the morning of the 25 April HMS *Antrim* and *Plymouth* commenced an accurate bombardment of two positions on the opposite side of King Edward Cove from the scientific station, Hestesletten and the slopes of Brown Mountain. This neutralised these areas and was followed by a landing of Royal Marines on Hestesletten who advanced across the cleared ground, through the whaling station, and ultimately across a minefield to King Edward Point. The bombardment was heard as far away as St Andrews Bay and had the effect of causing the Argentine garrison at King Edward Point to surrender without a shot being fired in their direction. They did not return fire and raised white flags before more British troops landed. Major G. Sheridan accepted their surrender at 14:15 and the signal 'Be pleased to inform Her Majesty that the White Ensign flies alongside the Union Jack on South Georgia. God save the Queen.' was despatched to the Ministry of Defence in London. King Edward Point was thus retaken with minimal use of force after 22 days enemy occupation. The Argentine prisoners were then supervised as they cleared mines, booby traps, and the large amount of rubbish they had deposited around the scientific station. The majority of their communications equipment, signals, orders, and classified documents were captured intact which subsequently proved very useful.

The 13 Argentine troops, commanded by Astiz, at Leith Harbour with 39 salvage workers were informed of these events by means of the captured communications equipment. The civilians left for Stromness whaling station, with a promise of safe conduct, on that evening and met a detachment of Royal Marines on the route. The Argentine forces were directed to remain at Leith Harbour overnight and ordered to move out to a specified hill to surrender early on the morning of 26 April. It was later found that much of the whaling station and a helicopter landing area designated on the football pitch had been heavily mined. Two weeks were required to detect and clear all the mines, booby traps, and other devices there, as well as generally clean up the accommodation etc.

The total number of Argentine troops taken prisoner was 156 and there were 39 civilian de-

Map showing the places mentioned in Chapter 9. (Author.)

1: King Edward Point, scientific and administrative station; invaded 3 April 1982, retaken 25 April. 2: Grytviken whaling station where the civilians from King Edward Point were taken prisoner. 3: Leith Harbour whaling station where *Bahía Buen Suceso* arrived unlawfully on 18 March 1982 and from where Argentine forces and salvage workers were removed on 26 April. 4: Observation post established on 21 March and manned by the British Antarctic Survey until the arrival of the Royal Marines. 5: Carlita Bay hut, the beginning of the 4-hour overland route to Leith Harbour. 6: Approximate position of the observation post established by helicopter from HMS *Endurance* to see the battle on 3 April. 7: Crashed enemy helicopter on the slopes of Brown Mountain (3 April). 8: Fortuna Glacier where an unsuccessful preliminary landing was made by the Special Air Service on 1 April. 9: Hound Bay where a concurrent landing was made by the Special Boat Service. 10: Grass Island where a second SAS position was established. 11: Hestesletten where, after an accurate patterned bombardment, British forces landed to advance to King Edward Point to accept the Argentine surrender on 25 April.

12: Approximate position from where HMS *Antrim* and *Plymouth* bombarded Hestesletten.

On the inset map: the sites with British personnel remaining after the invasion A: Bird Island station (4 persons). B: Schlieper Bay hut (2 persons) C: Lyell Glacier hut (4 persons). D: St Andrews Bay hut (5 persons).

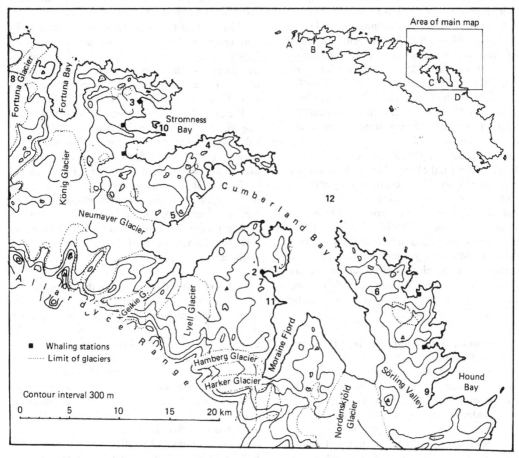

portees (the salvage workers). All were removed to Ascension Island aboard RFA *Tidespring* and then, with the exception of Captain Astiz, flown to Uruguay and released. Astiz was wanted for questioning by the French and Swedish Governments on charges involving torture and murder of their citizens after the 1976 revolution in Argentina. His status under the International Convention for the Treatment of Prisoners of War, of which the United Kingdom is a full signatory, permitted him to make no comment about these matters and, in due course, he was repatriated.

Only one man, an Argentine, was seriously injured during the retaking of South Georgia; he lost a leg following injury aboard *Santa Fé*. Another died on 26 April aboard the submarine, following confusion in orders while she was being moved to the Grytviken jetty. He was buried with full military honours at the Grytviken cemetery. *Santa Fé*, formerly USS *Catfish* built in March 1945, with a large amount of high explosive aboard, was later beached near Hestesletten and then prepared for subsequent disposal. An Argentine landing craft, *Fenix*, was also captured.

The British Antarctic Survey and other civilian personnel at the four field sites around South Georgia were relieved by the Royal Navy after the Argentine surrender. Some were able to visit King Edward Point before all were conveyed to Ascension Island and thence the United Kingdom. Many scientific specimens and other materials were taken out aboard HMS *Antelope*; these were lost when she was sunk by a bomb, dropped on her at the Falkland Islands, which detonated while being defused. Much other equipment and data were also salvaged by HMS *Endurance*.

The prisoners taken by the Argentines from South Georgia

The 13 British Antarctic Survey men and 22 Royal Marines captured at King Edward Point on 3 April were imprisoned aboard the *Bahia Paraiso* for 11 days under strong guard. Meals, several brief exercise periods, and movements in the prison quarters were all at machine-gun point. The ship reached Tierra del Fuego in 4 days where the wounded, including one Royal Marine, were taken ashore by helicopter. Then she took a position between Argentina and the Falkland Islands to support Argentine forces invading the Falkland Islands by becoming a landing station (Bravo I) to refuel helicopters otherwise unable to make the journey. After this was completed she proceeded to Puerto Belgrano where the prisoners were transferred to a hastily arranged prison camp in the dressing rooms of a conscripts swimming pool. Whilst held there, their possessions were thoroughly searched and items removed from them although most money and personal documents were held separately and properly accounted for. Two military and two civilian prisoners were taken before an Argentine naval tribunal to give evidence concerning the battle on 3 April and all were included on a television tape made in the prison camp.

After 4 days they were flown very late at night and early in the morning of the 20 April to Uruguay and released. During the 15 days' imprisonment the captives were held absolutely incommunicado and no news of their whereabouts was conveyed outside Argentina. Enquiries made by the Swiss Embassy on their behalf met with no response and, despite several requests, neither consular access nor prisoner of war communications were permitted. Conditions beyond these aspects and the active participation by *Bahia Paraiso* in the invasion while she had prisoners aboard were not incompatible with the requirements of the International Convention for the Treatment of Prisoners of War though Argentina is not a full signatory of this; the prisoners included civilians, and their legal status was not clarified.

Events after the removal of the Argentines

Following these events, South Georgia has been occupied by British forces. The great value of its many harbours was demonstrated during the campaign to free the Falkland Islands. HMS *Endurance* became the South Georgian flag ship and at one time controlled 25 ships in Cumberland Bay, transferring men and supplies for onward movement to the Falkland Islands. The largest ship ever to call at the island, *Queen Elizabeth II*, was anchored near Sappho Point. Other ships present from the reconquest of South Georgia to that of the Falkland Islands, apart from those already named, included: HMS *Ambuscade, Andromeda, Cordella, Farnella, Junella, Leeds Castle, Northella,* and *Pict*; RFS *Blue Rover, Pear Leaf, Regent, Sir Bedivere,* and

Stromness; SS *Canberra*, CS *Iris*, MV *Geestport*, *Nordic Ferry*, *Norland*, *Salvageman*, *Saxonia*, *Stena Inspector*, *Stena Seaspray*, *Typhoon*, *Wimpey Seahorse*, and *Yorkshireman*. Several Argentine aircraft flew over South Georgia and its vicinity during this period. One dropped a bomb aimed at a British ship, but fortunately this missed its intended target.

'M' company of the Royal Marines established themselves at King Edward Point, where damage to the buildings was mainly superficial, prepared defensive positions and dealt with enemy munitions. The garrison has since been relieved on regular occasions. Despite her defeat and Security Council Resolution 502, Argentina has yet to agree to the British proposal that an end to hostilities be declared and it is uncertain how long the garrison will remain on the island. In general, other than at King Edward Point, Grytviken, and Leith Harbour, together with the crashed helicopters and wreck of *Santa Fé*; South Georgia is largely unaffected by these events.

King Edward Point from the bridge of HMS *Hecate* in October 1982. (Author.)

In the Falkland Islands, the Argentine forces were defeated and surrendered on 14 June 1982 when the Government which the inhabitants desired was restored to them. The occupation of Southern Thule, another part of the Falkland Islands Dependencies, by an Argentine naval station, established on 7 November 1976, was terminated on 20 June 1982 by HMS *Endurance* which proceeded there from South Georgia accompanied by HMS *Yarmouth*, RFA *Olmeda* and the tug *Salvageman*. The station was closed and secured, its ten remaining men were removed to Argentina. Nearly all the technical equipment at the station had been destroyed by them a few days before HMS *Endurance* arrived; the naval station itself was destroyed in January 1983 following discovery of evidence of a visit by Argentine personnel after its closure.

In September 1982 the British Antarctic Survey despatched a three-man team to inspect King Edward Point and two men to reopen Bird Island

scientific station. A report of damage and losses together with many retrieved data and other items were obtained from the scientific station at King Edward Point. A large amount of scientific instruments, apparatus, and equipment was put into storage, pending future use. At Bird Island it was possible for normal operations to resume directly as the station had been closed by its scientific staff. The Survey has not resumed activities at King Edward Point; mines and other unexploded munitions, including those aboard *Santa Fé*, the amount of accommodation available, and the defence requirements have been some of the factors delaying this. None of the continuous research activity has yet been re-established except for some meteorological observations. Of the personal losses incurred by both civil and military personnel, some items have been recovered and a proportion of compensation paid for the rest.

Bird Island station operated virtually normally from the 1982/83 summer, with personnel of the Survey accompanied by two visiting United States scientists. King Edward Point remained occupied by the garrison. Another inspection of its facilities was made by the Survey in April 1983 and there have been later visits by the Survey's ships and personnel. Some officers of the Falkland Islands Government and the Civil Commissioner have also visited South Georgia.

Much political analysis and investigation have followed these events, especially with regard to the Falkland Islands. The British Government appointed a committee of Privy Counsellors, under the chairmanship of Lord Franks with the terms of reference 'To review the way in which the responsibilities of Government in relation to the Falkland Islands and their Dependencies were discharged in the period leading up to the Argentine Invasion of the Falkland Islands on 2 April 1982, taking account of all such factors in previous years as are relevant; and to report'. The Committee reported in January 1983 with a detailed investigation of relevant events. Many other publications have been issued following the invasion with various versions of events and differing interpretations.

The future for South Georgia after the war

The future of South Georgia after the events of 1982 is highly conjectural. Previously one had hoped that it might remain as one of the few almost undisturbed parts of the Earth, of great beauty and scientific interest. It had the rare distinction, until that year, of being one of the few parts of the world to have escaped war and had been effectively entirely a natural reserve from the time of the closure of its last whaling station. It is increasingly likely that it will experience another period of economic development with communications, administration, population and defence aspects much enlarged. This may not be for matters connected with South Georgia alone for, owing to its position, harbours, water supplies, and much else, it might again become 'the gateway to Antarctica' if exploitation occurs elsewhere in the region. The results of insufficiently controlled exploitation are abundantly demonstrated in the history of South Georgia. The lessons these provide may conceivably yield a better future. The establishment and maintenance of normal relationships with Argentina, when she acquires a stable and reliable government, would be of mutual advantage but it is very uncertain when that might be.

Two economic studies of the Falkland Islands and Dependencies were made in 1976 and 1982 with Lord Shackleton, son of Sir Ernest Shackleton, as chairman of both. Although the published reports principally concern the Falkland Islands, attention was given to South Georgia among the Dependencies. The reports described the circumstances of the island and made recommendations and suggested some courses for its future development. The terms of reference of the second report were to revise the first, following the Argentine invasion and the changes in the world's economic environment since 1976. Lord Shackleton who had visited South Georgia during the preparation of the 1976 report, stated in his introduction to the 1982 report that 'South Georgia may in the long run be of greater importance in the future development of the potential wealth of the South West Atlantic and Antarctic than the Falkland Islands'. The report describes possible exploitation of krill, fish, squid, seal, and minerals as well as conservation, the Antarctic Treaty, communications and existing circumstances of South Georgia.

The major recommendations concerned fisheries and included a 200 nautical mile fisheries limit around South Georgia, the South Sandwich

Islands and Shag Rocks should be established, and that consideration should be given to investigating, if possible, through a revived United Nations Southern Ocean Fishing Project, the technical possibilities of fishing both krill and fin fish in relatively small motor fishing vessels from South Georgia. The establishment of a 200 nautical mile (375 km) fishing limit, now general throughout the world, would permit effective measures for the long-term conservation of stocks of krill, fish, squid, and other animals to be applied. (The present fishing limit is the same as that of territorial waters, 3 nautical miles.) Already the fish stocks around South Georgia are significantly depleted by excessive catching. Sale of licences to foreign fishing fleets, or their exchange for other benefits (such as scientific information), subject to appropriate control, was proposed as a way of making this commercially viable. Such a licensing system operates over the fishing grounds off Kerguelen and raises much revenue for French Antarctic research and similar arrangements are made by the New Zealand and Australian governments for several southern islands under their control. The Falkland Islands might then become a forward base and communications centre for the investigation and development of these resources off South Georgia.

The re-establishment of sealing was also suggested. The South Georgian populations of fur seals have now increased enormously since the beginning of their very long period of total protection, while the elephant sealing industry was profitable throughout its second epoch. Both are suitable for scientific management to ensure that the industry is stable and the seal populations are in no danger of serious reduction.

The Shackleton Report referred also to hydrocarbon and other mineral exploitation around South Georgia but concluded that this is unlikely (see Chapter 7). Similarly, owing to changed economic circumstances, harvesting of kelp for alginate production is at present no longer viable. The revenue raised from Falkland Islands Dependencies postage stamps was referred to and the prospect of this increasing was considered extremely good owing to recent publicity about South Georgia. The report recommended that consideration should be given to establishing an airstrip on South Georgia and proposed two suitable sites. The report also referred to their military significance. Conservation of various aspects of South Georgia was regarded as very important; because of its geographical position and history the island possesses a range of natural resources and artifacts of industrial archaeology which makes it of special interest in a global context. It was emphasised that it is not only necessary to conserve the flora and fauna of the island with its surrounding waters for their natural and scientific value but also to ensure that resources are exploited in ways which are of maximum long-term economic benefit. The final recommendation with regard to South Georgia, in view of its industrial archaeological importance, was that there would be great benefit to be gained from a visit by an expert from a maritime museum or an industrial archaeologist to suggest what could and should be preserved, as well as how and where this may be done. In this respect it is fortunate that the British Antarctic Survey had recovered a large amount of documents relevant to this barely a month before *Bahía Buen Suceso* arrived at Leith Harbour.

An interesting and potentially important factor for any future commercial or other activity on South Georgia is the possibility of using hydroelectric power to obviate the need for expensive fuel imports. Grytviken whaling station obtained much of its electricity from this source and operated up to five turbines, such works are a Norwegian speciality. Several other areas have excellent potential for such schemes which have several advantages as sources of energy beyond cheapness alone.

There have been regular tourist visits to South Georgia since 1970 by *Lindblad Explorer* or *World Discoverer*. Previously, a return passage on *Darwin* or, much earlier, *Fleurus* permitted short visits from the Falkland Islands. For the future, it is probable that tourist visits will become more regular. Tourists may be able to remain ashore more than the usual day or less, as the attractions of the island have recently become far more widely known. There is no commercial accommodation ashore, hence such visits are necessarily ship-based at present. Provision of commercial accommodation has been suggested, using the facilities available at King Edward Point and suitable new construction. Proposals for an external studies centre for appropriate scientific disciplines, possibly combined with trekking and mountaineering, have also been made.

The British military forces at present stationed on South Georgia have come to appreciate it as a potentially excellent place for Polar training and exercises. They have demonstrated high regard for the wildlife in the operations so far conducted. The British Antarctic Survey has two principal requirements from the island: for land-based scientific stations, and for support of ship-based operations in the vicinity and at its stations elsewhere. The logistic value of the harbours of South Georgia for the military, the British Antarctic Survey and certain fishing fleets is already great; it will increase with future development. Christian Salvesen and Company are in an excellent position to become involved in any development of the resources of South Georgia as they hold the leases of large amounts of flat land at the heads of the five best harbours on the island. One hesitates to suggest, however, that a salvage operation to remove much dilapidated material not of an industrial archaeological nature from these sites might be economically worthwhile and would reduce some of the effects of commercial exploitation from the island.

The potential changes for South Georgia; from a very isolated island with a small scientific community, where a few ships call in summer and where during 8 months of the year there is only radio contact with the rest of the world; to whatever it may become after recent events may well be very great but, it is to be hoped, not uncontrolled. The prediction of the future is generally fraught with error. However, one can be sure that of South Georgia will be fascinating.

Appendices

Appendix 1

The sites leased to whaling companies on South Georgia

Site	Company (Country of registration)	Dates	Remarks
Grytviken	Compañia Argentina de Pesca (Argentina)	16 Nov. 1904 to 1960	The first whaling station in the Antarctic. The lease was formally granted from 1 Jan. 1906. Floating factories also operated there from 1909 to 1916.
	Albion Star (South Georgia) Ltd (F. Is.)	1960 to 1979	Albion Star finished whaling at the end of the 1961/62 season. The station was sub-leased to Kokusai Gyogyo Kabushike Kaisha of Japan for the 1963/64 and 1964/65 seasons. It closed on 4 Dec. 1964 and a caretaker remained to early 1971.
	Christian Salvesen Ltd (UK)	1979–current	
Jason Harbour	Compañia Argentina de Pesca (Argentina)	1 Jan. 1909 to 31 Dec. 1926	The site was not used for a whaling station but the whale catchers its lease permitted were deployed from Grytviken. Only one small hut was erected. The site reverted to the Government in 1927.
Stromness	Sandefjord Hvalfangerselskab (Norway)	1907/08 to 1919/20	Intended for a floating factory site in 1906/07 but the ship was wrecked. Another floating factory operated from 1907/08 and the lease was granted from 1 Jan. 1908. A shore station begun in 1912 started operating in March 1913. Station sub-leased to Southern Whaling and Sealing Co. in 1917/18 and 1918/19.
	Vestfold Hvalfangers A.S. (Norway)	1920/21 to 1945	A new company formed by amalgamation of Sandefjords Hvalfangerselskab and Hvalfangerselskab 'Ocean' used the station until it finished whaling on 12 Apr. 1931. From 1931–32 it was sub-leased to the South Georgia Co. as a ship repair yard.
	South Georgia Company (Chr. Salvesen Ltd) (UK)	1945–current	The station was closed after the 1960/61 season.
Husvik	Tønsbergs Hvalfangeri (Norway)	24 Dec. 1907 to 1959/60	Started as a floating factory site in 1907 which continued to 1913. The lease was formally granted from 1 Jan. 1908. Shore station started operating in 1910 and continued to 1930/31, resumed in 1945/46 to 1956/57 and in 1959/60.
	Albion Star (South Georgia) Ltd (F.Is.)	1960/61 to 1979	The station was partly dismantled in 1960/61 and its meat freezing plant moved to Grytviken.
	Christian Salvesen Ltd (UK)	1979–current	

Station	Company	Dates	Notes
Godthul	Bryde & Dahls Hvalfanger-selskab (Norway)	8 Jan. 1908 to 1928/29	This was only a floating factory site with a small shore base. The company purchased the South Georgia Exploration Company's lease and was granted a whaling licence from 16 Sept. 1908. The site was used from 1908/09 to 1916/17 and 1922/23 to 1928/29 (24 Apr.) after which the lease lapsed and the site reverted to the Government.
Leith Harbour	Christian Salvesen Ltd (UK)	13 Sept. 1909–current	Shore station from the beginning of operations. Lease formally granted on 1 Oct. 1909. Floating factories also operated from 1912 to 1918. The station did not operate in 1932/33, 1940/41, and from 1942/43 to 1944/45. The company's last whaling season was 1960/61. The station was sub-leased to Nippon Suisan Kaisha for 1963/64 to 1966/67 but closed in 15 Dec. 1965, the end of South Georgia whaling. Caretakers left in Jan. 1966.
Rosita Harbour (formerly Allardyce Harbour)	South Georgia Company (Chr. Salvesen Ltd) (UK)	1 Oct. 1909 to 1923	Nothing was done at the site but the whale catchers its lease permitted were deployed from Leith Harbour. The site reverted to the Government in 1924.
Ocean Harbour (formerly New Fortune Bay)	Hvalfangerselskab 'Ocean' (Norway)	10 Oct. 1909 to 1919/20	Shore station from the beginning of operations on 26 Oct. 1909. Following the company's amalgamation with Sandefjords Hvalfangerselskab the station was almost entirely removed to Stromness and the site reverted to the Government.
Prince Olav Harbour	Southern Whaling and Sealing Co. (Irvin & Johnson) (South Africa)	1 Aug. 1911 to Sept. 1919	Floating factory site to 1916. Shore station began in 1917.
	Southern Whaling and Sealing Co. (Lever Brothers) (UK)	Sept. 1919 to 31 Dec. 1934	Continued as a shore station until 14 Mar. 1931 and then closed. Site reverted to the Government from 1 Jan. 1935.
	South Georgia Company (Chr. Salvesen Ltd) (UK)	17 Sept. 1936–current	Station purchased for salvage.

Appendix 2

Whales and seals taken at South Georgia during the whaling period and the second epoch of sealing (maxima are underlined)[a]

Season	Number of			Number of whales taken							Whale oil produced (barrels)	Number of seals taken	Seal oil produced (barrels)
	Shore stations	Floating factories	Whale catchers	Blue	Fin	Hump back	Sei	Sperm	Other	Total			
1904/05	1	—	1	11	16	149	—	—	7	183	5 302	—	—
1905/06	1	—	2	27	68	288	—	—	16	399	12 002	80	168
1906/07	1	—	2	20	53	240	—	—	8	321	11 728	—	—
1907/08	1	2	7	4	4	1 281	—	—	_93_	1 382	39 660	—	—
1908/09	1	3	7	10	20	1 841	—	1	68	1 940	48 406	—	—
1909/10	2	7	17	26	58	3 391	—	4	37	3 516	104 316	595	1 082
1910/11	4	7	19	85	168	_6 197_	—	—	79	6 529	189 363	3 005	3 467
1911/12	4	_8_	21	298	516	5 635	—	4	82	6 535	212 262	2 059	4 031
1912/13	4	_8_	21	317	2 157	2 360	—	9	7	4 850	196 714	2 794	5 712
1913/14	5	3	21	940	1 716	512	94	21	66	3 349	176 487	4 881	7 804
1914/15	5	3	22	2 313	1 940	823	—	1	20	5 097	270 507	3 113	4 641
1915/16	5	4	28	3 026	2 744	1 578	—	1	12	7 361	346 270	2 016	2 537
1916/17	_6_	2	_32_	2 440	1 606	378	—	35	12	4 471	268 327	2 906	5 337
1917/18	_6_	1	_32_	1 871	1 144	60	49	37	35	3 196	202 503	3 018	5 297
1918/19	_6_	—	28	1 160	1 530	68	7	18	9	2 792	148 292	2 954	6 137
1919/20	_6_	—	26	987	1 673	79	71	8	14	2 832	147 029	1 230	1 650
1920/21	5	—	21	856	2 643	103	36	31	13	3 682	177 137	1 545	2 269
1921/22	5	—	20	2 570	710	9	103	3	—	3 395	249 042	1 114	1 660
1922/23	5	1	23	3 569	1 445	320	10	19	—	5 363	347 553	2 713	5 035
1923/24	5	1	23	1 927	1 378	130	191	49	—	3 675	247 463	2 994	6 375
1924/25	5	1	24	3 512	2 019	262	1	24	—	5 818	406 176	3 902	7 486
1925/26	5	1	23	1 855	_5 709_	236	13	12	—	_7 825_	404 457	3 801	6 891
1926/27	5	1	23	_3 689_	1 144	—	365	17	—	5 215	_417 292_	4 782	8 094
1927/28	5	2	23	2 125	1 357	15	95	60	—	3 637	303 480	5 515	10 033
1928/29	5	1	23	1 560	3 130	15	396	31	—	5 132	348 629	4 883	8 768
1929/30	5	—	27	488	3 396	46	216	39	1	4 186	247 963	5 102	9 224
1930/31	5	—	27	1 085	1 416	66	144	24	1	2 736	187 938	5 814	10 616
1931/32	2	—	12	438	1 735	6	16	10	—	2 205	122 205	5 929	11 580
1932/33	1	—	6	267	727	—	2	—	—	996	54 583	5 172	9 867
1933/34	2	—	12	536	1 728	92	—	7	—	2 363	132 187	5 987	12 252

Year													
1934/35	2	10	—	556	863	37	125	21	—	1 602	108 141	4 438	8 896
1935/36	2	10	—	1 221	520	41	—	3	—	1 785	143 185	5 838	11 892
1936/37	2	12	—	121	1 079	17	471	70	—	1 758	81 629	5 604	11 366
1937/38	2	12	—	97	1 552	40	155	43	—	1 887	90 266	6 000	12 295
1938/39	2	11	—	232	1 307	—	19	117	—	1 675	111 490	5 833	11 515
1939/40	2	12	—	88	937	—	80	85	1	1 191	64 782	6 000	12 154
1940/41	1	5	—	7	747	—	88	26	—	868	44 498	6 000	12 186
1941/42	2	12	—	59	1 189	16	52	109	—	1 425	77 819	5 831	11 786
1942/43	1	6	—	125	776	—	73	24	—	998	50 960	3 989	6 572
1943/44	1	7	—	28	632	4	197	101	—	962	50 001	5 927	11 167
1944/45	1	7	—	128	987	60	76	45	—	1 296	75 540	6 000	11 940
1945/46	3	16	—	80	1 456	238	82	57	—	1 913	78 877	5 382	10 382
1946/47	3	18	—	327	1 670	28	391	133	1	2 550	145 318	4 449	8 075
1947/48	3	21	—	46	2 142	24	609	128	—	2 949	163 398	6 000	11 994
1948/49	3	21	—	226	1 922	18	562	213	—	2 941	172 194	7 500	15 093
1949/50	3	21	—	14	1 999	26	1 183	157	—	3 379	147 121	6 876	13 358
1950/51	3	21	—	82	1 982	8	519	226	—	2 817	151 192	6 901	13 035
1951/52	3	21	—	6	2 007	10	498	141	—	2 662	144 375	7 877	14 608
1952/53	3	21	—	4	1 670	9	498	147	—	2 328	120 003	6 000	10 807
1953/54	3	21	—	13	2 673	11	778	179	—	3 654	184 836	6 000	11 475
1954/55	3	21	—	13	2 746	2	423	82	—	3 266	180 766	6 000	11 425
1955/56	3	21	—	3	2 669	—	284	93	—	3 049	172 363	6 000	12 068
1956/57	3	21	—	7	2 057	—	980	84	—	3 128	148 068	6 000	11 805
1957/58	2	20	—	6	2 251	—	924	225	—	3 406	171 432	5 408	11 020
1958/59	3	21	—	1	1 291	—	1 019	215	—	2 526	102 418	5 864	12 476
1959/60	3	21	—	9	1 160	—	1 075	89	—	2 333	97 546	5 787	12 562
1960/61	2	16	—	4	1 387	—	792	134	—	2 317	109 727	5 632	12 381
1961/62	1	8	—	—	661	—	447	86	—	1 194	49 815	4 765	9 666
1962/63	—	—	—	—	—	—	—	—	—	—	—	—	—
1963/64	2	16	—	—	552	—	409	60	—	1 021	41 282	3 998	7 156
1964/65	1	21	—	—	503	—	506	141	—	1 150	45 805	5 147	9 702
1965/66	1	10	—	—	218	—	4	17	—	239	9 964	—	—
Totals				41 515	87 555	26 754	15 128	3 716	582	175 250	9 360 084	260 950	498 870

"Floating factories include skrott processors. Other whales are Southern right whales except 4 Minke whales and 1 Bottlenose whale taken before 1914. Seals are Elephant seals except 755 Leopard seals taken before 1927, 97 Weddell seals taken before 1916, and 1 Fur seal taken in 1915. Sources: Captain Hodges' Report (1906), Command Paper 657 (1920), International Whaling Statistics (1931, 1964, & 1968), R.M. Laws (1960), R.W. Vaughan (1983), and various Magistrate's Reports. (There are some small discrepancies between these.)

Appendix 3

The postage stamps of South Georgia and the Falkland Islands Dependencies

Prior to 1944, Falkland Islands stamps were used in the Dependencies (and later for some higher values). In 1944 they were overprinted with 'South Georgia Dependency of' and likewise for the other parts of the Dependencies. From 1946 to 1955 and after 1980 they were designated 'Falkland Islands Dependencies' and between 1963 and 1969 'South Georgia'.

	Year of issue	Description and number in issue
1	1944	Falkland Islands overprinted stamps 'South Georgia', 8.
2	1946	Definitive issue of 'thick' maps, 8.
3	1946	Victory issue (omnibus), 2.
4	1948	Definitive issue of 'thin' maps, 8.
5	1948	Royal Silver Wedding Anniversary (omnibus), 2.
6	1949	Additional definitive 'thin' map $2\frac{1}{2}$d stamp, 1.
7	1949	Universal Postal Union 75th anniversary (omnibus), 4.
8	1953	Coronation of H.M. Queen Elizabeth II (omnibus), 1.
9	1954	Definitive issue of ships, 15.
10	1955	Trans-Antarctic Expedition 1956–1958 overprints, 4.
11	1963	Definitive issue including £1 Blue Whale, 15.
12	1969	Redesigned £1 definitive, King penguins, 1.
13	1971	Definitive issue overprinted for decimal currency 14.
14	1972	50th anniversary of the death of Sir Ernest Shackleton, 4.
15	1972	Royal Silver Wedding Anniversary (omnibus), 2.
16	1973	Wedding of Princess Anne (omnibus), 2.
17	1974	Centenary of the birth of Sir Winston Churchill (omnibus), 2.
18	1975	200th anniversary of Captain Cook's exploration and taking possession of South Georgia, 3.
19	1976	50th anniversary of the Discovery Investigations, 4.
20	1977	Silver Jubilee of the Succession of H.M. Queen Elizabeth II (omnibus), 3.
21	1978	Silver Jubilee of the Coronation of H.M. Queen Elizabeth II (omnibus), 3.
22	1979	Captain Cook's voyages, 4.
23	1980	Definitive issue of pictorials, 15.
24	1981	Plants, 6.
25	1981	Wedding of the Prince of Wales (omnibus), 3.
26	1982	Reindeer, 4.
27	1982	Insects, 6.
28	1982	21st birthday of the Princess of Wales (omnibus), 4.
29	1982	Rebuilding fund £1, surcharged £1, 1.
30	1983	200 years of manned flight, 4.
31	1984	Antarctic crustacea, 4.
	1984	Queen Elizabeth II pictorials, 13.
	1985	Volcanoes of South Sandwich Islands, 4.
	1985	Albatrosses, 4.
	1985	Life and times of Queen Mother (omnibus), 5.
	1985	Early naturalists, 4.
	1985	Queen Elizabeth II pictorials with imprint date '1985', 5.
	1986	60th birthday of Queen Elizabeth II (omnibus), 5.
	1986	Wedding of Prince Andrew (omnibus), 3.
	1987	Queen Elizabeth II, birds, 15.
	1987	30th anniversary of International Geophysical Year, 3.
	1988	Shells, 4.
	1988	Tercentenary Lloyd's of London 1688–1988 (omnibus), 4.
	1989	Glacier formation, 4
	1989	Combined Services Expedition 1964–65, 4.
	1990	Queen Elizabeth the Queen Mother – 90 glorious years (omnibus), 2.
	1990	Wrecks and hulks, 4.
	1991	65th birthday of Queen Elizabeth II and 70th birthday of Prince Philip (omnibus), 2.

Appendix 4

Summarised monthly meteorological tables from King Edward Point

King Edward Point: meteorological statistics 1951–80 (height of station 2.5 m).[a]

Month	Air temperatures (°C)			Mean wind speed (ms^{-1})	Mean Days of gale	Precipitation (Rain or water equivalent) mean (mm)	Mean days of		
	Mean	Maximum	Minimum				snow/ sleet	rain/ drizzle	fog
Jan.	+ 4.9	+ 22.3	− 4.8	4.2	1	112.5	10	22	1
Feb.	+ 5.6	+ 22.8	− 2.2	4.5	1	142.7	9	21	2
Mar.	+ 4.8	+ 22.2	− 4.3	4.8	2	158.6	11	23	2
Apr.	+ 2.7	+ 21.8	− 9.4	4.2	2	154.1	15	17	2
May	+ 0.3	+ 16.1	− 9.4	3.9	1	176.2	19	13	3
June	− 0.8	+ 15.6	− 12.1	4.1	1	148.8	21	10	2
July	− 1.5	+ 14.4	− 14.2	4.4	1	130.1	22	8	2
Aug.	− 1.2	+ 14.4	− 15.1	4.4	1	144.2	22	9	2
Sept.	+ 0.3	+ 14.6	− 11.7	4.5	1	133.3	18	11	2
Oct.	+ 2.1	+ 21.1	− 8.3	4.9	1	114.2	17	15	1
Nov.	+ 3.2	+ 19.7	− 6.3	4.7	1	96.3	16	16	1
Dec.	+ 3.9	+ 20.3	− 5.0	4.4	1	90.7	17	18	1
Year[b]	+ 2.0	+ 22.8	− 15.1	4.4	13	1601.6	198	183	20

[a]Copyright British Antarctic Survey.

[b]Annual means are not necessarily averaged monthly means.

Appendix 5

Systematic list of the native vascular flora of
South Georgia

Pteridophyta
Lycopodiaceae (Club mosses) 1
   Lycopodium magellanicum (Magellanic clubmoss)
Hymenophyllaceae (Filmy fern) 2
   Hymenophyllum falklandicum (Filmy-ferns)
Ophioglossaceae (Adder's tongues) 1
   Ophioglossum crotalophoroides (Adder's tongue)
Polypodiaceae 4
   Blechnum penna-marina (Small-fern)
   Cystopteris fragilis (Brittle bladder-fern)
   Polystichum mohrioides (Shield-fern)
   Grammitis poepigiana (Strap-fern)

Spermatophyta
Ranunculaceae (Buttercups) 1
   Ranunculus biternatus (Antarctic buttercup)
Caryophyllaceae 2
   Colobanthus quitensis (Antarctic pearlwort)
   Colobanthus subulatus (Sessile pearlwort)
Portulacaceae 1
   Montia fontana (Water blinks)

Rosaceae 2
   Acaena magellanica (Greater burnet)
   Acaena tenera (Lesser burnet)
   Acaena magellanica × tenera (Hybrid burnet)
Callitrichaceae 1
   Callitriche antarctica (Antarctic starwort)
Rubiaceae 1
   Galium antarcticum (Antarctic bedstraw)
Juncaceae (Rushes) 3
   Juncus scheuchzerioides (Great rush)
   Juncus inconspicuus (Lesser rush)
   Rostkovia magellanica (Brown rush)
Cyperaceae (Sedges) 1
   Uncinia meridensis
Gramineae (Grasses) 5
   Festuca contracta (Tufted fescue)
   Parodiochloa flabellata (Tussac, Tussock grass)
   Deschampsia antarctica (Antarctic hair-grass)
   Phleum alpinum (Alpine cat's-tail)
   Alopecurus antarcticus (Antarctic foxtail)

Appendix 6

The alien vascular flora of South Georgia (64 species)

Naturalised aliens (19)		Persistent aliens (15)	Transient aliens (30)
Widespread distribution (7)	Restricted distribution (12)		
Agrostis tenuis	*Achillea millefolium*	*Alium schoenoprasum*	*Alchemilla monticola*
Cerastium fontanum	*Achillea ptarmica*	*Anthriscus sylvestris*	*Alopecurus geniculatus*
Deschampsia caespitosa	*Agropyron repens*	*Cotula scariosa*	*Artemesia sp.*
Poa annua	*Agrostis canina*	*Deschampsia flexuosa*	*Avena fatua*
Poa pratensis	*Carex aquatilis*	*Empetrum rubrum*	*Brassica napus*
Rumex acetosella	*Carex nigra*	*Festuca ovina*	*Capsella bursa-pastoris*
Taraxacum officinale	*Festuca rubra*	*Lotus corniculatus*	*Carum carvi*
	Juncus filiformis	*Narcissus narcissus*	*Cerastium arvense*
	Pratia repens	*Nardus stricta*	*Daucus carota*
	Ranunculus repens	*Rorippa islandica*	*Lactuca sp.*
	Sagina procumbens	*Ranunculus acris*	*Lamium purpureum*
	Trifolium repens	*Rumex crispus*	*Lolium temulentum*
		Stellaria media	*Matricaria matricarioides*
		Trifolium hybridum	*Medicago sp.*
		Veronica serpyllifolia	*Phleum pratense*
			Pisum sativum
			Plantago major
			Plantago media
			Poa trivialis
			Raphanus sp.
			Rumex alpinus
			Senecio vulgaris
			Sinapis arvense
			Solanum tuberosum
			Sonchus olearaceus
			Stellaria graminea
			Thlapsi arvense
			Urtica cf dioica
			Urtica urens
			Veronica persica

Appendix 7

Fish recorded from the South Georgian region

Agnatha
Petromyzontes (Lampreys) 1
  Geotria australis

Chondrichthyes
Rajidae (Skates) 1
  Raja georgiana
Lamnidae (Mackerel sharks) 1
  Lamna nasus

Osteichthyes
Nototheniidae (Antarctic cods) 9
  Notothenia larseni
  Notothenia gibberifrons
  Notothenia nudifrons
  Notothenia neglecta
  Notothenia rossii
  Notothenia angustifrons
  Trematomus vicarius
  Trematomus hansoni
  Dissostichus eleginoides
Bathydraconidae (Dragon fish) 2
  Psilodraco breviceps
  Parachaenichthys georgianus
Harpagiferidae (Plunder fish) 2
  Harpagifer georgianus
  Artedidraco mirus
Channichthyidae (Ice fish) 3
  Pseudochaenichthys georgianus
  Chaenocephalus aceratus
  Champsocephalus gunnari

Liparidae (Snail fish) 2
  Careproctus georgianus
  Paraliparis gracilis
Zoarcidae (Eel-pouts) 1
  Melanostigma gelatinosum
Centrolophidae 1
  Pseudoicichthyes australis
Muraenolepidae (Eel cods) 1
  Muraenolepis microps
Bothidae (Flounders) 1
  Mancopsetta maculata
Paralepididae 1
  Notolepis coatsi
Scopelarchidae 1
  Neoscopelarchoides elongator
Trichiuridae (Cutlass fishes) 1
  Paradiplospinus glacialis
Myctophidae (Lantern fish) 13
  Kreffichthys anderssoni
  Protomyctophum tenisoni
  Protomyctophum normani
  Protomyctophum bolini
  Protomyctophum andriashevi
  Electrona antarctica
  Electrona carlsbergi
  Lampanyctis achirus
  Gymnoscopelus braueri
  Gymnoscopelus nicholsi
  Gymnoscopelus bolini
  Gymnoscopelus fraseri
  Gymnoscopelus hintonoides

Appendix 8

Birds recorded breeding on South Georgia

Spheniscidae (Penguins) 5
  Aptenodytes patagonicus (King penguin)
  Pygoscelis antarctica (Chinstrap or ringed penguin)
  Pygoscelis papua (Gentoo penguin)
  Eudyptes chrysolophus (Macaroni penguin)
  Eudyptes chrysocome (Rockhopper penguin)
Diomedeidae (Albatrosses) 4
  Diomedea exulans (Wandering albatross)
  Diomedea melanophrys (Black-browed albatross or mollymauk)
  Diomedea chrysostoma (Grey-headed albatross or mollymauk)
  Phoebetria palpebrata (Light-mantled sooty albatross)
Procellariidea (Petrels, prions, etc.) 8
  Macronectes giganteus (Southern giant petrel or fulmar)
  Macronectes halli (Northern giant petrel or fulmar)
  Daption capense (Cape pigeon, Pintado petrel)
  Pagodroma nivea (Snow petrel)
  Pachyptila desolata (Dove or Antarctic prion, Whale bird)
  Pachyptila turtur (Fairy prion, Snow bird)
  Halobaena caerulea (Blue petrel)
  Procellaria aequinoctialis (White-chinned petrel, Shoemaker)

Oceanitidae (Storm petrels) 3
  Oceanites oceanicus (Wilson's storm petrel)
  Fregetta tropica (Black-bellied storm petrel)
  Garrodia nereis (Grey-backed storm petrel)
Pelecanoididae (Diving petrels) 2
  Pelecanoides georgicus (South Georgia diving petrel)
  Pelecanoides urinatrix (Common, Kerguelen, or sub-Antarctic diving petrel)
Phalacrocoracidae (Shags) 1
  Phalacrocorax atriceps (Blue-eyed shag)
Anatidae (Ducks) 2
  Anas georgica (South Georgia pintail or teal)
  Anas flavirostris (Speckled or Yellow-billed teal)
Chionididae (Sheathbills) 1
  Chionis alba (Sheathbill, Kelp pigeon, Mutt)
Stercorariidae (Skuas) 1
  Catharacta lönnbergi (Brown or Southern skua)
Laridae (Gulls and terns) 2
  Larus dominicanus (Dominican, Southern black-backed or Kelp gull)
  Sterna vittata (Antarctic tern)
Motacillidae (Pipits) 1
  Anthus antarcticus (South Georgia pipit)

Appendix 9

Vagrant birds recorded at South Georgia

Spheniscidae (Penguins) 3
  Aptenodytes forsteri (Emperor penguin)
  Pygoscelis adeliae (Adelie penguin)
  Sphensicus magellanicus (Magellanic penguin)
Diomedeidae (Albatrosses) 2
  Diomedea salvini (Salvin's albatross)
  Phoebetria fusca (Sooty albatross)
Procellariidae (Petrels, prions, fulmars, etc.) 7
  Fulmarus glacialoides (Antarctic or Southern fulmar)
  Thalassoica antarctica (Antarctic petrel)
  Pachyptila belcheri (Narrow-billed prion)
  Pachyptila vittata (Broad-billed prion)
  Pterodroma brevirostris (Kerguelen petrel)
  Petrodroma mollis (Soft-plumaged petrel)
  Puffinus gravis (Great shearwater)
Areidae (Herons) 3
  Casmerodius albus (Great egret, White heron)
  Egretta thula (Snowy egret)
  Bubulcus ibis (Cattle egret, Tickbird)

Anatidae (Ducks) 2
  Anas discors (Blue-winged teal)
  Anas sibilatrix (Chiloé wigeon)
Rallidae (Rails) 1
  Porphyrula martinica (Purple gallinule)
Scolopacidae (Sandpipers) 4
  Tringa solitaria (Solitary sandpiper)
  Calidris minuta (Little stint)
  Calidris melanotus (Pectoral sandpiper)
  Calidris fuscicollis (White-rumped sandpiper)
Laridae (Gulls and terns) 2
  Larus atlanticus (Olrog's gull)
  Sterna paradisaea (Arctic tern)
Hirundinidae (Swallows) 2
  Hirundo rustica (Barn swallow)
  Delichon urbica (House martin)
Tyrannidae (Kingbirds) 1
  Tyrannus tyrannus (Eastern kingbird)

Appendix 10

Weights and measures referred to in this book

The Falkland Islands and Dependencies presently retain the Imperial system of weights and measures. Some conversion factors for this are given as it is becoming obsolete elsewhere, together with those for a variety of other units.

Système International conversions

League or lea: a nautical league of 3 nautical miles, 5.55 km.

Nautical mile: one second of longitude arc, 1.15 statute miles, 1.85 km.

Sazhen: Imperial Russian unit of depth, approximately 2 m.

Fathom: Imperial unit of depth, 6 feet, 1.83 m.

Ton: Imperial unit of weight, 1016 kg.

Ton: of registered or gross tonnage of a vessel is a unit of 100 cubic feet, 2.83 m^3.

Ton or Tun: of seal of whale oil is a unit of 6 barrels, approximately 1000 l.

Barrel: of seal or whale oil varies between 160 and 172 l, modern values are 40 Imperial gallons or 170 kg. (It is distinct from the barrel of petroleum or other mineral oil.)

Gallon: Imperial unit of volume, 4.55 l.

Pound: Imperial unit of weight, 454 g. (Also a currency unit.)

Foot and inch: Imperial units of length, 30.5 and 2.54 cm.

Temperature

Temperature is given in degrees Centigrade (or Celsius) throughout the book (except in the quotation from Captain Cook).

Time

Glass: a sand glass, generally of 1 hour.

Julian Calendar: is 12 days ahead of the Gregorian calendar near 1819.

Sea Day: begins at noon thus 12 hours are subtracted for a standard day.

Times given in the text are Falkland Island Dependencies time which is equivalent to Greenwich Mean Times minus 3 hours.

Currency

The Falkland Islands and Dependencies use a Pound (£) divided into 100 pennies (p). This is at par with the Pound Sterling. Prior to 15 February 1971 a non-decimal monetary system was in use with pounds, shillings, and pennies. 1 penny (1d) was 1/240 of a pound (£1), approximately 0.42 of a new penny (p). 1 shilling (1/-) was 1/20 of £1 or 5p. 2/6 represents 2 shillings and 6 pence which is 12½p.

Bibliography

A substantial amount of literature has been published concerning South Georgia but this is scattered and much is difficult to obtain. Below is a selection of generally available and more important works as well as those from which quotations are taken in this book. A comprehensive bibliography with 1344 references was published by the British Antarctic Survey in 1982 (Headland, 1982a) and is being updated. Most of the works listed are in English but much other material is available in Norwegian, German, Spanish, French, Russian, Japanese and some other languages (in approximate decending frequency). Many articles in newspapers and similar publications described events on South Georgia during 1982. Several subsequent works have been produced describing these events, some of which have given more attention to accuracy than to speed of reporting.

Maps are published by the Directorate of Overseas Surveys (DOS 610), Hydrographer of the Royal Navy (Charts; 3585, 3589, 3592, 3596, and 3597), and the Directorate of Military Survey (1501 SN24-9 1-GSGS) all of the British Government. A Gazetteer is published by Her Majesty's Stationary Office (HMSO), London.

Allen, J.A. (1899). Fur seal hunting in the Southern Hemisphere. In D.S. Jordan, *The fur seals and fur seal islands of the North Pacific Ocean*, document 2017. Washington, D.C.: Government Printer.
* Includes an account of South Georgia and other Antarctic sealing.
Andrews, J. (1974). *The Cancellations of the Falkland Islands and Dependencies and the Hand-struck Stamps; with Notes on the British, Argentine and Chilean Post Offices in the Antarctic*. London: Robson Lowe, 56 pp.
* Illustrated and with maps, South Georgia on pp. 30–7, much early history.
Barnes, R. (1972). *The Postal Services of the Falkland Islands, including the South Shetlands (1906–31) and South Georgia*. London: Robson Lowe, 96 pp. (Reprinted 1980.)

* Illustrated, contains much original research on administration and communications.
Beaglehole, J.C. (ed.) (1961). The Journals of Captain James Cook on his Voyage of Discovery. 2. The Voyage of the *Resolution* and *Adventure* 1772–1775. *Hakluyt Society* Cambridge No. 35 (extra series) 1021 pp.
* Includes Cook's account of his visit in 1775.
Bertrand, K.J. (1971). Americans in Antarctica 1775–1948. *Special publication*, No. 39. New York: American Geographical Society. 554 pp.
* Includes notes on early sealing and other visits to South Georgia.
Bogen, H.S.I. (1954 & 1955). Compañia Argentina de Pesca S.A. *Norsk Hvalfangst Tidende, Sandefjord*, **43** (10), 553–76; **44** (2), 88–95.
* Grytviken whaling station's history, catch statistics, etc. on its fiftieth anniversary. Second article has additions and rectifications. Several plates. In Norwegian and English.
Bonner, W.N. (1958). The introduced reindeer of South Georgia. *Falkland Islands Dependencies Survey Scientific Reports, London*, 22; 8 pp. and plates.
* History and detailed biology, plates and bibliography.
Bonner, W.N. (1968). The fur seal of South Georgia. *British Antarctic Survey Scientific Reports, London*, 56; 81 pp. and plates.
* History and detailed biology, plates and bibliography.
Bonner, W.N. (1980). *Whales*. Blandford Press, Poole. 278 pp.
* Contains much South Georgia material (history, personnel, etc.) and detailed information on whaling. Illustrated.
Bonner, W.N. (1982). *Seals and Man: A Study of Interactions*. Seattle & London: University of Washington Press. 170 pp.
* Includes much information about South Georgia sealing.
Boumphrey, R.S. (translator) (1967). A visit to South Georgia by H.W. Klutschak, 1877. *British Antarctic Survey Bulletin, London*, 12, pp. 85–92.

* Translation of Klutschak (1881) with notes, map and copies of woodcuts.

Brown, N.E. (1971). *Antarctic Housewife*. London & Melbourne: Hutchinson. 190 pp.
* Account of life at King Edward Point in 1955 by wife of radio operator, plates.

Burney, J.A. (1803–17). A chronological history of the voyages and discoveries in the South Sea or Pacific Ocean. In five volumes, 1803–17. London: G. & W. Nicol etc.
* Vol. 3, chapter 15 describes La Roche's visit in 1675. Vol. 5, chapter 6 describes *Leon's* visit in 1756.

Bush, W.M. (1983). *Antarctica and International Law: A Collection of Inter-State and National Documents*. 4 Volumes. New York: Oceana Publications.
* Includes a chronological arrangement of documents many of which concern South Georgia in the Argentine and British chapters.

Carse, D. (1959). The survey of South Georgia, 1951 to 1957. *Geographical Journal, London,* **125** (1), 20–37.
* Description of the conduct of the survey over four summer seasons, equipment, accommodation, fund raising, etc. Five plates and seven maps. Letter to Editor, next issue, also relevant.

Chaplin, J.M. (1932). Narrative of hydrographic survey operations in South Georgia and the South Shetland Islands 1926–1930. *Discovery Reports, Cambridge,* **3** 297–344.
* Survey of many harbours and anchorages, includes first charts and many photographs. Survey partly done from launch *Alert*.

Chapman, W. (ed.) (1965) *Antarctic Conquest*. New York: Bobbs-Merrill Company.
* Pages 15 and 16 give the account of South Georgia by Anthony de la Roché in 1675.

Christie, E.W.H. (1950). The supposed discovery of South Georgia by Amerigo Vespucci. *Polar Record, Cambridge,* **5** (40) 560–4.
* 'There is no reason what so ever for naming Amerigo Vespucci as the discoverer of South Georgia'.

Christie, E.W.H. (1951). *The Antarctic Problem*. London: George, Allen and Unwin. 336 pp.
* Historical introduction includes South Georgia. Argentine claims to it discussed. Bibliography and plates.

Clark, A.H. (1887). The Antarctic fur seal and sea elephant industries. In G.B. Goode, *The fisheries and fishing industries of the United States*, U.S. Commission on Fish and fisheries, **5** (2), 400–67. Washington, D.C.: Government Printer.
* Section about South Georgia on p. 412, details of early voyages, outfitting, conditions, etc. Atlas and plates.

Colnet, J. (1798). *A Voyage to the South Atlantic Ocean and Round Cape Horn into the Pacific Ocean for the Purpose of Extending the Spermacetti Whale Fisheries*. London: W. Bennett.
* Includes a note on whales near South Georgia in 1793.

Colonial Office, UK (1920). Report of the Interdepartmental Committee on Research and Development in the Dependencies of the Falkland Islands. *Command Paper*, No. 657. London: HMSO. 164 pp.
* Much information on early history, biology, sealing, survey, etc. The report led to the establishment of the *Discovery Committee*. Maps and bibliography included.

Cook, J. (1777). *A Voyage Towards the South Pole and Round the World Performed in HM Ships Resolution and Adventure in the Years 1772–1775*, two volumes. London: W. Strahan & T. Cadelle, 378 and 396 pp.
* Includes Captain Cook's account and taking possession of South Georgia, vol. 2, pp. 207–221. See also Beaglehole, J.C. 1961.

Dalrymple, A. (1771). *A Collection of Voyages Made to the Ocean Between Cape Horn and the Cape of Good Hope*, two volumes. London: published by the author.
* Contains an abstract of the journal of le Sieur Ducloz Guyot who sailed in the *Leon* past South Georgia in 1756.

Dautert, E. (1937). *Big Game in Antarctica*. Bristol: Arrowsmith. 254 pp.
* Translation from German edition of 1935. Whaling and sealing account, sixteen plates.

Deacon, G.E.R. (1975). Bicentenary of Captain Cook's landing on South Georgia. *Polar Record, Cambridge,* **17** (111), 692–4.
* Illustrated with plate of Possession Bay and map from Cook's original description.

Debenham, F. (ed.) (1945). The voyage of Captain Bellingshausen to the Antarctic Seas 1819–1821. *Hakluyt Society, London; series II*, (vol. 1) No. 91, 259 pp. and (vol. 2) No. 92, 474 pp.
* Translation of Bellingshausen's 1831 account published in St Petersberg. Pp. 86–92, vol. 1 concern South Georgia.

Destefani, L.H. (1982). *The Malvinas. The South Georgias and the South Sandwich Islands. The conflict with Britain*. Buenos Aires: Edipress S.A.
* Includes a description of the Argentine claim to sovereignty over South Georgia with some of the island's history.

Fanning, E. (1833). *Voyages and Discoveries in the South Seas, 1792–1832*. New York: Collins & Hannet, 512 pp. Also: London: O. Rich, 1834; Salem (Mass.): Marine Research Society (publication No. 6, 1924).
* Describes the island and sealing early in the nineteenth century, pp. 215–23.

Faustini, A. (1906). Di una carta nautica inedita della Georgia Austral. *Rivista Geografica Italiana, Firenze,* **13** (6), 343–51.
* Discusses and reproduces a map, possibly from Captain Pendleton 1802.

Filchner, W. (1922). Zum Sechsten Erdtheil die Zweite Deutsche Sudpolar Expedition. Berlin: Verlag Ullstein, 410 pp.
* Expedition in the *Deutschland* to the Weddell Sea. Visited South Georgia 1911–12. Account of whaling, history, etc. Many plates.

Fisher, M. & Fisher, J. (1957). *Shackleton*. London: Barrie, 559 pp.

* Detailed biography. Many sketches, plates, maps and a comprehensive bibliography.

Fitte, E.J. (1968). *La disputa con Gran Bretana por las islas del Atlantica Sur.* Buenos Aires: Emece Editores.
* Includes a description of the history and Argentine claims, to South Georgia.

Foreign Office, UK (1956). The British title to sovereignty in the Falkland Islands Dependencies. *Polar Record, Cambridge,* **8** (53), 125–51.
* Gives comprehensive statement of basis of British sovereignty, examines Argentine claims to South Georgia.

Forster, G. (1777). *A Voyage Round the World in His Britannic Majesty's Sloop, Resolution, Commanded by Captain James Cook during the Years 1772, 3, 4 and 5.* 2 vols. London: B. White, J. Robson, P. Emsley & G. Robinson, 602 and 607 pp.
* South Georgia described pp. 524–534, vol. 2 natural history and general account.

Franks, O.S. (Chairman) (1983). Falklands Islands Review, Report of a Committee of Privy Counsellors. *Command Paper,* 8787. London: HMSO, 106 pp.
* Includes an analysis of events leading to the Argentine invasion in 1982.

Fuchs, V.E. (1982). *Of Ice and Men.* Oswestry: Anthony Nelson, 383 pp.
* Includes account of Falkland Islands Dependencies Survey and British Antarctic Survey at South Georgia with several colour plates.

Greene, S.W. (1964). The vascular flora of South Georgia. *British Antarctic Survey Scientific Reports, London,* 45; 58 pp. with plates and map.
* Historical introduction, checklist of 51 species; 24 native and 5 naturalised ones described. Detailed bibliography. Many plates and diagrams.

Gressitt, J.L. (ed.) (1970). Subantarctic entomology, particularly of South Georgia and Heard Island. *Pacific Insects Monograph,* No. 23, Hawaii: Bishop Museum, 374 pp.
* Many papers on different groups of South Georgia's mites, spiders, and insects with a general account of the island.

Hardy, A.C. (1967), *Great Waters.* London: Collins, 542 pp.
* 'A voyage of natural history to study whales, plankton and the waters of the Southern Ocean'. Plates, watercolours, establishment of Discovery House and index of *Discovery* reports.

Hastings, M. & Jenkins, S. (1983). *The Battle for the Falklands.* London: Michael Joseph, 372 pp.
* Probably the best and most accurate account of the events of 1982.

Hattersley-Smith, G. (1980). The history of place names in the Falkland Islands Dependencies (South Georgia and the South Sandwich Islands). *British Antarctic Survey Scientific Reports, Cambridge,* 101; 112 pp. and map.
* Describes derivation of about 1200 South Georgia place-names. Historical introduction and extensive bibliography.

Headland, R.K. (1982a). South Georgia; a bibliography. *British Antarctic Survey, Data Report,* No. 7, 180 pp.
* Contains 1344 bibliographical references to the island.

Headland, R.K. (1982b). The German Station of the First International Polar Year, 1882–83, at South Georgia, Falkland Islands Dependencies. *Polar Record, Cambridge,* **21** (132), 287–301.

Headland, R.K. (1983). Military action in the Falkland Islands Dependencies April–June 1982. *Polar Record, Cambridge,* **21** (133), 394–5; (135) 549–558.

Hoare, M.E. (ed.) (1982). *The Resolution Journal of Johann Reinhold Forster.* 4 vol. London: The Hakluyt Society.
* Includes Forster's account of his visit to South Georgia in 1775, vol. 4, pp. 712–17.

Holtedahl, O. (ed.) (1929–61). *Scientific Results of the Norwegian Antarctic Expeditions 1927–1928, et sqq.* Instituted and financed by Consul Lars Christensen. 3 vols.: (parts 1–14); 2 (parts 15–27); 3 (parts 28–33). Oslo: Det Norsk Videnskaps Akademi, i Kommisjon Hos Jacob Dybwab.
* The series contains many papers relevant to South Georgia.

Hydrographer, R.N. (1974). *The Antarctic Pilot,* 4th ed. London: HMSO, 333 pp.
* Navigational and much general information concerning South Georgia with several plates. Earlier editions: 1st, 1930 (195 pp.); 2nd, 1948 (370 pp); 3rd, 1961 (448 pp) have much historical information. Supplements are issued periodically.

International Court of Justice (1956). *Antarctic Cases (United Kingdom v. Argentina, United Kingdom v. Chile); Pleadings, Oral Arguments, Documents.* Den Haag: International Court of Justice, 114 pp.
* Reproduces the documents submitted in the case, much concerns South Georgia.

Jackson, G. (1978). *The British Whaling Trade.* London: A. & C. Black, 325 pp.
* General history from 1600s to 1960s. Economic aspects discussed. Much material about Salvesens, Leith Harbour and Lever Bros., Prince Olav Harbour.

Jones, A.G.E. (1973). Voyages to South Georgia 1795–1820. *British Antarctic Survey Bulletin, London,* 32; pp. 15–22.
* Large number of early voyages described, map and plates from Cook. Bibliography and list of sources.

Jones, A.G.E. (1981). Three British Naval Antarctic Voyages, 1906–43. *Falkland Islands Journal, Stanley,* 1981; pp. 29–36.
* Gives details of *Sappho* and *Dartmouth* visits.

King, H.G.R. (1969). *The Antarctic.* London: Blandford Press, 276 pp.
* Includes general account of South Georgia, several plates.

Klutschack, H.W. (1881). Ein Besuch auf Sud Georgien. *Deutsche Rundschau fur Geographie und Statistics, Munchen,* **3** (11), 522–34.
* Account of a visit in 1877. See also Boumphrey (1967).

Kohl-Larsen, L. (1930). *An den Toren der Antarktis.* Stuttgart: Strecher & Schroder, xii & 288 pp.

* Historical introduction, account of whaling, sealing and natural history, exploration of the Kohl-Larsen Plateau. Expedition was on South Georgia in summer 1928/29. 39 plates and 3 maps.

Laws, R.M. (1960). The southern elephant seal (*Mirounga leonina*) at South Georgia. *Norsk Hvalfangst Tidende, Sandefjord*, **49** (10), 466–76; **49** (11), 520–42.

* Description of the sealing industry, production figures since 1910. Population, ages and biomass of the seals discussed. Map, plates and bibliography.

Laws, R.M. (1978). Ecological studies at South Georgia. *South African Journal of Antarctic Research, Pretoria*, **8**, 3–13.

* Historical introduction, discussion of environment and terrestrial, marine, sea bird and seal research. Large bibliography.

Laws, R.M. (ed) (1984). *Antarctic Ecology*. London: Academic Press.

* Includes nine papers relevant to South Georgia with a general introduction and comprehensive bibliography.

Leader-Williams, N. (1978). The history of the introduced reindeer of South Georgia. *Deer, Southampton*. **4** (5), 256–61.

* Map, photographs, tables of population and bibliography. Earlier work summarised.

Lönnberg, E. (1906). Contributions to the Fauna of South Georgia. *Kungliga Svenska Vetenskaps Akademiens Handlinger, Uppsala & Stockholm*, **40**, (5), 104 pp.

* Describes birds, whales, seals, fish, the rat, etc., several plates.

MacLaughlin, W.R.D. (1962). *Call to the South, A Story of British Whaling in the Antarctic*. London: George Harrap & Co., 188 pp.

* Narrative of period on a factory ship, includes a description of Leith Harbour and plates.

Marra, J. (1775). *Journal of the Resolution's Voyage, 1772–1775*. London: F. Newberry, 319 pp.

* Irregular journal of Captain Cook's voyage which surveyed part of South Georgia. Published anonymously.

Matthews, L.H. (1931). *South Georgia: The British Empire's Sub-Antarctic outpost*. Bristol: John Wright; and London: Simpkin Marshall, 163 pp.

* Detailed general account, quotations from early works, many photographs and drawings. Bibliography.

Matthews, L.H. (1951). *Wandering Albatross: Adventures Among the Albatrosses and Petrels in the Southern Ocean*. London: MacGibbon & Key, 131 pp. Also New York: Reinhardt & Evans.

* Popular account of investigation of procellariiformes on South Georgia.

Matthews, L.H. (1952). *Sea Elephant: The Life and Death of the Elephant Seal, illustrated by the author*. London: MacGibbon & Key, 190 pp.

* General account of South Georgia and account of sealing 1924–28, map and many illustrations.

Matthews, L.H. (1968). *The Whale*. London: Allen & Unwin, 287 pp.

* Detailed illustrated work on whales, much South Georgia material, large bibliography.

Matthews, L.H. (1977). *Penguin: Adventures Among the Birds, Beasts and Whalers of the Far South*. London: Peter Owen, 167 pp. Also New York: Universe Books.

* Popular illustrated account, mainly of South Georgia penguins, based on author's time at King Edward Point.

Morrell, B. (1832). *A Narrative of Four Voyages to the South Sea etc. From the Year 1822 to 1831*. New York: J. & J. Harper. Also New Jersey: Gregg Press, 1970.

* Page 58 describes a somewhat imaginative visit to South Georgia.

Murphy, R.C. (ed.) (1914). A report on the South Georgia expedition. *Science Bulletin, Museum of the Brooklyn Institute of Arts and Sciences, New York*, **2** (4), 41–102.

* Author visited South Georgia in 1912/13. Article includes general note and several papers by other authors.

Murphy, R.C. (1918). The status of sealing in the sub-Antarctic Atlantic. *Scientific Monthly, New York*, **7**, 112–19.

* Includes a historical description and contemporary account of early sealing at South Georgia with seven plates.

Murphy, R.C. (1922). An outpost of the Antarctic. *National Geographic Magazine, Washington*, **41** (4), 409–44.

* General account with map and many illustrations of South Georgia.

Murphy, R.C. (1947). *Log Book for Grace; Whaling Brig Daisy 1912–13*. New York: Macmillan & Co. 290 pp. Also, London: Robert Hale, 1948; facsimile edition: Time–Life Books, 1965.

* Describes the last old style sealing expedition, natural history and conditions on South Georgia in 1912/13.

Murphy, R.C. (1967). *A Dead Whale or a Stove Boat: Cruise of Daisy in the Atlantic Ocean, June 1912 to May 1913*. Boston: Houghton Mifflin, 177 pp.

* Illustrated companion volume to Logbook for Grace. One plate of South Georgia and many of whaling and sealing.

Neumayer, G. (ed) (1890 & 91) Die Internationale Polarforschung 1882–83. *Die Deutschen Expeditionen und Ihre Ergebnisse*. Berlin: A. Asher & Co. Two volumes, No. 1 (1891) 363 pp; No. 2 (1890) 574 pp.

* Scientific results of the Royal Bay expedition. Plates, plans, maps and list of publications included.

Neumayer, G. & Borgen, C. (1886). *Die Beobachtungs-Ergebnise der Deutschen Stationen, die Internationale Polarforschung 1882–83. Band II Sud Georgien*. Berlin: A. Asher & Co., 523 pp.

* General account of the expedition plans, maps and plates. Meteorological and geophysical data given and discussed.

Nordenskjöld, O., Andersson, J.G. & Larsen, C.A. (1905). *Antarctica, or Two Years Amongst the Ice of the South Pole*. London: Hurst & Blackett, 608 pp. Facsimile edition: London: C. Hurst (1977).

* Includes an account of the visit of *Antarctic* to South
Georgia in winter 1902. Naming of Maiviken and
Grytviken etc., many plates and some maps.

Ommanney, F.D. (1971). *Lost Leviathan, Whales and
Whaling*. London: Hutchinson, 280 pp.
* Describes the Discovery Investigations and whaling at
South Georgia.

Priestley, R., Adie, R.J. & Robin, G. de Q. (eds) (1964).
Antarctic Research. London: Butterworths, 360 pp.
* General account of Falkland Islands Dependencies
research, 21 papers, maps and many plates. Contains
much South Georgia information.

Prince, P.A. & Croxall, J.P. (1983) Birds of South Georgia:
new records and re-evaluation of status. *British
Antarctic Survey Bulletin, Cambridge*, 59, pp. 15–27.
* Gives an updated list of South Georgia birds to April
1982.

Rankin, N. (1951). *Antarctic Isle, Wildlife in South Georgia*.
London: Collins, 383 pp.
* Mainly ornithological account, descriptions of launch
journeys around the island. Includes 137 plates and a
map.

Roberts, B.B. (1958). Chronological list of Antarctic
expeditions. *Polar Record, Cambridge*, 9 (59), 97–134;
9 (60), 191–239.
* List includes many voyages to South Georgia.

Roberts, B.B. (1977). Conservation in the Antarctic.
*Philosophical Transactions of the Royal Society,
London, Series B*, 279, 97–104.
* Includes discussion of the 1975 conservation law for
South Georgia.

Rutter, O. (ed.) (1953). *A Voyage Round the World with
Captain James Cook in HMS Resolution*, by Anders
Sparrman. Translated by H. Beamish & A. Mackenzie-
Grieve. London: Robert Hale, 214 pp.
* Edited reprint of the 1785 work, briefly records a visit
to South Georgia (on pp. 196 & 197).

Salvesen, T.E. (1914) The whale fisheries of the Falkland
Island Dependencies. *Report of the Scientific Results of
the voyage of the SS 'Scotia', Edinburgh*, 4, 479–86.
* Describes operations, companies and catch of early South
Georgia whaling with several plates.

Shackleton, E.A.A. (1982). Falkland Islands economic study
1982. *Command Paper*, 8653. London: HMSO, 137 pp.
* Includes discussion of a potential future for South
Georgia after the events of 1982.

Shackleton, E.H. (1919). *South*. London: Heinemann, 380 pp.
* Account of the 1914–17 expedition, includes boat
journey to South Georgia and trek across the island.
Map and several South Georgia plates. Appendices:
Scientific work, J.M. Wordie; Southern whales and
whaling, R.S. Clarke also concern South Georgia.

Skottsberg, C. (1912). *The Wilds of Patagonia*. London:
Edward Arnold, 336 pp.
* Chapter 20: 'A winter trip to South Georgia'.

Smith, R.I.L. & Walton, D.W.H. (1975). South Georgia sub-
Antarctic. In *Structure and Function of tundra
ecosystems*, ed. T. Rosswall & O.W. Heal,

pp. 399–423. Stockholm: Swedish Natural Science
Research Council.
* Comprehensive botanical account, history,
environment, production, decomposition, nutrient
cycling, influence of man, etc. Large bibliography.

Smith, T.W. (1844). *A Narrative of the Life, Travels and
Sufferings of Thomas W. Smith, Comprising an Account
of his Early Life, Adoption by the Gypseys, his Travels
during Eighteen Voyages to Various Parts of the World,
during which he was Shipwrecked Five Times, thrice on
a Desolate Island near the South Pole, once on the
coasts of England and once on the coast of Africa...
Written by Himself*. Boston: W.C. Hill, 240 pp.
* Includes three visits to South Georgia, pp. 121–57.
Excellent descriptions of early sealing and conditions.

Stackpole, E.A. (1953) *The Sea-Hunters*. Philadelphia: J.B.
Lippincott. 510 pp.
* Includes details of earliest sealing voyages to South
Georgia.

Sutton, G. (1957). *Glacier Island, the official record of the
British South Georgia Expedition 1954–55*. London:
Chatto & Windus, 224 pp.
* Mountaineering, mainly in the southeast. Also
published in Travel Book Club (1958).

Tønnessen, J.N. & Johnsen, A.O. (1982). *The History of
Modern whaling*. Translation by R.I. Christopherson.
London: Hurst & Co., and Canberra: Australian
National University, 798 pp.
* Condensed translation of Johnsen (1959) and
Tønnesson (1967–8–9), *Den Moderne Hvalfangsts
Historie*.
Includes many details and a comprehensive
bibliography of South Georgia's whaling.

Vamplew, W. (1975). *Salvesen of Leith*. Edinburgh &
London: Scottish Academic Press, 311 pp.
* History of the company, including involvement in
South Georgia. Descriptions and plates of South
Georgia whaling, extensive shipping and production
tables.

Walton, D.W.H. (1982) The first South Georgia leases:
Compañia Argentina de Pesca and the South Georgia
Exploring Company Limited. *Polar Record,
Cambridge*, 21 (132), 231–40.
* Includes accounts of early lease applications, the two
companies, and the visit of HMS *Sappho*.

Walton, D.W.H. & Smith, R.I.L. (1973). Status of the alien
vascular flora of South Georgia. *British Antarctic
Survey Bulletin, London*, 36; pp. 79–97.
* Lists and discusses 20 transient, 16 persistent and 15
naturalised alien plants. Plates.

Watson, G.E. (1975). *Birds of the Antarctic and sub-Antarctic*.
Washington: American Geophysical Union, 350 pp.
* Includes description of South Georgia and its birds
with a general account of the region.

Weddell, J. (1825). *A Voyage Towards the South Pole
Performed in the Years 1822–1824*. London: Longman,
Hurst, Orme, Browne and Green, 276 pp. Facsimile
edition of the second edition (1827), 1970, David &

Charles Reprints, with a foreword by Sir V. Fuchs. 324 pp.
* Visit to South Georgia and sealing operations described pp. 50–60.
Wild, F. (1923). *Shackleton's Last Voyage; The Story of the Quest.* London: Cassell & Co., 372 pp.
* Records Shackleton's death at South Georgia and the work of the expedition.
Worsley, F.A. (1940). *Shackleton's Boat Journey.* London: Hodder & Stoughton, 147 pp. Also, the Folio Society (1974).
* Describes the journey from Elephant Island and trek across South Georgia.

Serial publications relevant to South Georgia

British Antarctic Survey Bulletin. London, numbers 1–38. Cambridge, number 39 onwards.
* Published several times a year since 1963. Contains many scientific and some other papers relevant to South Georgia and summarised meteorological data from King Edward Point.
British Antarctic Survey Scientific Reports. London to 1976 thence Cambridge.

* Published irregularly since 1963. A series of monographs, many relevant to South Georgia. Continues from the Falkland Islands Dependencies Survey Scientific Reports.
Discovery Reports. Discovery Committee, later the National Institute of Oceanography, Cambridge.
* Published irregularly since 1929. A series of monographs, many relevant to South Georgia.
Falkland Islands Gazette. Government Printer, Stanley.
* Published from 1891 onwards. Contains; laws, regulations, appointments, notices to mariners, meteorology, proclamations, general notices and other Governmental material concerning South Georgia.
Falkland Islands Journal. Port Stanley.
* Published annually since 1967. Contains occasional articles concerning South Georgia.
Polar Record. Scott Polar Research Institute, Cambridge.
* Published since 1931, presently three times a year. Includes many papers, notes, etc. concerning South Georgia.
Upland Goose. Falkland Islands Philatelic Study Group, Weston-Super-Mare.
* Published from 1960. Includes papers on South Georgia philately.

Index

Principal references are in bold characters. References
to or including illustrations are in italics. The place-
names of the map on pages 4 and 5 and names of
animals and plants in the appendices are not indexed.

Printed in the United States
By Bookmasters